ECOLOGICAL EFFECTS OF WASTEWATER

ECOLOGICAL EFFECTS OF WASTEWATER

Applied limnology and pollutant effects

E. B. Welch

Professor of Environmental Science
Department of Civil Engineering
University of Washington
Seattle, Washington, USA

With a contribution on Hydrographic Characteristics by

T. Lindell

Centre for Image Analysis
Uppsala University
Uppsala, Sweden

E & FN SPON
An Imprint of Chapman & Hall

London · Glasgow · New York · Tokyo · Melbourne · Madras

Published by Chapman & Hall, 2–6 Boundary Row, London SE1 8HN

Chapman & Hall, 2–6 Boundary Row, London SE1 8HN, UK

Blackie Academic & Professional, Wester Cleddens Road, Bishopbriggs, Glasgow G64 2NZ, UK

Chapman & Hall, 29 West 35th Street, New York NY 10001, USA

Chapman & Hall Japan, Thomson Publishing Japan, Hirakawacho Nemoto Building, 6F, 1-7-11 Hirakawa-cho, Chiyoda-ku, Tokyo 102, Japan

Chapman & Hall Australia, Thomas Nelson Australia, 102 Dodds Street, South Melbourne, Victoria 3205, Australia

Chapman & Hall India, R. Seshadri, 32 Second Main Road, CIT East, Madras 600 035, India

First published 1980 by the Press Syndicate of the University of Cambridge
Second edition 1992

© 1980 Cambridge University Press; 1992 E. B. Welch

Typeset in 10/12 Times by Excel Typesetter Company, Hong Kong
Printed in Great Britain by St. Edmundsbury Press, Bury St. Edmunds, Suffolk

ISBN 0 412 34940 X

A catalogue record for this book is available from the British Library

Library of Congress Cataloging-in-Publication data
Welch, E. B. (Eugene B.)
    Ecological effects of waste water : pollution in fresh water
systems / E. B. Welch, T. Lindell. – 2nd ed.
        p. cm.
    Includes bibliographical references and index.
    ISBN 0-412-34940-X
    1. Sewage disposal in rivers, lakes, etc. – Environmental aspects.
    2. Water – Pollution – Environmental aspects.   I. Lindell, T.
    II. Title.
    QH545.S49W44   1992
    628.1'68–dc20
                                                                    92–16060
                                                                        CIP

CONTENTS

PREFACE

The purpose of this book is to convey a broad, but in some respects still detailed, account of the effects of several important substances in wastewater on aquatic ecosystems. To cope with expanding knowledge, there is an increasing need for synthesis and integration of pertinent information in many areas. This is especially true with effects of wastewater in aquatic ecosystems which is a prime example of an area requiring broad, interdisciplinary study. However, many pollution problems may each require a separate treatise in order to cover the recent literature fully. Although much detail must be sacrificed in a book such as this, its value is in introducing the reader to the basic concepts involved with several pertinent problems all under one cover. The emphasis is on conveying an understanding of cause and effect in assessment rather than an assessment of the national or world-wide state of the various problems.

Material is presented at a level that hopefully appeals to scientists and engineers with and without a biological background. In this regard, a course, which has the same title as this book, has been successful in drawing from upper division undergraduate and graduate students from several different disciplines at the University of Washington for twenty years and is offered conjointly between Civil Engineering and Fisheries. Many of the concepts presented have been formulated over that time and this second edition is an update adding pertinent literature published over the past 12 years, as well as a rewriting and restructuring of most of the original content. Scientists and engineers with pollution enforcement and resource agencies, and in industries involved in pollution assessment/ mitigation, should find this new edition worthwhile. The reader will find a much improved organization, with division into general concepts, standing waters and running waters.

Although much progress has occurred in the control of water pollution and improvement in water quality in developed countries over the past two decades, the degradation of water resources by waste substances continues in developed and developing countries alike. Eutrophication, the problem to which much of this book is devoted, has been controlled in several, well-publicized point-source problems in larger lakes. But degradation continues in countless small and medium sized lakes in developed countries through non-point sources of nutrients such as urban and agricultural runoff, and the control of excess algal biomass in streams is poorly understood. In that regard, the first three-quarters of the book could be called applied limnology. While thermal effects due to waste heat from power plants never reached a problem of large magnitude due to the insistence on cooling facilities, it threatens to reappear through global warming. Acidification of freshwaters continues to be a near global problem although reductions in emissions may occur on a broad base. Toxicants in the environment currently receive the major attention and although groundwater is most impacted by seepage from landfills, some surface waters are also affected. Problem emphasis changes, but the response of ecosystems to various levels of wastewater substances does not change. For instance, trout will respond in a predictable way to increased temperature whether from a power plant effluent or from future global warming. Thus, there is need for people in environmental protection in general and water quality control in particular to understand the basic concepts regarding the assessment of aquatic ecosystem degradation. It is to this end that I have revised and rewritten this book.

I am indebted to many for encouragement in this endeavour, including the following: countless suggestions and assistance from students; reviews by Gunnel Ahlgren (temperature effects on phytoplankton), Gertrude Nürnberg (eutrophication), Dimitris Spyridakis (phosphorus and carbon cycles), Jean Jacoby (periphyton) and James Buckley (toxicants and toxicity); many discussions with colleagues, such as Ingemar Ahlgren, Barry Biggs, Dennis Cooke, John Quinn, Rich Horner, Brian Mar and Dimitris Spyridakis; and finally the contribution by Tommy Lindell on hydrographical characteristics.

E. B. Welch
Seattle,
Washington, USA

Part One

General Concepts of Aquatic Ecosystems

ECOSYSTEM FUNCTION AND MANAGEMENT

An ecosystem can be described as some unit of the biosphere, or the entire biosphere itself, within which chemical substances are cycled and recycled while the energy transported as part of those substances continually passes through the system. Although every ecosystem must ultimately obey the laws of thermodynamics and degrade to complete randomness (wind down), consistent with the universe, energy may be accumulated momentarily in an ecosystem. However, without a continuously renewed input of energy, the accumulation would be exhausted and the system would ultimately wind down.

The single continuous input to the world's ecosystems is from solar energy, and the conversion of that electromagnetic energy into chemical energy and then into work is what allows an ecosystem and the organisms that make up that system to function.

Because the processes of energy flow and nutrient cycling are quite variable, the assignment of definite boundaries between what might otherwise seem like clearly separate systems is not easy. However, for the sake of practicality and manageability, boundaries are cast that allow ease of study and process measurement. Thus, a stream and its immediate watershed, as well as a lake and its watershed inputs, may be considered ecosystems. The system could be considered closed under some conditions if one is describing chemical nutrients, but never if one is referring to energy, because there is always an input to and a loss from the system. Because an ecosystem responds to inputs as an integrated system, the study of whole systems is useful for management.

1.1 ECOSYSTEM COMPOSITION AND ENERGY SOURCES

Each ecosystem has a structure that determines how it functions in the transfer of energy and the cycling of nutrients. This structure can be

TABLE 1.1.   Comparison of energy sources in two ecosystems

| | Energy (g carbon m^{-2} $year^{-1}$) | | | |
| | Deciduous forest | | | Coniferous forest |
Energy source	Forest	Streams	Lake	Lake
Autochthonous	941	1	88	4.8
Allochthonous	3	615	18	9.4

Source: Data on deciduous forest from G. E. Likens, personal communication; data on coniferous forest from Wissmar *et al.* (1977)

[handwritten annotations: food from outside an aquatic community from detritus; food from inside an envir. in form of photosyn or autrophic production]

thought of as the organization of the internal groupings of chemical nutrients and energy through which the functioning occurs. This matter is living and dead, the living being represented best by the trophic levels of organisms leading from algae and/or rooted plants to fourth-level carnivores, for instance, in some grazing food webs. Outside organic matter is processed and decomposed by insects, fungi and bacteria through to the carnivores in detritus-based food webs.

The watershed has come to be regarded as an integral and inseparable part of the ecosystem (Borman and Likens, 1967; Likens and Borman, 1974). The character of the watershed determines whether the stream's energy source is mostly autochthonous (produced within) or allochthonous (produced outside). In a comparative study of areas in the northeastern USA, Likens (personal communication) has shown that an autochthonous-producing forest results in the domination of the drainage stream by allochthonous inputs, whereas the watershed lake is autochthonous. Wissmar *et al.* (1977) found a high mountain lake in a coniferous forest to be dominated by allochthonous inputs (Table 1.1). Lakes and streams in poorly vegetated watersheds (Minshall, 1978) and relatively large lakes tend to be dominated by autochthonous sources. According to the river continuum concept, low order streams in forested watersheds, as well as high order lower stretches of often turbid rivers, tend to be heterotrophic or detritus based, while intermediate order streams are autotrophic (Vannote *et al.*, 1980). Chapter 11 includes further discussion of stream productivity.

The relative proportions of allochthonous to authochthonous energy sources in ecosystems is useful for management decisions affecting aquatic ecosystems. To protect and use aquatic systems effectively, one must know how systems use and respond to different amounts and varying compositions of natural energy.

AQUATIC ECOSYSTEM

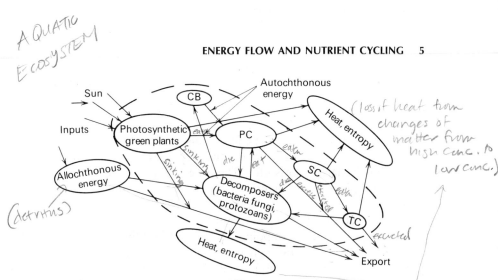

Figure 1.1. Energy flow and nutrient cycling in an aquatic ecosystem (Boundary indicated by dashed line). CB, chemosynthetic bacteria; PC, primary consumers; SC, secondary consumers; TC, tertiary consumers

any arrow going from ⊝ to decomposers must also go to heat entropy

1.2 ENERGY FLOW AND NUTRIENT CYCLING

If one could measure all the organisms and their main processes in an ecosystem and if their energy-consuming and energy-processing characteristics fell neatly into separate levels, one would be able to arrange a flow diagram as shown in Figure 1.1. Such a diagram shows several important points: (1) energy flows through the system and does not return, because by the second law of thermodynamics matter moves toward randomness, from states of high concentration to states of low concentration, and when that happens the energy contained becomes less (note entropic heat loss); (2) loss in energy as heat occurs at each step in the transfer process; and (3) allochthonous energy moves through the heterotrophic microorganisms or decomposers. A significant role in net production of usable energy for consumers is attributed to these microorganisms (note arrow from decomposers to consumers) in addition to decomposition (Pomeroy, 1974).

Even for allochthonous sources from surrounding forests or from cultural input of organic waste, the ultimate source must be either photosynthesis or to a lesser extent chemosynthesis.

The process of photosynthesis includes two phases: (1) a light reaction that traps solar energy, releases molecular O_2 and produces ATP (adenosine triphosphate) through photophosphorylation; and (2) a dark reaction (light not needed) that utilizes the trapped energy as ATP and fixes CO_2 or HCO_3^- into reduced organic cell material. The process yields energy and synthesizes new cells and can be summarized by:

$$6CO_2 + 12H_2O + \text{light} + (\text{chlorophyll } a \text{ and accessory pigments})$$
$$\rightarrow C_6H_{12}O_6 + 6O_2 + 6H_2O$$

The organic compound is glucose – the compound that holds the sun's energy trapped by photosynthetic organisms and allows the maintenance of the biosphere (that thin film of atmosphere, water and land around the planet Earth that supports life). The process is performed by green plants and a few pigmented bacteria. Only the green plants release O_2 as a byproduct.

Chemosynthesis is the other process through which organisms are totally self-sufficient in trapping energy and building cell material. This process yields energy through the oxidation of reduced inorganic compounds and thus requires no previous biological mediation by the various bacteria utilizing this process. The primitive Earth was rich in reduced inorganic compounds that are currently used by some bacteria for energy. An example is nitrification:

$$2NH_3 + 3O_2 \rightarrow 2HNO_2 + 2H_2O + \text{energy} \; (\text{small})$$

This process is less important to ecosystems as an energy yielder and more important as a material recycler. Chemosynthetic processes by bacteria are heavily involved in the cycling of such nutrients as nitrogen and sulphur.

Nutrients, such as N, C, P and S, follow the same pathways as energy in the system; they are consumed by autotrophs (green plants and chemosynthetic bacteria) from inorganic pools and fixed into organic compounds. These nutrients are then transferred through trophic levels just as energy is, because the reduced organic compounds, of which organisms are composed, are the carrier of the entrapped chemical energy. For example, CO_2 is the most oxidized state of C, but when fixed into glucose through photosynthesis, 1 mole ($C_6H_{12}O_6$) contains 674 kilocalories (kcal) of energy releasable through respiration by plants, animals, or decomposer microorganisms by the same biochemical pathways. The energy content of whole organisms in the various trophic levels ranges from 4 to 6 kcal g^{-1} dry weight.

The principal difference between energy and nutrient transport is that once organisms have utilized the energy from complex compounds and oxidized them completely (e.g. glucose to CO_2 and H_2O), the compounds are recycled through the inorganic pool(s) and are almost totally reusable by the community. There is no permanent loss, but, practically, a certain fraction may be lost to the sediments and require tectonic uplift for recycling. The recycling process is shown in Figure 1.2 (note heavy arrow from decomposers to inorganic pool). There are also chemical–physical processes involved in nutrient cycling. Chapter 4 includes further discussion of nutrient cycling.

ANY
ECOSYSTEM

efficient E'   net prod. trophic level
transfer   =   net producti.
                      available for
                      consumption

recycling   heat entropy?

Export

Figure 1.2. Transport and recycling of nutrients through an ecosystem (boundaries indicated by dashed line). CB, chemosynthetic bacteria; PC, primary consumers; SC, secondary consumers; TC, tertiary consumers

1.3   EFFICIENCY OF ENERGY AND NUTRIENT USE

The efficiency of the transfer of energy and/or nutrients is usually measured by the ratio of net productivity of a trophic level to the net productivity available for its consumption (Russell-Hunter, 1970, pp. 25, 191). These values usually fall between 10 and 20%. The level of this efficiency depends greatly upon the structure of the food web through which the materials are moving.

Structure in ecosystems can be thought of as the organization of species populations into appropriate trophic levels. However, trophic level is rather artificial and few organisms conform to a single trophic level throughout their entire life cycle. Nevertheless, populations organize, in time, in such a way that energy usage is optimized. Ecosystems that are physically stable, such as tropical rain forests and coral reefs, maximize organization and complexity and remain rather constant in biomass, productivity, species diversity and, consequently, efficiency. Very simply, the greater the variety of energy users in an ecosystem (alternative pathways), the greater the chance that a quantity of energy packaged in a particular way will be intercepted and used before it leaves the system. Instability in ecosystem structure brought about by natural or culturally caused variability in the physical–chemical environment should result in decreased efficiency and instability in energy and nutrient usage, and nutrient recycling is, therefore, much 'looser'. Where systems approach steady state, that is, inputs equal outputs, nutrient recycling tends to be 'tighter' and 'more complete' (Borman and Likens, 1967; Bahr *et al.*, 1972).

Odum (1969) has illustrated the effect of maturity on the characteristics and processes of ecosystems (Table 1.2 and Figure 1.3). He proposed

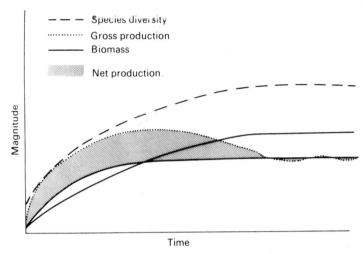

Figure 1.3. Conceptual plot demonstrating successional changes in species diversity, gross production (respiration + net production) and biomass. The shaded area representing net production approaches zero as the ecosystem matures (Bahr *et al.*, 1972, modified from Odum, 1969)

TABLE 1.2. Characteristic properties of ecosystems

	Type of ecosystem	
Property	*Immature*	*Mature*
Net production (yield)	High	Low
Food chains	Linear	Weblike
Nutrient exchange	Rapid	Slow
Nutrient conservation	Poor	Good
Species diversity	Low	High
Stability-resistance to perturbations	Low	High
Entropy	High	Low

Source: (Odum, 1969), with permission of the American Association for the Advancement of Science

that diversity tends to be low in immature systems and high in mature systems. Stability is often suggested to increase with diversity, but a definition of stability and whether it is directly related to diversity is a matter of controversy. Washington (1984) reviewed the problem and cited Goodman (1975) who suggested that stability has generally included one or more of the following: a constancy of numbers, a resistance to perturbation, or ability of a system to return to a previous state after displacement. According to Figure 1.3, net production tends to be inversely related to diversity. Margalef (1969) suggested that at low diversity, fluctuations in production per unit biomass are quite large and diversity changes are small, indicating instability, whereas at high diversity, fluctuations in production per unit biomass tend to be small with considerable changes in diversity, indicating stability. Figure 1.3 suggests that with time ecosystems tend to develop into a rather steady-state condition, with little net productivity and maximized structure. If production is low and associated fluctuations are small in a mature eco-system, it follows that nutrients are conserved and cycled more slowly and energy thus moves through the system at a slower rate and entropy is less – the system is more efficient in its functioning.

Examples of efficiency in ecosystems give some idea of their maturity. Fisher and Likens (1972) showed that Bear Brook, a northeastern forest stream, was only 34% efficient (total respiratory loss/total energy input). Thus, 66% of the energy left the system downstream unused, indicative of immaturity. Lindeman's (1942) study of Cedar Bog Lake showed it to be 54.5% efficient with 45.5% lost to sediments. Odum's (1956) results from Silver Springs in Florida showed only 7.5% lost from the system downstream with 92.5% being utilized.

1.4 MANAGEMENT OF ECOSYSTEMS

What do humans want from aquatic ecosystems? Obviously they do not want to destroy them, but they are highly inclined to manipulate them to their purpose whether it be to increase or decrease their productivity, keep them natural, or have them accept (assimilate) wastes with a minimum adverse effect on other uses. The problem is that these uses are conflicting. For example, to produce more and bigger sport fish would usually require stimulation of primary production, which, in turn, may result in decreased structure (species diversity) with possibly less stability. The yield of particular game-fish populations may increase, but at the expense of the overall efficiency of energy-handling and nutrient-recycling capabilities of the system. That would be indicated by the occurrence of a large biomass of inedible algae along with a wider range in oxygen, pH and even toxic ammonia. Fish may be larger and more abundant, but they could also be subject to high mortality times. In the same sense, ecosystems cannot assimilate wastes without some cost to their structure and stability, with consequent reduced efficiencies in nutrient cycling and energy utilization.

In reality, stable communities are probably not more able to resist change from waste input than unstable ones. Bahr et al. (1972) argue that if this were true, species diversity would decrease with waste input

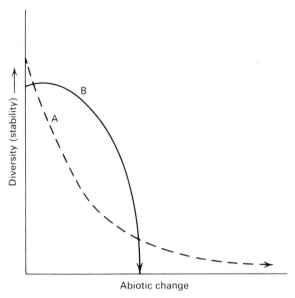

Figure 1.4. Ecosystem stability as a function of intensity in a stressful abiotic change. The two curves are discussed in the text (Bahr *et al.*, 1972)

(abiotic change) at an increasing rate (curve B in Figure 1.4), rather than at a decreasing rate as is usually observed (Figure 1.4, curve A). Highly diverse communities may produce stable behaviour in ecosystems, but such a state is very sensitive in itself to change in the environment. Any change, or manipulation, often tends to lower that stability and decrease the efficiency of ecosystem functioning. Humans probably cannot have their cake and eat it too. That is, highly structured and diverse ecosystems that maximize recycling and energy-using efficiency are probably not highly resistant to disturbance. The use without abuse idea is discussed further, e.g. in Chapter 11.

High diversity → stable ecosystem → easily changed be environ. → low diversity

Low diversity is a result of an unstable environment

STANDING CROP, PRODUCTIVITY AND GROWTH LIMITATION

Before population growth and control are discussed further, some important terms should be understood. Standing crop is often synonymous with biomass in that it represents the quantity per volume or area that is present and can be measured. The units of measure are wet weight, dry weight, numbers, or the quantity of some elemental constituent
- (N, P, C). Standing crop is the more general term; biomass is appropriate, of course, when the unit is mass.

Productivity is the rate of biomass formation and is a dynamic process. Productivity is often proportional to biomass, but in the case of fast-growing and efficiently grazed populations, biomass may be relatively small, whereas daily productivity is quite high. To understand that phenomenon, the concept of turnover rate is useful and can be defined as:

$$\left(\frac{1}{t}\right) \text{Turnover rate} = \frac{\text{Productivity } (m/l^3 - t)}{\text{Biomass } (m/l^3)}$$

where m = mass, l = length and t = time.

Turnover rate is thus the number of times that the biomass is theoretically replaced per unit time. Because they have the same units $(1/t)$, turnover rate equals growth rate but only when biomass remains constant. When biomass is not constant from one time to another, the growth rate over that period is determined by the increase or decrease in biomass as well as the instantaneous measurement of productivity.

2.1 PYRAMID OF BIOMASS

Productivity is directly proportional to biomass in a food chain only if the turnover rate at each trophic level is constant. A hypothetical example of this is shown in Table 2.1. Because energy is lost at each transfer in the

TABLE 2.1.   Hypothetical relation of biomass and productivity in a three-tiered food chain. Units are in mass (m), length (l) and time (t)

Trophic level	Biomass $(m\,l^{-2})$	Productivity $(m\,l^{-2}\,t^{-1})$	Turnover rate (t^{-1})
Secondary consumer	1	10	10
Primary consumer	10	100	10
Primary producer	100	1000	10

food chain, productivity must always decrease from primary producer to secondary consumer. In this instance, productivity decreases by a factor of 10 at each level. There is no reason to believe turnover rate would remain constant. Biomass may be greater at consumer than producer levels because producers are smaller organisms with a higher ratio of cell surface to volume and therefore have more rapid turnover rates. In such a case the biomass pyramid may be inverted as cited by Odum (1959, p. 64). In Long Island Sound, zooplankton and bottom fauna biomass were double that of phytoplankton. In the English Channel the ratio was 3:1. Thus, from this theoretical standpoint, productivity may not, in fact, be very tightly related to biomass in nature, particularly in open water systems. If the biomass in the examples from Odum were multiplied by turnover rate, the pyramids of biomass would be pyramids of productivity (and energy flow) and become upright.

2.2   PRODUCTIVITY

Primary production is the rate of biomass formation of the autotrophs, which are primarily the chlorophyll-bearing micro- and macroorganisms. Net production rate is measured as the net amount of organic matter fixed (e.g. carbon assimilated) and transferrable to the next trophic level. Gross production equals total assimilation including respiration. These quantities can be represented as shown in Figure 2.1.

In an ecosystem with plants only, net production by primary producers equals net ecosystem production (NEP) and is actually the realized increase in biomass available for harvest or grazing by primary consumers (secondary producers). With consumers present, net ecosystem production (NEP) is the sum of the fraction of net production (NP) not consumed plus the realized growth of consumers (secondary production). The ratio of gross production to respiration (P:R) in these hypothetical examples is 2 with plants only (assuming R = 0.5 GP) and between 1 and 2 with consumers present. Tropical rain forests have P:R ratios near 1 due to high rates of decomposition.

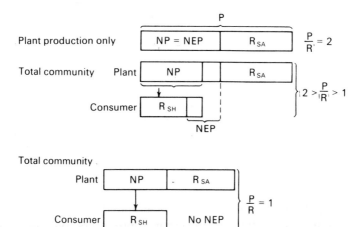

Figure 2.1. Hypothetical diagram of a simplified ecosystem energy budget. P, Gross production; NP, net production; R_{SA}, respiration autotrophs; R_{SH}, respiration heterotrophs; NEP, net ecosystem production (modified from Woodwell, 1970)

In heterotrophic, or allochthorous-fed ecosystems, P:R ratios are typically less than 1. The use of ratios of heterotrophic:autotrophic components is discussed further in Chapter 11.

2.3   EFFICIENCY OF ENERGY TRANSFER

The energy efficiency of ecosystems has direct management implications, as pointed out earlier. At the first trophic transfer, photosynthesis by chlorophyll-bearing organisms is rather inefficient. For terrestrial systems, efficiency of light utilization is about 1%, but in water it is much lower: 0.1–0.4% (Kormondy, 1969). The lower efficiency in water results from the absorptive and scattering effects of light by water and its particles.

Utilization of primary productivity through secondary production is nominally about 10%, but may range from about 5–20%. The 90% that is lost at each trophic transfer is largely through respiration. The steps involved in handling energy at the grazer level are shown in Figure 2.2. Food energy assimilated is equal to that ingested minus excretion. Respiration and any non-trophic growth (e.g. clam shells) are the other losses that subsequently result with net production remaining for transfer to the next level.

In essence, the efficiency of energy utilization at each trophic level depends upon how individual consumers handle the energy ingested. Trophic transfer is relatively efficient or inefficient because of the individual populations that are present. If the structure of an ecosystem is disrupted by waste input such that many of its efficient energy utilizers

Figure 2.2. Energy budget of individual animal. PB, plant biomass; TI, total energy ingested; TA, total energy assimilated; NA, non-assimilation (excretion); RA, respired assimilation (maintenance and activity demand); NRA, non-respired assimilation; NTG, non-trophic growth; NP, net production (after Russell-Hunter, 1970)

are eliminated, then overall efficiency in the system would naturally decrease. Large populations of unconsumed producers could indicate the degrading effect on the dynamics of the system.

2.4 POPULATION GROWTH AND LIMITATION

2.4.1 Growth

Growth of any population of living organisms can be described as exponential in an environment where resources are unlimited. With exponential growth, individuals are added at an increasing rate as the population becomes larger. This unlimited growth can be described by the following first-order, differential equation:

$$dN/dt = rN \qquad (2.1)$$

where r is the growth rate constant and N is population size.

The integrated form of this equation is:

$$N_t = N_0 e^{-rt} \qquad (2.2)$$

where N_t is population size at some time t after growth begins, N_0 is the initial size of the population and e is the base of natural logarithms.

Exponential growth of a population operates like compound interest. That is, the interest gained on an investment after some time interval is analogous to individuals added to a population and is dependent on the size of the investment (or population). By multiplying the interest rate by the new balance after each interval of time, the rate of increase in the total investment increases with time. This is shown graphically in Figure 2.3 and in Equation 2.1; as N increases, dN/dt increases.

The growth rate constant, r, has a maximum for each population that is consistent with its genetic capacity in an optimum environment. If the environment is less than optimum, r will not reach the genetically fixed maximum. Thus, an evaluation of r will indicate how optimum an en-

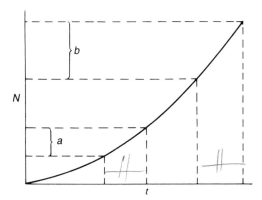

Figure 2.3. Unlimited population growth curve. Note that population increment b is much greater than a for an equal time increment, which is due to the larger population size. This is a result of the 'compound interest' phenomenon characteristic of populations

vironment is for a given population. In multicellular sexually reproducing organisms, r is affected by the age structure, or that fraction of the population that is reproductively active. If the age structure remains constant, r can be evaluated with respect to environmental variables. However, much of the change in r could be the result of altered age structure. This is not a problem in the analysis of unicellular asexually reproducing populations because theoretically all have an equal chance of reproduction. In both types of populations, r is the difference between birth and death rates, although with phytoplankton, as will be discussed later (Chapter 6), the growth rate is treated separately from loss (or death) rates. Typical growth rates range from $0.01\,\text{year}^{-1}$ ($1\%\,\text{y}^{-1}$) for a human population to $1.0\,\text{day}^{-1}$ ($100\%\,\text{d}^{-1}$) for algae and about $3.0\,\text{h}^{-1}$ for bacteria.

In reality, population density cannot continue to increase exponentially because some resource will eventually limit further increase. The resource that exhausts first will slow and ultimately stop the process as its supply diminishes. An example of populations that grow rapidly, seemingly to 'explode', are spring phytoplankton blooms. They tend to exhaust the limiting resource and reach their maximum attainable biomass all at once, which is followed by a crash. This type of growth is often represented by J-shaped curves. On the other hand, if environmental resistance is gradual, the growth curve resembles an S-shape, in which specific stages in growth can be defined. Initially, population increase is slow as the organisms adjust to their environment (Figure 2.4). This is followed by exponential growth when resources are unlimited. When resources become limiting, population increase slows and ultimately stops if resource supply is

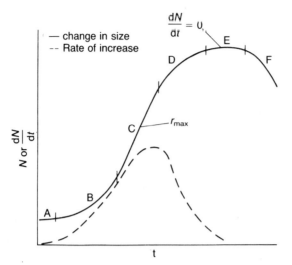

Figure 2.4. Population growth curve, environmentally limited, showing changes in size and rate of increase. Growth curve phases are: A, lag; B, accelerated growth; C, maximum rate; D, decelerated growth; E, stationary; F, death

exhausted. If additional resource is not forthcoming, the population will die and decomposition may recycle the limiting resource resulting in the start of a new growth curve.

The growth curve in Figure 2.4 is referred to as logistic and can be described, at least through the stable population phase, by the following equation;

$$dN/dt = rN[(K - N)/K] \qquad (2.3)$$

where K is the maximum attainable population size and r and N are as defined previously. Note that dN/dt, which is net productivity, reaches its peak consistent with the maximum slope (r) of the population curve (Figure 2.4).

The integrated form of Equation 2.3 is:

$$N_t = \frac{K}{1 + [(K - N_0)/N_0]e^{-rt}} \qquad (2.4)$$

2.4.2 Limitation

The concept of resource limitation will be briefly considered with respect to nutrients. Justus Liebig's law of the minimum is fundamental to understanding the control of population growth. It states that the yield of a crop will be limited by the essential nutrient that is most scarce in the

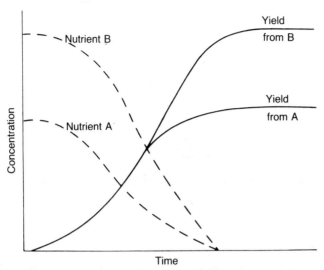

Figure 2.5. Growth curves for a microbial population showing the maximum yield as a function of initial nutrient (limiting) concentration

environment relative to the needs of the organism. Relation to need is an important part of that definition. The element lowest in concentration may not be the limiting nutrient. For example, nutrient A may be twice as abundant in the environment as nutrients B and C, but if plant growth requires three times the amount of A compared to B or C, further growth will first be limited by a reduction in A. Growth can also be limited by more than one nutrient simultaneously, if all required nutrients are reduced proportionately, but remain in the required ratio for growth.

The yield limitation of a crop of microorganisms by the scarcity of a critical nutrient is illustrated in Figure 2.5. In this example, the growth rate (denoted by the maximum slope of the line; C in Figure 2.4) can vary with temperature, light intensity, nutrient concentration or species used in the test, but the maximum yield attained (in the Liebig sense) depends on the abundance of the most limiting nutrient. Note that in the case of a limiting nutrient, K in Equation 2.3 is analogous to the quantity of nutrient available for assimilation and converted to biomass. This concept is the basis for the algal growth potential (AGP) test (Skulberg, 1965; Miller *et al.*, 1978). AGP tests are also sensitive to inhibitors and have been used to evaluate toxicity of wastewater and hazardous waste sites (Peterson *et al.*, 1985).

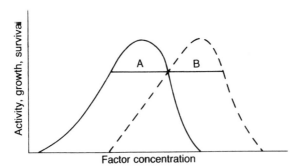

Figure 2.6. Hypotherical response of two populations to a required factor

2.4.3   Tolerance

Shelford's law of tolerance includes the concept of too much of a resource, as well as too little (Odum, 1959). The growth and survival of a population can be controlled by qualitative or quantitative deficiency or excess of any one of several environmental factors that may approach the limits of tolerance of that population. Aquatic environmental variables may include all or one of the following: toxic inhibitors, quality or quantity of food resources, micro- and macro-inorganic nutrients, temperature, dissolved solids (salinity), sediment, light, current velocity, dissolved oxygen, hydrogen ion (pH), etc.

2.4.4   Adaptation

Populations can be considered to have 'built-in', or genetic tolerances for environmental variables. Such heavy metals as Cu and Zn, which are usually thought of as only being toxic, are also required at low concentrations for metabolism. As such, they are referred to as micronutrients. However, at some higher concentration (depending on the population), they will become toxic.

This variation in genetically based tolerance is simply illustrated in Figure 2.6. An organism whose optimum for activity, growth and survival is denoted by the solid line would probably be lost from the system and replaced by a species whose optimum is denoted by the broken line if the factor increased in concentration range from A to B. The optimum range of environmental factors that describe the niche for a population include biotic as well as abiotic factors. However, physicochemical factors of water quality will be emphasized here.

Each species is adapted through genetic selection to a set of environmental conditions. Further genetic adaptation to changed environmental conditions proceeds slowly through several generations. However, within the limits of genetic tolerance, a species can adapt physiologically. For example, fish adapt to higher temperatures in summer, taking their cue from increased temperature and photoperiod. Algae adapt to low and high light by adjusting cell chlorophyll content. Overall, the species that has the greatest capacity to maintain the most growth to offset respiratory demands on energy, in the face of a changing environment beyond its optimum range, should be able to maintain the most stable population.

3
HYDROGRAPHIC CHARACTERISTICS

The effects of wastewater in the aquatic environment are largely determined by the physical characteristics of the receiving water. For example,
the initial condition for the mixing process can be modified by adding the
wastewater at variable depths or using different types of diffusers. The
final concentration of a contaminant in the receiving water is thus determined by these initial manipulations plus the physical characteristics of
the wastewater itself and the receiving water body. One type of environment, effective in dispersing the contaminant, may produce low concentrations; another environment, capable of very little dispersion may
produce high concentrations. However, the best solution for water quality
protection is not always to produce effective dispersion and low concentrations of a contaminant. Sedimentation represents a loss of pollutants
from the overlying water, and it is primarily a function of the pollutant
concentration, as well as the physical characteristics of the water body.
Thus, the movement of water and its entrained particles is fundamental to
the understanding of the processes that control water quality.

3.1 TYPES OF FLOW

Depending upon the movements of water particles, the flow may be
either laminar or turbulent. Within a laminar flow the water movement is
regular and horizontal, that is, the separate layers flow on top of each
other without extensive mixing. The mixing that does occur is on the
molecular level. Laminar flow, however, is a rare phenomenon in nature.

Water movement that is irregular, causing mixing eddies to develop on
a macroscopic scale, is called turbulent. Within a turbulent water mass
the water velocity varies considerably in both vertical and horizontal
directions. The movement is characterized by erratic motions of fluid
particles and considerable lateral interchange of momentum. This type of
movement is, therefore, most easily described statistically. Turbulent
motion totally dominates in a natural environment.

When flow is constant in both time and space, it is termed uniform and is characterized by parallel stream lines. Even though a uniform flow is also very rare in nature, the term is often useful for purposes of modelling pollutant transport and distribution. The opposite to a uniform flow is called non-uniform.

If the flow varies with time, it is termed unsteady; a flow that is at every point constant with respect to time is called steady. The steady-state condition is also a very useful generalization in modelling processes, as will be illustrated later with phytoplankton growth and lake phosphorus content. A laminar flow occurs where the water velocity is low and/or close to an underlying solid, smooth surface. In this case the flow is sometimes both steady and uniform. Steady and uniform conditions are also rare in nature, but a situation close to this state is often found in the near-bottom layers of lakes. The laminar, uniform flow is easily transformed into a turbulent flow with only a small change in velocity. When one layer of fluid slides over another, the friction between the layers gives rise to a shear stress that, with increasing velocity, results in turbulent flow. This state is reached faster if the underlying bottom is rough. The shear stress for laminar flow is expressed as:

$$\tau = \mu \frac{du}{dz} \tag{3.1}$$

where τ = tangential force per unit area, μ = coefficient of viscosity, and du/dz = velocity gradient perpendicular to the direction of motion.

For turbulent motion, eddy viscosity is much larger than the molecular viscosity so the shear stress is often written as:

$$\tau = \varepsilon \frac{d\bar{u}}{dz} \tag{3.2}$$

where ε = eddy viscosity.

The magnitude of physical processes in water bodies greatly depends on the movement of the water. If the water is moving fast enough, as in some rivers, it is often impossible for a (gravity) wave to travel upstream. The flow in this case is termed shooting. The velocity of propagation (c) of a gravity wave in this case is given by:

$$c = \sqrt{(gh)} \tag{3.3}$$

where g = gravity and h = depth of water.

If the flow is slower than the propagation of a gravity wave, the term tranquil (or streaming) is used. In this case a gravity wave could move upstream. The different regimes of flow can be separated according to water velocity and distance from a solid river bottom (Figure 3.1). There is a transition zone between a turbulent and a laminar flow within which the type of flow is essentially dependent on the water temperature.

Figure 3.1. Types of water flow in a wide, open channel. See text for explanation (Sundborg, 1956)

3.2 BOUNDARY LAYERS

Figure 3.2 shows schematically the flow conditions in a stream above the first few centimetres from the bottom. However, closer to the bottom there is a thin layer where flow conditions undergo rapid changes. The water in immediate contact with the bottom tends to adhere to it, and there is, therefore, a sharp velocity gradient in this zone. This layer, in which the flow is appreciably retarded by the viscosity of the water and the friction against the bottom and within which there are shear stresses in the fluid, is called the boundary layer. Periphyton and benthic invertebrates inhabit this reduced-velocity environment. The flow in the boundary layer is either laminar or turbulent. In rivers the flow is so turbulent that the boundary layer often extends throughout the entire water column. In that case friction against the bottom is very predominant and velocity gradients exist throughout the entire depth. The boundary layer could be laminar in channels with a very slow flow, or in lakes.

The initiation of turbulence in the boundary layer is primarily a function of the water velocity and the stability of the flow (see the following discussion of density effects). The importance of the boundary layer and the effect of velocity and shear stress on nutrient availability to periphyton will be discussed later (Chapter 11).

In natural water courses, the boundary layer may be upset by the character of the bottom; for example, if the bottom is convex (e.g.

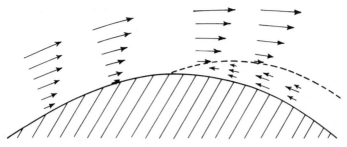

Figure 3.2. Boundary layer separation. See text for explanation

surface of a boulder), the flow on the downstream side of the convex surface tends to create a reverse velocity component close to the bottom and causes a boundary layer separation (Figure 3.2). The boundary layer separation, as well as other effects of bottom irregularities and shear stresses between adjacent water layers that exist with normal velocity patterns, suggest that nearly all natural water movements are turbulent.

The interaction between the bottom and the overlying water is very important from the standpoint of water quality. The processes of erosion, deposition and resuspension of sediment are dependent upon the variations in the state of turbulence, the viscous shear and the form resistance of the bottom.

The velocity patterns and the state of flow with regard to the processes of deposition and resuspension are just as important as the dispersion and mixing of a pollutant entering a receiving water. For example, within a slow, near-laminar running stream a pollutant can accompany the stream for several kilometres without mixing on a macro scale; however, within a fast-moving turbulent stream the mixing may be practically complete in as little as 10 metres or so downstream, depending on stream size, manner of pollutant introduction, and so on.

3.3 PHYSICAL FACTORS CONTROLLING WASTEWATER DISTRIBUTION

There are several very important factors that control the responses of a lake to a waste input, e.g. size, shape, depth, residence time/ flushing rate, temperature conditions (degree of stratification), hydro-meteorological conditions and the hydrological regime. Any or all of these factors may influence the effects of a certain waste discharge in different ways and some of these effects will be discussed in detail below.

Water body size is important in defining the type of phenomena that will occur, e.g. a large enough lake will be dominated by the Coriolis force and thus circulate in a particular manner. A very deep lake, compared to a shallow lake, will show much less influence from the bottom, particularly with regard to short-term changes in water quality, i.e. there is none or little resuspension of particulate matter. A deeper lake, furthermore, generally has a long residence time or a low flushing rate, which is probably the most important single factor in governing physical processes in lakes. Variation in density with depth as a function of temperature, is also an important factor that can limit the water volumes involved in mixing and dispersion processes.

Meteorological, hydrological and hydro-meteorological factors strongly modify and govern the physical behaviour of water bodies and sometimes completely screen a positive or negative human impact. A period of sudden precipitation in spring (within a temperate climate) may cause a substantial nutrient load, which can completely hide and destroy the long-term effects of a wastewater treatment programme. Climatological effects (e.g. amount of solar energy) are also important for the overall functioning of lakes.

The importance of density effects in general will be discussed below before focusing on the effect of temperature on density. Finally, some physical properties of large lakes will be discussed, which are of importance in understanding the behaviour of pollutants in such lakes.

3.4 DENSITY EFFECTS

The previous discussion assumes a water volume of equal density. However, conditions could be markedly changed by density differences within the water mass. Vertical density differences are very common features of both lakes and estuaries and sometimes of rivers, and therefore the causes and effects of variations in density must be understood.

Density gradients in a water mass are caused by differences in water temperature and in dissolved or suspended material in the water. In freshwater lakes, temperature is the dominating cause of density stratification. Freshwater reaches its maximum density at about 4°C, while density decreases above and below that temperature (Figure 3.3). In estuaries, dissolved substances (salinity) are often the principal cause for stratification. In the rivers and mainstream impounds, suspended material may influence stratification, although not as predictably as temperature and salinity.

Density usually increases towards the bottom of a water column, sometimes gradually over the entire vertical distance and sometimes rapidly within a certain depth interval. The relative effect of a given Δt/depth on stratification, as temperature increases will be discussed in relation to phytoplankton, light and mixing in Chapter 6. When a strong vertical density gradient is present, which is common during summer in lakes, the turbulent velocity fluctuations are subdued or prevented and the turbulent exchange of water between contiguous layers becomes

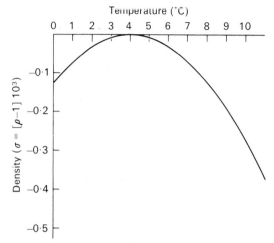

Figure 3.3. The relation between temperature and density in lake water

[FRESH]

weaker or ceases. This occurs because the energy needed to move lighter layers downward and heavier layers upward increases with the density differences. Not only will the turbulent exchange be affected but also the vertical velocity distribution, the shear stress, and the transition from laminar to turbulent motion will be suppressed. Furthermore, if the stratification in a lake is sufficiently strong, it may even be useful to consider the lake as two bodies of water separated by the layer of greatest change in density with depth. This vertical distribution of water masses is quite similar to that of air masses above the earth.

The type of flow is often determined by the density gradient. A dimensionless parameter that describes the stability of the flow can be derived if such a parameter is called the Richardson number and is calculated as follows:

$$R_i = -\frac{g}{\rho}\frac{d\rho}{dz} \Big/ \left(\frac{d\bar{u}}{dz}\right)^2 \tag{3.4}$$

where ρ = density of the liquid, $\bar{u}$ = mean water velocity, and g = acceleration resulting from gravity. When the liquid is homogeneous R_i = 0, the stratification may be called neutral. If $R_i > 0$, the stratification is considered stable; for $R_i < 0$, it is unstable. One can see that stratification increases as the density gradient increases.

There are some contradictory opinions on the minimum value of R_i for a transition from laminar to turbulent flow. However, far from boundary surfaces, a value of 0.25 (Prandtl, 1952) is usually accepted. Much lower values have been observed near a solid boundary.

3.5   DENSITY CURRENTS

Density stratification can greatly influence the distribution of an incoming mass of water that could contain pollutants. The incoming volume of water may flow over, under, or through the main water mass as a result of gravity and the density characteristics of the respective water masses. Such a phenomenon is referred to as a *density current*. This phenomenon was described early in the classic work by Forel (1895). Although not often easily observed directly, density currents are nevertheless frequently present in lakes. For example, streams entering lakes often have different density characteristics than the lakes themselves. Density differences between a lake or reservoir and its inflowing water may be enhanced by man to create a density current.

The persistence of a density current in a lake is mainly dependent upon the state of flow which is indicated by the Richardson number. The density current will gradually decrease and disappear, but more rapid disappearance of the current will result with mixing across the interface if the Richardson value is low or a disturbance develops, such as a wavelike phenomenon similar to surface waves. If this wave travels with increasing amplitude, it finally breaks and extensive mixing occurs (Knapp, 1943). Small-scale mixing occurs when the density current leaves the water surface and continues as an undercurrent or when an obstacle on the bottom disturbs the current. If the high density of a current is mainly a function of suspended matter (usually associated with high turbulence), the phenomenon is called a turbidity current. Turbidity currents are most common in estuarine environments and in reservoirs on large rivers. They may also occur as artificially produced currents, for example, as a result of dredging.

In an estuary, a heavy suspended load causes a very rapid current, which is characteristic for most turbidity currents. These rapid currents are likely to be erosive, as evidenced by the large canyons formed in nearshore waters of the ocean that receive the world's largest river estuaries. When velocity of the inflow decreases, due to the topography of the lake or ocean bottom, the turbulence decreases and the suspended material settles out.

Rapidly moving turbidity currents are rare in lakes, but the slow-moving density displacements of volumes of water are common features.

3.6   THERMAL PROPERTIES OF LAKES

In addition to causing density stratification in freshwater lakes, temperature also affects the chemical and biological processes. Factors determining lake temperature are primarily latitude, altitude and continental location. Within low-lying equatorial areas (between about 15°S and 15°N latitude), lakes are warm (27–30°C) all the year with essentially no yearly, but sometimes a daily, variation. Seasonal changes may occur in the subtropical regions (about 15–30°S and 15–30°N), however, especially in mountainous areas. Thornton (1987a) cites lakes throughout the tropics (30°S to 30°N) that undergo seasonal stratification patterns although less pronounced than in temperate regions. Near polar areas, lake temperature is low all year but usually with greater yearly than daily variations. Some lakes are frozen throughout the year.

[margin note: Typical T's °C]

Within the midlatitudes, lake temperature is strongly variable during most of the year, and variable air temperature also creates considerable vertical gradients in lakes. The sun is the principal dominating source of energy causing the seasonal variation. Incoming radiation, assuming a sufficiently large angle of incidence, is usually absorbed rather quickly within the upper water layers. The visible-range wavelengths are almost totally absorbed within the upper 2 m, and of the total radiation energy (about twice the visible portion) entering the lake, about 50% is absorbed within the first 2 m. The attenuation of visible light in relation to phytoplankton photosynthesis is discussed in detail in Chapter 6.

Wind is the distributing force for the heated water within the basin. Variations in other factors – evaporation, inorganic material in the water, or loss of long-wave radiation from the water surface at night – also contribute to the distribution; but only to a small extent compared to the wind. The effect of wind and solar heating in determining the depth of mixing will be discussed later. Water's high specific heat is another important thermal property in lakes. This property of water allows for a large storage of energy and results in a much slower rate of heating and cooling of the lake than the surrounding land surface. When water cools, its density is increased. Most lakes are shallow and, to at least 100 m depth, the pressure effect on density caused by water depth can be ignored.

[margin note: wind distributes heat & diffusion ??]

3.7 ORIGIN OF LAKES

The origin of a lake is an important key to understanding its water quality and response to different pollutant inputs. The origin is also an important clue to the lake's form and its sediment characteristics. These factors will in turn determine the lake's detention time and affect its nutrient content. A detailed discussion of the classification of lakes according to their origin is found in classical textbooks (e.g. Hutchinson, 1957), but will only be discussed briefly here.

Lakes caused by tectonic movements (e.g. Lake Tahoe, Baikal, Tanganyika and Nyassa) and vulcanism (e.g. Lake Kivu and Crater Lake) are often deep with steep shores and frequently have a long detention time. Lakes formed by glacial activities, like most lakes in Canada,

Scandinavia, the Great Lakes area, may appear in many different shapes due to different forces involved. However, they are mostly shallow and nutrient-poor, with rocky shallow shores and swash-zones. They often have medium to long detention times due to a complex morainic landscape.

Glacial lakes in flat environments which formerly were sea bottoms, or lakes caused by fluvial action (ox-bow, levee or coastal lagoon lakes), generally have high nutrient levels, high sediment loads and are shallow.

Lakes caused by chemical solution are often very complex and sediment-poor and the opposite is applied to deflation lakes caused by action of the wind.

Reservoirs are man-made, created by the damming of rivers. They are usually nutrient- and sediment-rich and have short detention times. Natural basins have usually developed slowly and continue to change very gradually in shape and physical characteristics, whereas reservoirs fill up rapidly and may continue to undergo marked changes in their characteristics.

3.8   THERMAL CYCLE IN A TYPICAL MID-LATITUDE LAKE

Most lakes are located within the mid-latitudes (glacially sculptured or remoulded) and consequently undergo strong temperature variations during the year. Thus, discussion about temperature in lakes should begin with those in the mid-latitudes.

Consider a lake with ice cover in winter. When sunlight begins to have an effect in early spring, the ice cover melts. This process is very often accelerated in the final phases by wind, rain and darkening of the ice. The actual breaking up of the ice cover is often very fast, and usually occurs in a few hours, most often in the month of April. In the mid-latitudes, the spring flood often occurs at the same time and a period of strong vertical mixing and turbulence is initiated in lakes that have high flushing rates. When the spring flood ceases and the radiation from the sun has increased, the upper water layer begins to warm (generally in May). That process accelerates through mid-summer. Until mid-May, when the water is heated to about 6–8°C, little energy is usually required to mix the whole volume of water. The length of this circulation period is dependent upon several factors: the surface area, depth, topographic location, presence of winds, and so on. Those factors also determine the temperature the lake will reach before the vertical stratification begins. During calm and clear periods, stratification strengthens. The density gradients rapidly increase, which means that the wind is increasingly retarded in working its mixing effect through the water mass. The upper layer is mixed to a homogeneous temperature by turbulence, and below this surface layer, the vertical exchange of water is very limited, as previously discussed.

The depth of the mixed layer (thermocline, see below) occurs where the force by the wind mixing the water downward is balanced by the resistance to that mixing. Resistance to mixing is due to buoyancy of the water, caused by heating and consequent density reduction.

The result of this process is the creation of three relatively well-defined separate layers in the water mass developed in the summer: an upper homogeneous layer (the epilimnion), a lower fairly homogeneous cold layer (the hypolimnion), and an intermediate layer characterized by a strong temperature gradient. The definition and description of the intermediate layer has always been a matter of controversy. The most commonly used terms are thermocline and metalimnion. However, this layer has little physical or chemical uniformity. The use of the term metalimnion, which implies a uniformity similar to the epi- and hypolimnions, may be misleading. Therefore, throughout this chapter the term thermocline is used.

A lake with a typical summer stratification has a vertical distribution of temperature as shown in Figure 3.4. The late-summer lowering of the thermocline is essentially an effect of wind and reduced solar heating. Although epilimnetic water is gradually heated until the latter part of July, the thermocline depth increases throughout the summer until it finally breaks up completely in the autumn (Figure 3.5). The initiation of autumn circulation (complete mixing) at that time is a direct function of the frequency and force of winds together with a decrease in temperature. The autumn circulation usually starts in September. The initiation, development, and termination of summer stratification is often very different from year to year (Figure 3.5).

Figure 3.5 also shows that the temperature of the hypolimnion gradually increases through the summer. The reason for this increase is still not satisfactorily explained. Solar radiation, turbulent transfer of heat and extensive mixing during upwelling events, as well as biological decomposition, certainly contribute heat. The concept of upwelling is best described as a situation in which the hypolimnetic water reaches the water surface because of strong wind effects that disturb the thermocline (discussed below). Hutchinson (1957) proposed that density currents are a main source of energy transferred to the hypolimnion (later adapted by Wetzel, 1975). This process has never clearly been established and extensive sampling in Swedish lakes with the aid of very sensitive instrumentation has so far not provided clear evidence for the importance of this process. There is, however, some indication that vertical entrainment occurs in summer and may be responsible for transporting phosphorus vertically from anoxic hypolimnia into the photic zone. This will be discussed in Chapter 4.

As solar radiation decreases in late summer, income of energy to the lake, and consequently its temperature, decreases. After the circulation

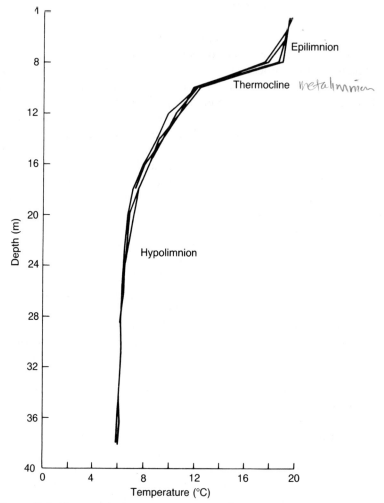

Figure 3.4. Five vertical temperature soundings along transects (Figure 3.17) from a typical dimictic lake, Lake Ekoln, Sweden, June 1969 (Kvarnäs and Lindell, 1970)

in September, the water mass usually has a temperature of about 10°C. The temperature decreases further during total circulation reaching 4°C or lower. As the density decreases at temperatures below 4°C and during a calm, cold night (usually in December) an ice cover will typically occur. The ice cover will continue to increase during winter, with a resulting vertical distribution of temperature in late March similar to that shown in Figure 3.6. The water temperature is close to 0°C in the surface layer and close to that of maximum density (4°C) in the bottom layers. The increase

Figure 3.5. Yearly distribution of temperature as isotherms, Lake Ekoln, Sweden. Dashed line is approximated (Kvarnäs and Lindell, 1970)

at depth in this case was caused by heating from the sediments plus inflowing water. The apparent slight instability indicated by the temperature difference between 8 and 22 m (i.e. denser water overlying lighter water) was actually compensated by a salinity gradient, which resulted in a stable water mass.

3.9 CLASSIFICATION OF LAKES BY TEMPERATURE REGIME

Much of the nomenclature of lakes is based on their thermal characteristics, which may be peculiar to a particular region and greatly affect the lake's physical, chemical and biological processes as shown in Chapters 4 and 6. Therefore, the different lake types are graphically described in Figures 3.7 and 3.8.

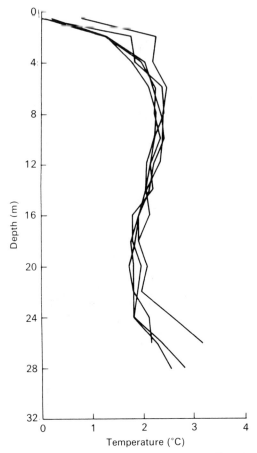

Figure 3.6. Five vertical temperature soundings from a dimictic lake, Lake Ekoln, Sweden, March 1970 (Kvarnäs and Lindell, 1970)

The mid-latitude lake discussed above is called dimictic, indicating the two circulation periods per year. Dimictic lakes also exist at high elevations within subtropical regions. The requirement of two circulation periods per year is fulfilled in the mid-latitudes as long as the lake is continentally located. Within maritime areas in the mid-latitudes, the winter temperature does not reach a low enough level for the lake to freeze. Thus, only one circulation period occurs, resulting in a warm monomictic lake. The temperature during the winter is usually 4–7°C in a monomictic lake. Within cold regions, a corresponding lake type exists, with its one circulation period occurring during the summer when the temperature reaches 4°C. That lake is referred to as cold monomictic. If the climate is so extreme that the lake is always frozen, it

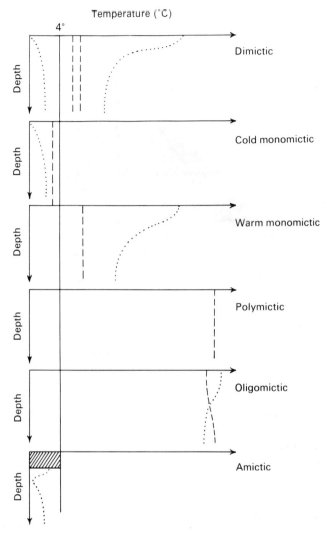

Figure 3.7. Lake types classified according to temperature criteria. Dotted lines indicate winter or summer stratification. Dashed lines indicate circulation temperatures. Shaded area indicates ice cover

is called amictic. Its opposite, a polymictic lake, is found within tropical areas where the yearly variation is insignificant, although short-term variations create weak stratification and periods of circulation in between. Polymixis can also occur in shallow-temperate lakes in summer and that feature will be shown later to greatly influence nutrient cycling (Chapter 4). Cold polymictic lakes also exist within high-altitude tropical areas with

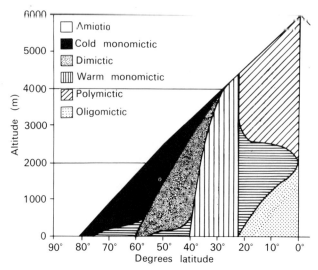

Figure 3.8. Classification of lake types, based on latitude and altitude. Horizontal lines in the figure are transitional zones (Wetzel, 1975)

strong winds. Within the tropics, some lakes, called oligomictic, circulate only sporadically.

In the above types of lakes, temperature variation is the main cause of stratification and, when circulation occurs, the whole vertical water column is included. A complete vertical circulation may in certain cases be suppressed by a permanent or a quasi-permanent chemical stratification (e.g. a salinity gradient) in a portion of the water mass. In those rare cases, lakes are termed meromictic and a very atypical temperature gradient occurs.

3.10   WATER MOVEMENTS

The water movement of a river is caused by gravity. The water surface usually has a substantial inclination (gradient). Lakes normally have no gradient, so that factor is unimportant. The gradient is considered the cause of a lake current in only elongated, flow-through type lakes with short detention times.

The wind causes waves on the surface of the lake due to the friction against the water and the currents, which rapidly decrease with depth. An asymmetrical distribution is frequently created when water is transported through the lake. A continuous water transport in one direction at the surface is compensated by a transport in the opposite direction, which often occurs in deeper parts of the lake. This transport of water is especially interesting during the summer when the thermocline is inclined and the waters of the epilimnion and hypolimnion have essentially separate circulations because of the small amounts of friction between the two layers.

If the wind suddenly ceases, the thermocline will oscillate back and forth like a seesaw until its resting position in the horizontal plane is achieved. The rhythmic movement of the thermocline is called an internal seiche, and the corresponding movement on the water surface itself is referred to as a surface seiche. The oscillations of the surface seiches are of considerably less magnitude.

3.11 SURFACE WAVES

Surface waves are ecologically interesting primarily in near-shore areas. In the deep parts of lakes the waves are non-transporting and in general each water particle describes an orbital movement, the circle (wave amplitude) diameter of which is halved for each change in depth of λ/g, where λ = wavelength. The wave height is usually about 1/20 of the wavelength. The wave height is a function of the distance along the water surface in the direction of the wind (= fetch) and a function of wind speed and duration. In the nearshore zone where the wave breaks, the orbital movement is transferred to a horizontal water movement. The wave generates an erosive character that causes the bottom substrate to be unstable and highly turbid. Large waves in oceanic coastal zones or in large lakes influence the bottom down to great depths; however, the velocities resulting from the waves usually die out at a depth of approximately $\lambda/2$ below the water surface. In small lakes the wave action is limited, but even the leeward sides of large lakes are often shallow and heavily vegetated, whereas the windward sides are deeper and biologically bare, which is a result of the stronger wave action on the windward side.

3.12 WIND-DRIVEN CURRENTS

The direction and velocity of wind-driven currents are determined by the classical Ekman spiral. To be able to apply this law, however, one must assume a free water surface (no bottom and lateral friction). The Ekman current implies a surface current directed 45° to the right of the wind and a mean water transport 90° to the right of the wind because of the Coriolis force (Figure 3.9). In lakes, both bottom and lateral friction exist to a variable degree, and there is no generally accepted method of computing the currents from wind data in small lakes. Computations have been attempted based on the equations of continuity and motion, but they vary considerably in their final form.

For large lakes, the three-dimensional models for the Great Lakes by

Figure 3.9. The distribution of water movements in a wind-driven ocean (the Ekman current) (Hellström, 1941)

Simons (1973), based on classical oceanographic models but also incorporating the topography of the lake, have been very successful. Rather simple relations have also been proposed, as for Lake Ladoga (Witting, 1909), where a current velocity was calculated according to

$$v = 0.48/\sqrt{w} \tag{3.5}$$

where w = wind velocity. Witting observed a direction of the current slightly to the right of the wind. For some small Swedish lakes the following formula has been used:

$$v = chw \tag{3.6}$$

where $c = 2 \times 10^{-5}$ (for no stratification), h = mean depth, and $c = 1.5 \times 10^{-5}$ (for h = mean depth of the thermocline). As a rough estimate of surface currents in lakes, it is possible to consider them parallel to the wind and in the order of 2% of the wind velocity. In the deeper parts of lakes the water movements are slow, generally only a few centimetres per second. This presents severe problems in recording water movements because the velocities are lower than the activating velocity of most current meters.

3.13 SEICHES

When the wind blows, it causes a piling up of the water on the windward side of the lake. The inclination of the water surface resulting from the wind is usually in the order of a few decimetres in large lakes and a few centimetres in small lakes. Morphometric conditions in lakes are, however, strongly modifying and, consequently, general rules are difficult to postulate. For a rectangular-shaped lake, Hellström (1941) suggested the inclination to be given by

$$s = \frac{3.2 \times 10^{-6}}{g\bar{z}} w^2 l \qquad (3.7)$$

where s is expressed as the total difference in height between the windward side and the leeside of the lake, g = acceleration resulting from gravity, $\bar{z}$ = mean depth, w = wind velocity and l = length of the basin. Thus, seiches should be greater in longer, shallower lakes that are exposed to the wind.

Oscillation of the water surface after the wind has ceased can be determined by the classical formula of Chrystal (1904):

$$t_n = \frac{l}{n} \frac{2l}{\sqrt{g\bar{z}}} \qquad (3.8)$$

where the ratio of the periodicities for the uninodal, binodal and trinodal seiches is $t_1 : t_2 : t_3 = 100 : 50 : 33$, with respective nodes at $x = 0$ (uninodal), $x = \pm 0.25\,l$ (binodal), and $x = \pm 0.33\,l$ (trinodal); l = length of the basin, g = acceleration resulting from gravity, $\bar{z}$ = mean depth, and n = number of nodes.

This general formula is derived from a rectangular basin with constant depth where the seiche can be considered a shallow water wave, but it has been shown to be applicable to most types of lakes. Surface seiches generally have a fairly small impact on the variability of water-quality characteristics. The internal seiche could, however, cause strong fluctuations in water quality both in time and space as the thermocline is separating two water masses that could be of very different quality. The internal seiche has an amplitude of a magnitude 100–1000 times larger than the surface seiche and is primarily determined by the density differences between the epilimnion and hypolimnion. The periodic motion of the internal seiche can be determined (Mortimer, 1952) by:

$$S_n = \frac{2l}{\left[g(\rho_h - \rho_c)/\left(\frac{\rho_h}{h_h} + \frac{\rho_c}{h_c} \right) \right]^{0.5}} \qquad (3.9)$$

where h_c = thickness of epilimnion, h_h = thickness of hypolimnion, ρ = density and l = length of the basin.

The period of the internal seiche is often in the order of hours, whereas the surface seiche is calculated in minutes. Slow water movements in lakes, as mentioned previously, are difficult to record physically, but the slow velocities associated with the internal seiche can be roughly estimated (Mortimer, 1952) as:

$$v_c = \frac{sl^2}{2Th_c} \tag{3.10}$$

$$v_h = \frac{sl^2}{2TH_h} \tag{3.11}$$

where s = maximum slope of the thermocline, l = length of the basin, t = period, and h_c, h_h = depth of epilimnion and hypolimnion respectively.

3.14 TIDAL EFFECTS

The tide is caused mainly by the gravitational pull of the moon. The effect of the sun's gravitation can be observed in connection with extreme high or low tide. However, tides probably have little effect on the water quality in lakes. Very little knowledge has been generated so far in this field, but it is believed that tides may cause waves of the order of a few centimetres in large lakes.

In oceanic coastal zones the tides are highly variable because of the general distribution of the land masses and the local morphometry of the coastal zone. In some areas the tides could be close to zero, whereas in other places, especially in funnel-shaped estuaries, the tides may be of the order of several metres. The tides have two maxima and two minima per day for a lunar day that is 24 h 50 min. An estuary with a large tidal range, therefore, has an exceptionally good transport mechanism as flushing occurs twice a day. See Chapter 6 for the effect of tidal prism thickness on phytoplankton.

3.15 WATER MOVEMENTS IN LARGE LAKES

The effect of the earth's rotation may be detectable in the inflow to small lakes, but almost never in the internal water movements. In large lakes and coastal waters the Coriolis force effect is visible with all water movements and may totally dominate the current patterns (Figure 3.10). The meaning of 'large' must be defined in relation to the latitude, and the Coriolis parameter, f, is defined as:

$$f = 2\Omega \sin \phi \tag{3.12}$$

where Ω = angular velocity of the earth and ϕ = latitude.

The Coriolis force is directed perpendicular to the right of the motion in the Northern Hemisphere and to the left in the Southern Hemisphere. The

Figure 3.10. Distribution of Glomma River water in the Skagerrak Sea

Figure 3.11. Geostrophic circulation of a large lake (Lake Vänern, Sweden). Along a cross section (a) of the lake the temperature pattern (b) is dome-shaped (c) The resulting dynamic pattern of height anomalies in centimetres (Lindell, 1975)

effect of the Coriolis force increases with latitude and is zero at the equator.

By using the definition of the Coriolis parameter, it is possible to calculate the radius of the inertia circle:

$$r = \frac{u}{f} \tag{3.13}$$

where u = water velocity. For a typical current velocity of 10 cm s^{-1} at 45° latitude, for example, $r \approx 1$ km. If the width of the lake is five times the inertia circle radius, the effects of the earth's rotation can be visible, but if the width is 20 times the radius or more, the Coriolis force will dominate the currents.

A very typical feature of large lakes is rapid water movement in nearshore areas known as 'coastal jets'. When a wind initiates Ekman drift in the centre of a lake, the Coriolis force balances the wind stress. In a narrow coastal bank, however, the wind stress accelerates the water

Figure 3.12. The temperature distribution in Lake Ontario (surface and longitudinal section) during 7–10 June 1965, showing the position of a thermal bar (Rodgers, 1966)

alongshore. This mechanism transports water to the windward side of the lake, and when the wind ceases, a return flow develops that is again confined to the nearshore bands. It is now an accepted fact that the coastal jets are sometimes connected to a type of deformed, so-called Kelvin wave (a seiche dominated by the Coriolis force), resulting in a very asymmetrical wave that progresses counterclockwise with a very rapid countercurrent (clockwise) on its rear side.

A persistent geostrophic counterclockwise circulation dominates during summer on the lakeward side of coastal jets (Figure 3.11). This circulation pattern is caused by a combination of wind, Coriolis force and temperature contrasts and may be so strong that the common and substantial vertical shifting of the thermocline resulting from wind never overcomes the effect of the circulation.

3.16   THERMAL BAR

A certain type of circulation common in large dimictic lakes during spring (and theoretically also in the autumn), caused by the uneven heating of the water, is called a thermal bar. If the temperature of the whole lake volume is well below 4°C and the heating during the spring is rapid, the rate of heating will be most pronounced in the nearshore shallow areas. The deep pelagic areas of the lake remain below or near 4°C. In the transitional zone between the high temperatures of the coastal areas and the cold open water there is a zone with a strong horizontal temperature gradient (Figure 3.12).

This zone moves slowly from the nearshore areas toward the open waters. The duration of this process varies from days to months, depending on the differences in heating and size of the lake. The typical thermal bar phenomenon is transformed to the temperature pattern typical for geostrophic water movements. The currents associated with these phenomena are characterized by a predominant alongshore counterclockwise circulation in the nearshore stratified zone. Modelling of the thermal bar has shown downward movements on the lakeside of the bar and upwelling in the warm water. Its development from early spring to the summer circulation is not yet satisfactorily explained.

The thermal bar has relevance for the dispersion of wastewater because the nearshore warm waters are trapped close to the coastline. The resulting high nutrient concentrations could in turn trigger algal blooms. Such a nearshore effect has been suggested in the Great Lakes (Beeton and Edmondson, 1972).

3.17 MEASUREMENT OF WATER MOVEMENTS

Water velocities in natural bodies vary from several $m s^{-1}$ in rapidly running rivers through several $dm s^{-1}$ in coastal areas to commonly less than $30 cm s^{-1}$ in lakes. To record horizontal moderately high velocities in water is fairly straightforward and could be performed by current meters, drifters or dye. However, water movements can be complex in nature and especially slow in lakes, even slower than the starting velocity of common current meters. In that case special vanes have been used, where the inclination of the vane compared to the vertical direction is recorded.

The most commonly used instrumentation is the current meter which usually includes an impeller, placed in the water in certain transects at defined depths and the water movement is then integrated over several transects.

Alternatives to current meters are drifters or drogues which could be placed at different depths and often have a surface float. The float can be followed by optical instruments from land, indicating direction and distance, or simply followed by radar.

Dyes have frequently been used for studies of diffusion in lakes and slowly moving rivers. Most practical from the measuring point of view is the use of fluorescent and radioactive tracers, but those may cause environmental hazards and permission may be complicated. A new technique of using remote sensing has been widely applied in large and medium scale systems, e.g. to study coastal circulation, or for detecting and defining natural inputs or waste discharges (suspended matter or temperature) in lakes or coastal waters. By using near infrared bands from satellites or infrared films from aircraft, the movements of algae-laden water have been estimated. An example of the use of satellite imagery in tracing pulp mill discharges into the Baltic Sea is shown in Figure 3.13. It should be noted that the observed distribution patterns of a satellite image do not directly indicate the present water movement, but is a result of many earlier long- or short-term variations (cf. Spigel and Imberger, 1987).

3.18 DISPERSAL PROCESSES

In principle it is possible to separate the dispersion processes among small-scale and large-scale phenomena. The small-scale diffusion processes appear in the neighbourhood of the sources of pollutants or water courses, whereas the large-scale processes are closely associated with the pelagic areas of lakes. The parameters that influence the dispersion processes are water velocity and eddy diffusivity in the horizontal (K_y) and vertical (K_z) direction. K_y and K_z are dependent upon the character of the turbulence, which in turn is dependent upon water velocity and stability, as has been mentioned earlier. They are also dependent on the scale of the phenomenon.

3.18.1 Horizontal diffusion

The horizontal eddy diffusion along the horizontal plane (x) may be defined as

$$K_y = ku\frac{dw^2}{dx} \tag{3.14}$$

where u = water velocity, in the x direction; w = width of the plume, in the y direction; and k is a constant. Csanady (1970) has proposed a value of 0.03 for the constant in the equation.

Horizontal diffusion is of interest in the dispersal of a wastewater plume into a lake. The diffusivity increases rapidly with size of the plume. The increase in the width of the plume exceeds that of a simple linear increase. Although resulting from turbulence, the dispersion process at sufficient distances from the source becomes more similar to molecular diffusion. This development is the consequence of the growing plume, causing an increase in the size of turbulent eddies, which become increasingly important. The size of the turbulent eddies does have an upper limit, which is why this type of diffusion eventually becomes similar to that of molecular diffusion. Although the processes are non-linear with complicated interrelationships, a linear diffusion model is often sufficiently good for most purposes. A typical development of a plume from a dye injection will be related to distance as shown in Figure 3.14. Csanady (1970) defines plume diameter as $\approx$ 10% of the width (w) of the

Figure 3.13. The relative turbidity of the water along the coastline of Gävle, 6 October 1981 from Landsat-2 MSS

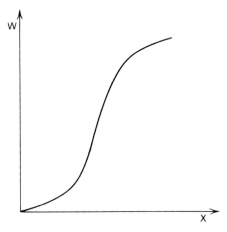

Figure 3.14. Horizontal diffusion. Changes in the width of a plume (W) in the direction of the length axis (X)

cross section at the surface. The horizontal diffusion coefficient in lakes is usually in the order of $100-1000\,\text{cm}^2\,\text{s}^{-1}$.

3.18.2   Vertical diffusion

Vertical diffusion has certain similarities with horizontal diffusion, but the magnitude of vertical diffusion can be affected by many important variables, such as density, water velocity and current direction, which can change rapidly with depth. Vertical diffusion is especially different in the presence of a thermocline. As noted earlier in discussing density, it is sometimes useful to consider a lake as two lakes with the epilimnion as the 'upper lake'. The lower limit of the epilimnion acts as a floor in the process of diffusion. The two main types of vertical diffusion may, therefore, be generalized according to Figure 3.15. Typical vertical diffusion values are $\approx 5\,\text{cm}^2\,\text{s}^{-1}$, but initially much higher values (up to $30\,\text{cm}^2\,\text{s}^{-1}$) are common (Csanady, 1970). Compared to horizontal diffusion, vertical diffusion is much smaller in magnitude because of the differences in the horizontal and vertical limitations to turbulence.

3.18.3   Large-scale phenomena

The most important large-scale diffusion process for most sizes of lakes is the meandering type that follows a large inflow. Meander diffusion is similar in appearance to the meandering of rivers; that is, the movement is sinusoidal in plan view with a helicoidal component. There are no general rules for predicting the occurrence and frequency of meandering in lakes.

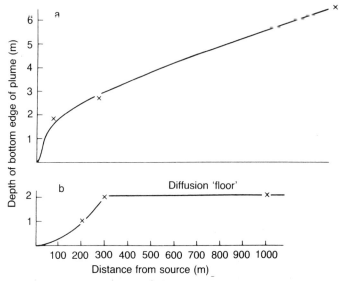

Figure 3.15. Vertical diffusion. Typical patterns of diffusion in Lake Huron in the absence (a) and presence (b) of a thermocline (from Csanady, 1970)

In large lakes where thermal bars and coastal jets exist, the dispersion of pollutants often leads to concentrated levels in the coastal zones, thus restricting horizontal diffusion. It should be noted, however, that the jets and thermal bars vary substantially with time and the diffusivities, therefore, could be very different from one day to the next.

3.18.4   Typical dispersion patterns of plumes

The following types of dispersion may occur.

1. Complete mixing;
2. The pollutant usually disperses horizontally;
   (a) At the surface;
   (b) At an intermediate level;
   (c) Along the bottom;
3. Combinations of 1 and 2.

Complete mixing is almost entirely a function of the velocity of the receiving water because diffusion coefficients depend on velocity. A rapidly flowing stream always has a high degree of turbulence and, consequently, an input of wastewater from a point or non-point source can be homogeneously distributed within the whole river width usually after only a few hundred metres. However, the distance will depend on how the pollutant enters the stream and the stream size. It is easy to be misled by this fact when considering the water quality of a river system.

To determine the water quality in a downstream lake, in addition to observing pollutant effects in the stream near the input, the total transport per unit time of the pollutant should be calculated. To calculate the transport of a substance requires knowledge of the quantity of water as well as the average concentration. In a strongly turbulent stream there is little interest in whether the pollutant is dispersed as a concentrated jet or by the diffusion process. The turbulence of the stream will dominate the dispersion of material immediately and the resulting cross-sectional concentration may become homogeneous a few hundred metres downstream in a moderate-width stream. Consequently, adverse effects are noted in turbulent waters downstream from the mixing zone only at relatively higher pollutant-loading rates, whereas in a non-turbulent stream adverse effects (or 'hot spots') may occur.

If the densities of the receiving stream and added wastewater are different and the stream velocity is low, the pollutant will mix into the stream very slowly. High concentrations of the discharged pollutant often remain unmixed very far downstream, sometimes several kilometres, if the degree of turbulence is low, the density gradients are strong and the stream width is substantial. Considerable variability exists in the mixing processes because of density gradients and bottom topography. Another form of dispersion occurs, however, when the usually denser wastewater progresses downstream close to the bottom. This is the result of high dissolved solids content causing the incoming wastewater to be extremely dense. In that case the temperature has little effect on density and, hence, on dispersion.

When the density differences are considerably less, the situation is equally critical. The incoming water mass tends to hold to the entering side of the river for a long distance, still assuming that fairly low river water velocities exist. This process can frequently be observed, for instance, at the confluence of two tributaries carrying very different loads of suspended sediment.

In streams with this type of incomplete mixing of wastewater, the ecological effects may be severe on one side of the river and non-existent on the other. The monitoring of stream water quality seldom includes samples from the entire river width. An incomplete mixing could, therefore, easily lead to non-representative water-quality sampling and local samples would not provide spatially averaged concentrations, which would occur if the pollutant were uniformly distributed over the cross section.

A wide variety of different patterns of dispersion that are mainly a function of water velocity and, to some extent, density gradients, also

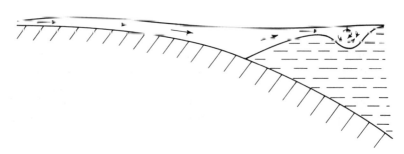

Figure 3.16. Surface inflow in a lake

occur. To accurately monitor stream water quality usually requires a careful study of the patterns of turbulence during different conditions of stream flow. This can be accomplished with the aid of tracer studies.

Although dispersion and mixing in estuaries and lakes are more re-stricted by large differences in density, the mixing processes, on the other hand, are accelerated by wind and currents in the open water. The diffusion process in a lake or an estuary is fairly simple to define from a number of theoretical linear models, but the complete modelling and forecasting of the dispersion of a wastewater plume is very complicated, not only because of the complexity of the model itself, but also because of the stochastic nature of the varying meteorological and hydrological con-ditions. However, a few descriptive cases of mixing will be discussed.

An estuary[1] in a lake, formed at the mouth of an entering stream, is similar to a point-source input to a stream. The main types of mixing are complete mixing and mixing that is density-dependent at the surface, bottom, or intermediate inflow level.

The basic condition governing complete mixing is a non-stratified water mass (spring and autumn for a dimictic lake). This is most often an adequate assumption because the density of the stream seldom deviates much from that of the lake. Only very dense incoming water, close to the density maximum of 4°C, or a flow of dense suspended or dissolved material, will dive as a density current toward the bottom of the lake and remain there for some time. The best opportunity to obtain a low con-centration of a pollutant in a lake occurs during the normal spring–autumn mixing periods.

During summer stratification, essentially three types of dispersion and mixing occur. The inflow of water stays at the surface of the lake if the density of the incoming water is lower than that of the lake water. That

---

[1] Estuary in this freshwater case is defined as a drowned river mouth, rather than where seawater is measurably diluted with freshwater

process is most simply illustrated in Figure 3.16. Early summer is usually the time when this condition first occurs. During that part of the year, the incoming water is warmer than the lake water (unless the river water is meltwater) because the land mass warm up more rapidly than the water and the volume of incoming water is usually small relative to the lake volume. This phase often starts at the end of the spring flood in dimictic lakes and ends as the incoming water begins to dive later in the summer. In the beginning of this phase, the initial mixing is quite strong because of the high velocity of the incoming water. The stream water gradually becomes more concentrated as evaporation processes become significant. Stream velocities gradually decrease and the inflowing water dives to a depth of equal density, usually the upper level of the thermocline. This inflow at the top of the thermocline is generally very distinct with little initial mixing due to low velocities. Dispersion of the inflowing water is mostly in the horizontal and longitudinal directions. In some lakes it is possible to follow an inflowing stream for a long distance. A heavily polluted river is entering the lake close to section 01 (Figure 3.17). The high specific conductance of the river could be traced on top of the thermocline all through the lake to the outlet in the south-eastern part.

The remaining type of dispersion/mixing, in which the inflow dives to the near-bottom layers of a lake, usually exists in winter. This type is analogous to the inflow moving along the bottom in rivers as described earlier. Occasionally, when small differences in temperature exist between the lake and inflowing water, only a slight increase in dissolved substances is sufficient to cause the inflow to move to the bottom (Figure 3.18). Because river discharge is low in winter, there is little initial mixing in dimictic lakes. Ice cover also protects the lake from direct wind effects minimizing mixing. It is a common feature in dimictic lakes in winter for inflowing water to accumulate in the deeper parts of the lake and not mix with the rest of the lake water until the spring flood occurs.

In low-elevation estuaries there is often an interaction between the water masses of the sea or the lake and the incoming water. The initial mixing zone could vary greatly because of differences in tides, discharges and wind effects. Lake water, for example, may be forced far upstream from the estuary under a strong onshore wind and the initial mixing thus occurs not at the river mouth but several hundred metres upstream (Figure 3.19).

In an estuary facing the sea, low-density inflowing water flows on top of the salt water. The initial mixing zone, as well as the degree of mixing, is dependent primarily upon the ratio of tidal prism volume and volume of river inflow per tidal cycle. A salt-water estuary tends to be more highly stratified when the river inflow is large compared to the tidal prism and hence mixing associated with tidal energy is not so large. In very exposed shallow estuaries, however, the wind is as important a factor as the tide.

Figure 3.17.

Figure 3.17. Cross-sectional recordings of specific conductance in the River Fyris within Lake Ekoln, Sweden, August 1970

Figure 3.18. Bottom inflow in a lake

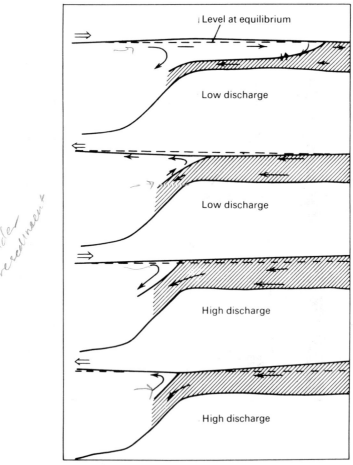

Figure 3.19. Interaction between river water and lake water at different flow conditions and different wind directions. Open arrows indicate wind, solid arrows indicate flow of water. The river water is shaded

3.19   SEDIMENTATION

In a system in equilibrium between erosion, transport and sedimentation, the plant and animal life has adapted to the stable conditions. When one or more of these processes is altered by natural or human causes, life in the system will be directly influenced. Sedimentation processes may be particularly affected by increased sheet erosion of nutrient rich soils from watersheds or resuspension of bottom material with high concentrations of nutrients, pesticides, metals or other pollutants, having significant ecological effects. These effects will be dealt with later, but a few facts concerning the physical conditions governing the sedimentological processes in lakes will be mentioned here.

There are essentially two main areas where sedimentological processes are important in lakes: the littoral and pelagic zones. The innermost part of the littoral is affected by a large transport of sediment caused by wave erosion and is called the swash-zone. River deltas are also important in that sense.

Deltas and the swash-zone are characterized by coarse sediment while the deeper central parts of the lakes are depositional areas, usually containing very fine sediments. This non-uniform distribution of sediments, where fine material is continually removed from shallow areas and transported to deep areas is referred to as focusing. The usual differences in physical processes between small and large lakes will also govern the characteristics of the sediments. In large lakes, the previously discussed dominating forces affecting sedimentation are wind and wind-induced currents, but in small lakes, inflows of rivers may dominate the process. The common requirement for sedimentation of very fine material, however, is a period in which the water-mass is completely calm (e.g. during the ice-covered winter for dimictic lakes) or if physiochemical conditions favour flocculation of material.

fine - low energy
coarse - high energy

4
NUTRIENT CYCLES

4.1 PHOSPHORUS

A discussion of the cycles of nutrient elements will begin with phosphorus (P), because it is the most limiting nutrient, or key nutrient, in fresh water. The annual productivity of freshwater lakes is controlled more often by P than by any other nutrient or environmental factor. The forms of P and a general cycle will be described first and the more important processes will be indicated. Because of the importance of the P store in the sediments as a long-term source to the overlying water in lakes, even after external inputs have been curtailed, the reactions at the sediment–water interface will be discussed in detail. The principal processes involved in P cycling will be applied to laboratory and open-lake conditions by reviewing pertinent experimental results.

4.1.1 Forms of P

P is separated into several forms in the analytical process through primarily mechanical filtration. The only form that can be determined with the acid molybdate method for P analysis (APHA, 1985) is orthophosphate. Orthophosphate is the form available for uptake and is determined in the filtrate after a portion of the water sample is passed through a $0.45\,\mu m$ pore size filter. Because colloidal particles with sorbed P may pass through such filters, the analysis may actually measure some of the sorbed P and hence, more than orthophosphate. Thus, the results from the undigested filtrate fraction is referred to most commonly as SRP (soluble reactive P) and is analogous to DIP (dissolved inorganic P). Total P (TP) is determined by complete digestion of the unfiltered water with persulphate, or other strong oxidant, and autoclaved. This process releases all sorbed and complexed inorganic and organic P. If digestion is also performed on the filtrate, then the difference between SRP and TP in the filtrate (which is TDP) is dissolved organic P (DOP) and the difference

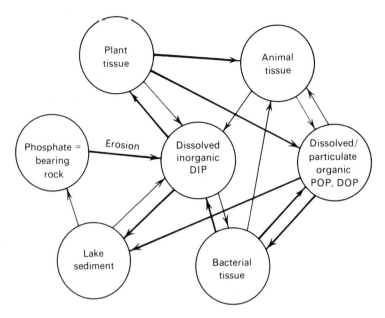

Figure 4.1. The aquatic phosphorus cycle with the transfers of greatest magnitude emphasized

between TP in the unfiltered sample and TDP is total particulate P (TPP). TP and SRP are usually the only two determinations made routinely and the difference between them can be taken as an estimate of TPP, because DOP is usually in relatively low concentration. TP ranges from $\approx 5\,\mathrm{mg\,l^{-1}}$ in sewage effluent (most of which is SRP) to as little as several $\mu\mathrm{g\,l^{-1}}$ in remote oligotrophic lakes. SRP varies greatly with abundance and demands of the algae and recycling activity, often occurring at levels that stretch the capability of the methods. A detection level of at least $2\,\mu\mathrm{g\,l^{-1}}$ is mandatory in lake research and requires a 10 cm cell length for best results.

4.1.2 General cycle for P

Pools and transfers of P are depicted generally in Figure 4.1. The ultimate source of P to aquatic ecosystems is from phosphate rock, whether through erosion from developed or undeveloped land or through wastewater inputs. P is initially utilized through plant and microbial uptake of dissolved inorganic P (DIP) associated with the processes of photosynthesis, chemosynthesis and decomposition. All organisms require P for metabolism and structure. Photosynthesis by green plants, which are the phytoplankton in the open water of lakes and estuaries, is most responsible for the uptake of DIP. Macrophytes in the shallow areas and bacteria in the water column and in the surficial sediments can also be effective in removing DIP from the water. Subsequently, phytoplankton and bacteria are consumed by animal grazers, which are in turn consumed by predators. A fraction of the DIP taken up can enter the dissolved and particulate organic P pools (DOP, POP) through, respectively, excretion and death. DOP and POP can be recycled to DIP by bacteria through decomposition of organic matter. DIP can be released through autolysis (Golterman, 1972) without microorganisms and through excretion by animals.

The remaining processes include sedimentation of PIP and POP and sediment release of DIP. Sedimentation of detrital (dead) and living phytoplankton and zooplankton fecal pellets are important processes in the loss of P from the open water. Much of the sedimented P is refractory and becomes part of the permanent sediments. However, depending on the physical–chemical conditions at the sediment–water interface and in the surficial sediments, a portion of sedimented P can be released back into the overlying water as DIP.

4.1.3   Sediment–water interface processes

The exchange of P between sediment and water in a lake can depend on several factors acting separately or together. The more important ones are: oxidation–reduction potential (redox), which is dependent on the oxygen concentration; pH; water exchange, as it affects diffusion and transport; temperature, as it affects microbial activity; and the relative fractions of P in the sediment that are bound with iron/aluminium, calcium and organic matter, which can vary greatly from lake to lake. The relative importance of these processes will vary with depth and degree of thermal stratification. The annual cycle in a typical temperate, stratified, anoxic (hypolimnion) lake will be considered first.

At the onset of thermal stratification, oxygen (DO) declines in the hypolimnion due to microbial decomposition of organic matter in the water column and in the surficial sediments. As DO approaches zero, reducing conditions prevail and iron in the surficial sediments is reduced from its ferric form (Fe^{3+}) to its ferrous form (Fe^{2+}). P, that was bound to the hydroxy complexes of ferric iron (e.g. $FeOOH-H_3PO_4$), is now solubilized and released into the interstitial pore water of the sediments and is available for diffusion into the overlying, anoxic water. The rate of diffusion is a function of the concentration gradient in SRP between the interstitial pore water and the overlying water. Adequate exchange of low P water overlying the sediment can maintain a high concentration gradient and enhance the release rate. As a result of a reasonably constant rate of release, the hypolimnetic P content increases more or

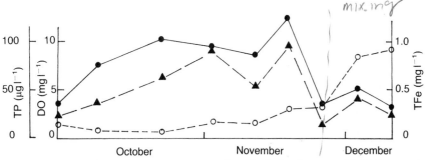

Figure 4.2. Hypolimnetic mean concentrations in Lake Sammamish before and after autumn turnover (Rock, 1974). Closed circles = TP; triangles = TFe; open circles = DO.

less linearly throughout the stratified period. Rates of release attributed to the iron redox process are quite variable and values as high as $52\,mg\,m^{-2}\,day^{-1}$ have been reported (Löfgren, 1987), but normally they are lower; Nurenberg (1984) has reported a range from 1.5 to $34\,mg\,m^{-2}\,day^{-1}$ with a mean of $16\,mg\,m^{-2}\,day^{-1}$ for 21 lakes.

While P is increasing in the hypolimnion, it is decreasing in the epilimnion through uptake and sedimentation by phytoplankton cells and zooplankton fecal pellets. While some of the P increase in the hypolimnion may be due to microbial decomposition of settling phytoplankton cells and detritus, results by Gächter and Mares (1985) show that SRP is readily sorbed by particulate matter as it settles through the hypolimnion regardless of the lake's trophic state and that the source of increased SRP in the hypolimnion is the sediments. They conclude this because in the upper hypolimnion P is not released in proportion to O_2 loss, while in the deeper portion, P increased in excess of the expected stoichiometric gain of one mole P for each 138 mole loss in O_2.

When the lake destratifies in the autumn, the whole water column and surficial sediments are replenished with DO. As that occurs, ferrous iron is oxidized to the ferric state and P is once again sorbed to its hydroxy complexes and returned to the sediment. Birch (1976) showed that 90% of the P released during the stratified anaerobic period in monomictic Lake Sammamish, Washington was quickly resedimented in November following turnover. The net result of these processes can be seen in the seasonal changes in DO, TFe and TP concentrations in the hypolimnion of Lake Sammamish (Figure 4.2).

P may again increase during the winter in dimictic lakes if DO reaches low levels under the ice. However, if the water overlying the surficial sediments remains oxic, then P will not be released, even if the deeper sediments are highly reducing and the lake is eutrophic. Figure 4.3 shows the redox potential in sediments and immediately overlying water during winter in Esthwaite water and Ennerdale, two English lakes (Mortimer,

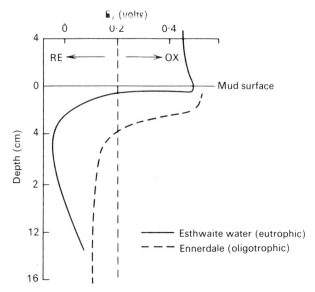

Figure 4.3. Vertical distribution of oxidized and reduced conditions in the sediments of two English lakes in winter (modified from Mortimer, 1942 with permission of Blackwell Scientific Publishers Ltd)

1941, 1942). Although of widely different trophy, as indicated by the redox level in the sediments, the overlying water was well oxidized in both lakes. Under reducing conditions in the sediment, P can diffuse upward to replace P that was previously released from surficial sediments, but if an oxidized layer exists, as shown in Figure 4.2, P will be readily sorbed in the surficial layer.

The iron redox cycle in lake sediments and the implications to P exchange with the overlying water were described initially by Einsele (1936), but have been an intriguing and important topic among limnologists ever since. Using fractionation procedures from soil chemistry, researchers have characterized sediments according to the P fractions that are easily desorbed, iron/aluminium bound, calcium bound and incorporated with organic matter (Williams *et al.*, 1967; Boström *et al.*, 1982). Sediments that have high iron/aluminium bound fractions can be expected to react readily to DO and pH changes. Sediments that are dominated more by the loosely bound and organic fractions are often from lakes that have been enriched by sewage effluent and can be expected to be more sensitive to temperature and microbial activity and less to pH and DO (Boström, 1984). However, lake sediments are very complicated chemically and iron redox may not be controlling P exchange, as, for example, in sediments where high concentrations of live *Microcystis* in

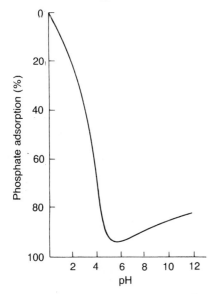

Figure 4.4. Relative removal of PO_4^{3-} from solution with $Fe(OH)_3$ versus pH (Ohle, 1953)

the surficial sediments and high carbon content in deeper sediment favour the reduction of carbon instead of iron (Davison *et al.*, 1985). Gächter *et al.* (1988) found that one half the dry matter in suficial sediments of Lake Sempoch, Switzerland was contained in bacterial cells. Laboratory and field experiments indicated that uptake of P by bacteria during oxic conditions and release during anoxia is an important process in the exchange of P between sediments and water. They also suggested that there tends to be an uncoupling of Fe and P when those elements are released following the onset of anoxia in several lakes (e.g. Figure 4.2). Nevertheless, there still seems to be general agreement that iron redox is the principal mechanism controlling P exchange in the deep-water sediments of most stratified lakes.

But what are the principal mechanisms operating in shallow, unstratified, apparently oxic lakes and littoral zones of deep lakes? Numerous investigations have shown that P from sediments in shallow oxic lakes can represent a large fraction of the total P loading (Lennox, 1984; Jacoby *et al.*, 1982; Riley and Prepas, 1984; Ryding, 1985). Under oxic conditions, the solubility of iron is controlled by pH (Stumm and Leckie, 1971). At a pH of ≈ 6, which is rather typical for the sediments of soft-water lakes, the solubility of ferric iron is minimal and therefore P can be effectively sorbed and removed from the water column (Ohle, 1953; Figure 4.4). With increasing pH the solubility of iron increases and

low pH = low ferric Fe solubility = low TP

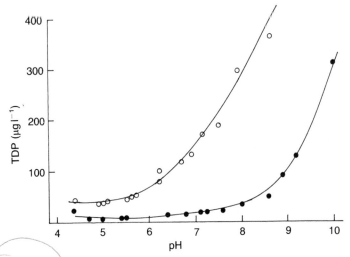

Figure 4.5. Relationship between equilibrium total dissolved P and pH in surficial sediments at two sites in Long Lake, WA (Jacoby *et al.*, 1982). ○, Midlake (3.5 m); ●, Lilies (1 m)

P that was sorbed to ferric iron hydroxy complexes is released (Figure 4.5). The high photosynthetic rates in nutrient-rich lakes can result in mid-day pH values of 10, which could produce very high rates of P release from sediments in littoral regions in deep lakes and throughout in shallow lakes (Jacoby *et al.*, 1982; Andersen, 1975). However, it is not clear that such high pH levels could persist near the well-buffered sediments, which normally have a pH ≈ 6. Photosynthetically caused high pH could, nevertheless, maintain P solubilized in the water column and available for algal uptake (Löfgren, 1987; Ryding, 1985).

Temperature can be an important driving force in the release of P especially from organic sediments. Kamp-Nielsen (1974) has shown that P release from Lake Esrom sediments was directly related to an increase in temperature. Also, Ryding (1985) found the rise in water temperature to be a key factor in the release of P in 16 shallow, eutrophic Swedish lakes. The role of temperature is related to the stimulation of bacterial activity, which in turn creates anoxic conditions and the reduction of iron. Bacterial activity can increase dramatically in response to normally moderate increases in temperature as shown by Boström *et al.* (1985) for highly organic Lake Vallentunasjön sediments. They found a Q_{10} of 10.6 over a temperature range of 4 to 20°C. Water near the sediment surface would undergo such temperature increases only in shallow unstratified lakes and in littoral regions. But wind mixing apparently maintains nearly saturated DO conditions throughout the water column, which should not favour the iron redox mechanism as an explanation for P release in such environments.

So what are the mechanisms for 'aerobic' P exchange with sediments in shallow environments? Although oxygen-temperature profiles in shallow environments normally show largely uniform temperature and saturated or supersaturated DO concentrations throughout the water column, these conditions do not persist at the sediment–water interface. Slight thermal stratification can occur temporarily under quiescent conditions permitting DO to deplete in the water layer immediately overlying the sediment (Stefan and Hanson, 1981; Riley and Prepas, 1984; Löfgren, 1987). Even if no thermal stratification is present, DO can deplete in the immediate overlying water under quiescent conditions (Welch *et al.*, 1988a). Moreover, iron has been observed to increase before DO in the overlying water is completely depleted (Mortimer, 1941, 1942; Davison *et al.*, 1985). Thus, large release rates of P from sediments during summer in unstratified lakes can be attributed to the iron redox mechanism. In addition to bacteria causing the reduction of iron indirectly through DO depletion via their metabolic activity, evidence exists that bacteria may reduce iron directly (Jones *et al.*, 1983). This is supported by strong positive correlations between iron and TP during the summer in shallow, eutrophic Lake Finjasjon in Sweden (Löfgren, 1987).

Increased wind mixing, following the quiescent period, would then entrain the layer overlying the sediment and transport the P throughout the water column. Although the entrained iron would oxidize upon mixing, the entrained P could remain in the water column under conditions of photosynthetically caused high pH. Ahlgren (1980) has shown a positive relationship between the wind-fetch distance per unit mean depth and the summer amplitude in TP in some shallow, eutrophic Swedish lakes. In a similar fashion, Osgood (1988a) found that internal loading in lakes around the Twin Cities, Minnesota, increased when the mean depth/area$^{0.5}$ (m/km) was less than about seven. The relative thermal resistance to mixing, and to a lesser extent flushing rate, was inversely correlated ($r^2 = 0.92$) with internal loading over a 9-year period in Moses Lake (mean depth 5.6 m), Washington (Jones and Welch, 1990).

Also, it is clear that P released from sediments in enriched, shallow, polymictic lakes would immediately be available in the lighted zone for uptake by phytoplankton. Rather than sedimented P and sediment-released P being trapped in the hypolimnion until turnover, as is usually the case in deep stratified lakes, P can be continually recycled throughout the summer in unstratified lakes.

A rather more extreme case can be used to illustrate the importance of sediment–water exchange of P in shallow, eutrophic, polymictic lakes. Lake Trummen, in Växjö, Sweden, has gained international fame as the first, large sediment dredging project to restore a lake (Björk, 1972). The

TABLE 4.1. Release of N and P in containers with water and sediment under aerobic and anaerobic conditions (mgm^{-2} day^{-1})

	Aerobic		Anaerobic	
	PO_4-P	NH_4-N	PO_4-P	NH_4-N
Black gytta deposited in 1960s	1.7	0.0	14.0	73.0
Brown gytta deposited 1000 yr BP, depth >40 cm	0.0	0.0	1.5	0.0
After dredging[a]			1.24	

Source: Björk *et al.* (1972); [a] Bengtsson *et al.* (1975)

lake's maximum depth was only 2 m prior to dredging and core studies showed that although the sedimentation rate had decreased over its 10 000 year history, it had recently increased from 0.2 to 8 mm year^{-1}, in response to the discharge of sewage effluent from the city. The sediment P content correspondingly increased during 1920 to 1958, from 50 to 800 mg l^{-1} ($\approx$0.5–8 mg g^{-1} dry matter, assuming 10% water content). Zn and Cu increased as well. Sediment release experiments showed that the most recently deposited sediments were obviously significant sources for nutrients, compared to the deeper, historical sediments (Table 4.1). In an effort to recover the lake, the sewage effluent was diverted, but no improvement was noted over the subsequent 10-year period. Removal of the top 1 m of rich sediment quickly resulted in the recovery of the lake. Thus, the sediment–water exchange of P was sufficient to maintain the high productivity and poor water quality in spite of the removal of the sewage input. Following removal of the rich layer of sediment, the release rate of P declined (Table 4.1). More details about this project will be covered in Chapter 10.

Although sediment-released P in unstratified lakes is readily available to phytoplankton in the lighted zone, this is not usually the case in stratified lakes. Sediment-released P in the hypolimnion can become available to phytoplankton in the lighted epilimnion through erosion of the thermocline and vertical entrainment of hypolimnetic water as a result of summer storms. This has been shown for several lakes. For example, an August storm resulted in disruption of the thermocline in Lake Ballinger (Z_{max} = 12 m) near Seattle, apparently causing an increase in SRP, which stimulated a phytoplankton increase (Figure 4.6). This phenomenon was first demonstrated by Stauffer and Lee (1973), who showed that a thermocline sinking of 1 m in Lake Mendota, Wisconsin resulted in a doubling of the eplimnetic P content and contributed significantly to summer algal blooms. Analysis of continuous temperature data from Lake Ekoln, Sweden showed seven events of thermocline

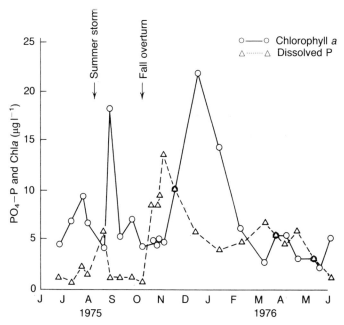

Figure 4.6. Dissolved P increase in the epilimnion of Lake Ballinger, Washington, as a result of an August storm with a subsequent response in plankton chlorophyll

erosion during the stratified period in 1970. An example is shown in Figure 4.7, when the TP content in the epilimnion was calculated to have increased by 15%. However, the best evidence for thermocline migration and entrainment of hypolimnetic P comes from Larsen *et al.* (1981). They showed that for four wind-mixing events during summer in Shagawa Lake, Minnesota the epilimnetic P increased from 74 to $130\,\mu g\,l^{-1}$ and that coincidentally there was nearly as great a decline in the hypolimnetic P content. In addition to entrainment, they also estimated that diffusion could account for a vertical transport of 4.5 to $9\,mg\,m^{-2}\,day^{-1}$ of P into the epilimnion.

There are other processes that can account for the transport of sediment P to lake water. Gas ebullition from sediments can serve as a transport process for interstitial SRP if the surficial sediments are reduced (Enell and Löfgren, 1988). Bioturbation of the surficial sediments by benthic invertebrates may also be an important transport process (Boström *et al.*, 1982). Excretion by carp feeding on detritus and sedimented algae can be substantial (Shapiro *et al.*, 1975). Death and decomposition of rooted macrophytes, which depend largely on sediments for their P supply, is a process by which sediment P can be transported to lake water (Schultz and Malueg, 1971; Carpenter and Adams, 1977; Landers, 1982; Rorslett

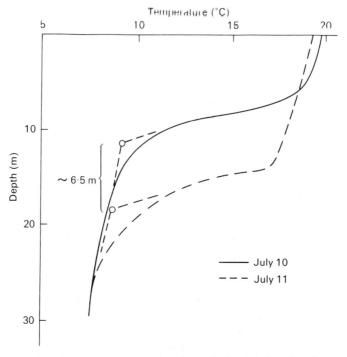

Figure 4.7. Temperature profiles from Lake Ekoln, Sweden preceding and following one of seven thermocline disturbances during the summer of 1970. Thermocline position estimated by extrapolation to approximate the overall thermocline displacement (6.5 m). Temperature was continuously monitored at 1 m intervals (data from National Swedish Environmental Protection Board)

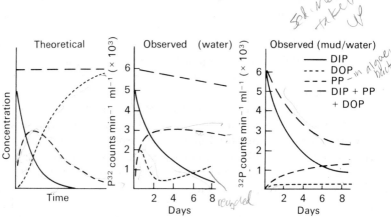

Figure 4.8. Distribution of ³²P among three forms after addition to microcosms. DIP, dissolved inorganic P; DOP, dissolved organic P; PP, particulate P (Hays and Phillips, 1958)

et al., 1986). The relative significance of that source will be discussed further in Chapter 9. Vertical migration of buoyant and mobile algae is yet another transport process that has been given recent attention. Trimbee and Harris (1984) found that the blue green algae *Microcystis aeruginosa*, *Gomphosphaeria lacustris* and *Lyngbya birgei* may have originated largely from sediments in Guelph Reservoir, Ontario because downward facing traps at 10 m captured 2–4% of the bloom population that could have been produced by only five or six divisions of the 2–4% inoculum. TP was also shown to migrate with the algae. Transport by algae within the water column has been shown for vacuolate *Oscillatoria* (Klemer *et al.*, 1985) and for a motile alga, *Cryptomonas Marssonii* by Salonen *et al.* (1984). See 'Succession' for further discussion of this process (Section 6.5.6).

4.1.4 Recycling and the fate of added P

Lakes tend to reestablish equilibrium levels of P in the water in a relatively short time following manipulation of the inputs. Permanently increasing the P concentration in lakes is a very difficult process; many frequent and small additions are necessary in order to raise and maintain a higher concentration, as has been shown in experimental enrichment of lakes (Schindler, 1974). The rapid cycling of P has been demonstrated in laboratory microcosms using radiophosphorus (^{32}P). The principal results of this early work by Hays and Phillips (1958) were as follows (Figure 4.8).

1. DIP was rapidly removed from the water phase at an exponential rate, but the reaction series did not reach completion as theoretically predicted, i.e. all DIP was not converted to DOP.
2. Bacteria took up and held much of the DIP as PP, which was recycled back to the water phase as DIP.
3. The loss of total P was much greater with mud, which competed with bacteria for DIP and was much slower to release DIP.
4. An equilibrium was established between water and solid phases in which total ^{32}P was only ≈0.1 of the initial level.

The experiment was repeated in a small Nova Scotia lake to which 11 mCi of ^{32}P was added as $KH_2{}^{32}PO_4$. The level of ^{32}P decreased to 80% of the initial after 2 weeks. The concentration peaked in plankton after 4–16 h and in fish after 3–4 days (Hays and Phillips, 1958).

Similar experiments on the fate of P were performed in Crecy Lake (20 ha, mean depth 2.4 m) in New Brunswick, but with inorganic fertilizer. Nevertheless, results were comparable with the ^{32}P work. The fertilization theoretically raised P to 390 μg l^{-1}, N to 210 μg l^{-1} and K to

Figure 4.9. Effect of fertilizing Crecy Lake with P, N, and K. Concentrations of dissolved inorganic and organic phosphorus (DIP, OP) during years of fertilizing and succeeding years are shown (after Smith, 1969)

$270\,\mu\text{g}\,\text{l}^{-1}$. The results from fertilizations in 1951 and 1959 and successive years are shown in Figure 4.9 (Smith, 1969).

From such experiments, the turnover rates can be estimated by assuming first-order kinetics and calculating the rate of disappearance of P in the various phases following fertilization. The loss rate for DIP was between 5–6% day^{-1} for the three fertilizations (Table 4.2). The turnover times (time required for loss from a phase) could then be calculated as the ratio of the amount in a particular phase to the loss rate for that phase. P in the solid phase underwent a 10 to 20-fold slower turnover than did P in the dissolved phase, which illustrates how P was conserved. Nevertheless, by the next year there was no evidence of the previous year's fertilization, thus supporting the earlier statement that permanently increasing the P content in lakes is a difficult process.

Phosphorus pool size and transfer among pools within the water column can be modelled for a system if enough data are available. The definitions of pools and transfer rates are often grossly oversimplified for the sake of practicality and lack of knowledge. However, transfer rates are variable and can be highly dependent on such factors as light, tem-

TABLE 4.2.   Turnover time of P in water and solids of Crecy Lake estimated from observed loss rates

Fertilization	Loss rate (% day^{-1})	Turnover time (days)	
		Water	Solids
First	5.9	17	176
Second	5.6	18	248
Third	5.3	19	394

Turnover time (days) = (amount in phase)/[loss rate (per day)]. *Source:* Smith (1969)

perature and species composition of the trophic level. As a result, dynamic models of lake ecosystems are more useful for investigating how the system works than for prediction for management purposes.

To illustrate how a model could be constructed, Stumm and Leckie (1971) developed a hypothetical, steady-state model for the water column of a stratified lake (Figure 4.10). Any substance can be modelled, but usually either C, N or P is used. Conversions from one element to another are made using an appropriate elemental ratio, which is another simplification. P is modelled in Figure 4.10 and the transfer rates are expressed in $\mu g\,l^{-1}\,day^{-1}$ and pool sizes in $\mu g\,l^{-1}$. The model suggests that $\approx 30\%$ of the uptake demand (1.1/3.5) is supplied by bacterial decomposition in the epilimnion, whereas 20% is diffused from the hypolimnion and the remainder comes from the inflow. The hypolimnetic source is reasonable considering the earlier discussion of thermocline erosion due to storms. About one third of the P loss from the plankton rains into the hypolimnion via sinking fecal pellets and dead phytoplankton cells. There are also losses from the soluble phase to the sediments due to inorganic precipitation. Chemical regenerative processes from the sediment account for only $\approx 5\%$ of the biological contribution to the soluble P pool (0.1 versus $2.0\,\mu g\,l^{-1}\,day^{-1}$). As indicated earlier, and to be discussed further in Chapter 7, contributions of P from chemical exchange accross the sediment–water interface in shallow stratified and unstratified lakes alike can be much more significant.

Some omissions in Figure 4.10 are separate pathways for regeneration from dead cells via autolysis, which is not a function of bacterial activity, and excretion of DIP direct from phytoplankton and grazing zooplankton, the latter being one of the more important regenerative processes. These processes were indicated separately in Figure 4.1.

Figure 4.10. Hypothetical steady-state model of phosphorus cycling in a lake (after Stumm and Leckie 1971)

Another important process often omitted in lake models of the water column and sediments is the interchange with the littoral zone. Although contributions from the littoral zone have been emphasized recently, there are pertinent results from an early study conducted by Hutchinson (1957, p. 748; Hutchinson and Bowen, 1950) on Linsley Pond, Connecticut. ^{32}P was added on 21 June 1946 and results were observed during August 1–15 by noting the change in the amount of tracer in the various pools. This whole-lake experiment illustrates the interaction in the transfer of P among the epilimnion, littoral region and the hypolimnion. The transfer rates of P in kg week^{-1} were as follows for the epilimnion: epilimnetic increase, 0.26; loss to hypolimnion, 1.55; transport from littoral to epilimnion (sum), 1.81; and for the hypolimnium: hypolimnion increase,

3.75; gain from epilimnion, 1.55; gain from sediment (difference), 2.20. The hypolimnion was assumed to act as a sink in this balance analysis. Some of the epilimnetic gain may have come from the hypolimnion due to processes mentioned earlier. However, the littoral was probably the major contributor as indicated.

Rich and Wetzel (1978) have stressed the importance of the littoral region in supplying energy and materials to the economy of lakes. Recent studies have confirmed the importance of the littoral in supplying P and other substances to the pelagic region. Cooke *et al.* (1978, 1982) showed by mass balance calculations that internal loading of P was reduced but not eliminated in West Twin Lake, Ohio after an alum treatment of the anoxic hypolimnetic sediments. Internal loading in East Twin, the control lake, remained about double that of West Twin after treatment. The conclusion was that the continued internal supply in treated West Twin must have come from the littoral region, because the alum floc had blocked P release from the sediments.

The littoral processes could include macrophyte decomposition, bioturbation or aerobic release from sediments. Decomposition of *Myriophyllum spicatum* in the littoral zone was found to be a significant recycling process in Lake Wingra, Wisconsin, contributing an amount equal to $\approx 60\%$ of the external P load and 30% of the total load (Carpenter and Adams, 1977; Prentki *et al.*, 1979; Carpenter, 1980a). Loss of P from healthy shoots was considered nil, but the net (minus uptake) release from the littoral through senescence was estimated at $2\,\mathrm{g\,m^{-2}\,year^{-1}}$, which is $17\,\mathrm{mg\,m^{-2}\,day^{-1}}$ (Smith and Adams, 1986, see Chapter 9). For comparison, Nurnberg (1984) reported that the release rate from anoxic hypolimnia of 21 lakes ranged from 1.5 to $34\,\mathrm{mg\,m^{-2}\,day^{-1}}$ with a mean of 16. Thus, lakes with extensive beds of rooted macrophytes (at least *M. spicatum*) may have a significant portion of internal loading coming from the littoral zone.

4.1.5   Sediments as a source or sink for P

The question as to when and to what extent sediments are a source or sink for P is important for understanding lake productivity as well as for the management of lake quality. Most of the processes involved in the sediment–water exchange of P have been covered in the previous discussion. However, knowledge of the relative contribution of each of those processes in a given lake is lacking and may be impossible to obtain. Some general summary comments may help clarify the issue.

1. Lakes are nearly always a sink for P in the long term, whether they are shallow, deep, oxic or anoxic. Unless a fraction of the input has been recently removed, the input mass of P will usually be greater than the

usually,
input > output

output on a year-to-year basis. The fraction retained is usually greater than 0.5, but can vary widely (Vollenweider, 1975). Another way to verify this is to compare the retention times of P and water; they are, respectively, 0.77 and 2.0 years in Lake Sammamish, calculated by the ratio of lake quantity to inflow rate.

sources of P

2. Although sediments act as an ultimate sink, they can act as significant sources during a portion of the year. The most significant process determining when and to what extent sediments will act as a source is iron redox, whether the lake is stratified, unstratified, oxic or anoxic. Stratified, oxic lakes generally yield little P from sediments because iron is always oxidized and effectively sorbs P. However, in stratified lakes that have anoxic hypolimnia and in unstratified lakes that experience alternating quiescent and windy periods during the summer, the release of sediment P, due to iron reduction, results in lake output of P greatly exceeding the input. Photosynthetically caused high pH and recycling processes in the water column tend to retard the resettlement of P. The other processes that supply P from sediments, such as macrophyte decomposition, bioturbation, excretion by carp, gas ebullition, algae migration, etc., have not been demonstrated to be as significant as iron redox generally.

3. Whether the sediment-released P in stratified lakes actually reaches the photic zone and is available for algal uptake is a significant issue, but not with respect to the source/sink question. So long as P reaches the hypolimnion, even if unavailable to algae, the sediments are technically a source for that P. However, there is ample evidence, as indicated earlier, that some of that hypolimnetic P can be transported into the epilimnion.

4.2 NITROGEN

The general cycle of N in an aquatic ecosystem is shown in Figure 4.11. The cycle is very complex, and many of the transfer processes and associated pool sizes are important not only to aquatic productivity but also to environmental quality in general and human health. Two large biological source and sink processes occur with N that do not occur with P, which greatly influences the significance of N in the control of productivity and waste treatment. These processes involve the transfer of N from atmospheric N_2 through microbial fixation and its return to the atmosphere via N_2O and N_2 through denitrification. The processes of nitrification (oxidation of ammonia) and denitrification are reactions that would proceed without biological mediation but at a slow rate. Microorganisms greatly speed the rate of reaction and at the same time capture the energy available in the reduced compounds through an ordered series of cellular enzyme catalysed reactions. Because the energy sources are

N_2 fixation $: N_2 \rightarrow NO_3$

denitrification $: N_2O \rightarrow N_2$ atmosphere

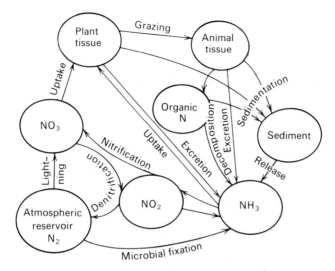

Figure 4.11. The aquatic nitrogen cycle

inorganic, the organisms are referred to as chemolithotrophs (Brock, 1970).

4.2.1 Pool concentrations

Nitrogen is most abundant as N_2, compromising nearly 80% of the atmospheric constituents. However, before this form is available to plants it must be fixed either biologically or through a light-energized reaction. Nitrate (NO_3^-) is the most common form that can be used by aquatic plants, and its content ranges from a trace when productivity is high, and all available N has been absorbed into plant tissue, to concentrations usually between 500 and $1000\,\mu g\,l^{-1}$ during periods of no utilization. Concentrations well over $1\,mg\,l^{-1}$ can occur but are usually associated with inputs. Ammonia (NH_3) or ammonium (NH_4^+), which is the principal form in water, becomes abundant in the absence of oxygen or in very enriched waters, but it is usually less abundant than NO_3^-. Because NH_4^+ is a more reduced form, it is often preferred over NO_3^- by plants. NO_2^- is not used and in fact can be rather toxic. Because it readily oxidizes to NO_3^-, it is usually not present in appreciable amounts.

4.2.2 Nitrification

Nitrification is the process by which NH_3 is transformed into NO_2^- and finally into NO_3^-. The process occurs only under aerobic conditions.

Organisms that commonly perform the transformations are *Nitrosomonas* and *Nitrobacter*. Although the processes are energy yielding, as shown below, the yield is rather low compared to other transformations in the cycle (Delwiche, 1970). The reactions in nitrification are:

$$2NH_4^+ + 3O_2 \rightarrow 2NO_2^- + 2H_2O + 4H^+ + energy$$

$$2NO_2^- + O_2 \rightarrow 2NO_3^- + energy$$

4.2.3 Denitrification *Anoxic process*

This process occurs only in the absence, or near absence, of oxygen. A common denitrifying organism is *Thiobacillus denitrificans* which is a chemolithotroph and the associated reaction is:

$$5S^{2-} + 6NO_3^- + 2H_2O \rightarrow 5SO_4^{2-} + 3N_2 + 4H^+ + energy$$

Heterotrophic bacteria, such as *Micrococcus, Serratia, Pseudomonas* and *Achromobacter*, are also denitrifiers when oxygen concentrations are low. As they are facultative anaerobes, they can also carry on aerobic respiration and are part of the normal flora in sewage (Christensen and Harremoes, 1972).

In denitrification, the reverse process of nitrification, the bacteria reduce NO_3^- to NO_2^- and in turn form molecular gaseous N_2 or N_2O (nitrous oxide). This represents the mechanism of N loss from ecosystems and can be used as a treatment method for the removal of N from wastewater. However, the environment must be aerobic for nitrification and anaerobic for denitrification. The necessary alternation of aerobiosis/ anaerobiosis in order for denitrification to represent a significant N loss has implications in the N balance and nutrient limitation in lakes as well as in wastewater treatment. The more productive a lake, the greater is the chance for oxygen to reach low levels, which would allow denitrification. For wastewater treatment, an aerobic period must precede the anaerobic period in order for NH_4^+ to be converted to NO_3^-, which can then be denitrified without oxygen. N reduction to NH_4^+ from NO_3^- also occurs.

4.2.4 N fixation

Nitrogen fixation is an energy-consuming aerobic process carried on in aquatic environments by such bacteria as *Azotobacter* and *Clostridium* and by the blue-green algae (Cyanobacteria) *Nostoc, Anabaenopsis, Gleotrichia* and *Aphanizomenon*. N-fixing blue-green algae usually dominate in enriched lakes in summer. Nitrogen fixation can represent a significant input of N to an ecosystem. Measured rates for *Anabaena* range from 0.04 to $72\,\mu g\,l^{-1}$ day^{-1}. Such rates could represent a several-

TABLE 4.3.   N balance from Lake Mendota, Wisconsin

Income		Loss	
Source	(%)	*Sink*	(%)
Waste water	8.1	Outflow	16.4
Surface water	14.7	Denitrification	11.1
Precipitation	17.5	Fish catch	4.5
Groundwater	45.0	Weeds	1.3
Fixation	14.4	Sedimentation	66.7
Total	100	Total	100

Source: modified from Brezonik and Lee (1968)

fold turnover in dissolved inorganic N in enriched systems, because NO_3^- is usually depleted to very low levels or is undetectable when blue-green fixers appear. Horne and Goldman (1972) have shown that blue-green fixation contributed 43% of the annual N input to Clear Lake, California.

N fixation is performed in the heterocysts of blue-green algae. The abundance of these extra large cells in the filaments or chains of the above N-fixing blue-greens usually increases with NO_3^- depletion (Horne and Goldman, 1972). Because fixation is an energy-demanding process, it becomes advantageous only when NO_3^- and NH_4^+ are no longer available.

This source of N for the N fixers also becomes available for other algal species as the cells of the N fixers decompose. In many eutrophic lakes *Aphanizomenon* and *Microcystis* usually do not occur together in large masses but often seem to alternate in abundance. N availability may be the cause for that alternation. Also, the availability of N can be a cause for the succession to blue-green algae, particularly the heterocystous blue-greens (see Chapter 6 on succession).

Just as N loss through denitrification can increase in significance as productivity increases (because of oxygen depletion), fixation also increases with productivity due to depletion of NO_3^-. Fixation rates of the planktonic blue-greens in enriched Lake Erie have been determined to range from 4.2 to 230 nmol acetylene $mg^{-1} N h^{-1}$ and from 0.69 to 25 nmol $mg^{-1} h^{-1}$ in less rich Lake Michigan (Howard *et al.*, 1970). Horne and Goldman (1972) have measured a maximum rate of several hundred nmol $l^{-1} h^{-1}$ in highly enriched Clear Lake.

Complete budgets for N in lakes are constructed much less frequently than for P. Brezonik and Lee (1968) developed a complete budget of N income and loss in Lake Mendota (Table 4.3). In that case the gain from fixation was very similar to that lost by denitrification, although neither was the most important source or sink. Sedimentation was estimated by differences between loss and income. Another source and sink, respectively, were marsh drainage and groundwater recharge for which they had no values. Messer and Brezonik (1983) cite several N budgets that show much larger fractions (>50%) of incoming N lost through denitrification.

4.2.5   Implications to nutrient limitation

The availability of N and P to the primary producers in freshwater ecosystems varies markedly because of unique processes in the N cycle, together with the differing chemical behaviour of N and P. Reiteration of some of the contrasts between the N and P cycles should give better insight as to which nutrient is apt to be most limiting.

1. In a freshwater lake the residence time for an incoming quantity of N is longer than that for P, because the strong sorptive capacity of inorganic metal complexes and organic particulate matter for PO_4^{3-} tends to remove P to the sediments. Although some sedimented P can be recycled via several processes (e.g. iron redox, macrophyte decomposition, etc.) in most cases the overall efficiency of such recyling does not seem to be very high, i.e. even anoxic lakes are still net sinks for P. N, on the other hand, in the form of NO_3^- and NH_4^+ is much more soluble and is not readily sorbed by inorganic complexes as is PO_4^{3-}. Besides NH_3 can also be released from anaerobic sediments.

2. Although significant fractions of N income can be lost from aquatic systems via denitrification, this occurs only in anaerobic and thus, highly enriched waters. As a result, the process contributes to N limitation in highly enriched lakes, but should not greatly contribute to an N shortage in waters of low or moderate enrichment.

3. The atmosphere provides a ready source of N to N-depleted systems through the biological fixation of N_2. This occurs in aerobic environments and requires only the presence of heterocystous blue-green algae. When available N is depleted, these organisms can dominate and thus provide a very large part of the N supply to enriched systems. In low or moderately productive systems, however, inorganic N has not declined enough to offer an advantage to N fixation. Because it is an energy-consuming process, fixation is advantageous only if NO_3^- and NH_4^+ are scarce. Another factor that restricts the supply of N through fixation is a rather low maximum rate of N replacement in cells, i.e. growth rate, of 0.05 day^{-1} (Horne and Goldman, 1972).

4. Yet another source of N that does not occur in the P cycle is via precipitation. Although rainwater contains P and has been shown to be important in lakes in phosphorus-poor watersheds (Schindler, 1974), NO_3^- in rainfall is a rather constant phenomenon, having been trans-

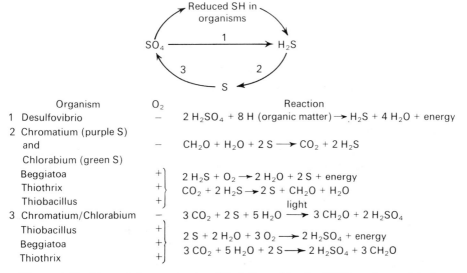

Figure 4.12. The sulphur cycle (modified from Brock, 1970) and associated bacteria-mediated reactions (Klein, 1962)

formed from the N_2 reservoir in the atmosphere by lightning. In a temperate area the N in rain is about $6\,kg\,ha^{-1}\,year^{-1}$ for a 75 cm level of precipitation (Hutchinson, 1957).

Overall, there are obviously fewer sources for P than for N and sedimentation is probably a more efficient remover of P than N in most aquatic ecosystems. Therefore, P should usually be less available than N and considering the potential for P recycling and loss of N through denitrification in anoxic environments, N should probably limit in highly enriched systems. Nutrient limitation will be discussed further in Chapter 6.

4.3 SULPHUR

The sulphur cycle is interesting from the standpoint that several significant water-quality changes result from waste inputs and involve processes in the S cycle. However, S itself is almost never a limiting nutrient in aquatic ecosystems. The normal levels as SO_4^{2-} are more than adequate to meet plant needs. The cycle is shown in Figure 4.12 with some of the bacteria and associated processes that are responsible for the indicated (numbered) transformations.

Odorous conditions are easily created when waters are overloaded with organic waste to the point that O_2 is removed. Then SO_4^{2-} is the electron acceptor often used for the breakdown of organic matter. Step 1 in the cycle shows the production of H_2S, which has a rotten-egg smell. If NO_3^- is available, N-reducing bacteria will dominate and odours will be minimal. The production of H_2S, which is toxic, may create a problem for fish. However, it does not persist long in the presence of O_2.

Thiobacillus ferroxidans contributes to acid mine water by oxidizing FeS (ferrous sulphide), resulting in H_2SO_4 and a pH that may be as low as 1.0. Bacterial activity may cause 80% of such acidity. The same process occurs with *T. thiooxidans* and is responsible for pipe corrosion (Brock, 1970).

Of considerable interest is the occurrence of 'plates' of sulphur bacteria in lakes (Brock, 1970), composed of either photosynthetic, facultative anaerobic purple or green sulphur bacteria or aerobic, colour-less sulphur bacteria. Regardless of type, they are rather restricted to intermediate layers, usually the metalimnion. Photosynthetic bacteria need light as well as H_2S. Although H_2S may be abundant under anoxic conditions at depth, light is not abundant. The abundant light in the surface layer is of no help to these organisms because H_2S is unstable in the presence of O_2. So these organisms occur at a rather restricted depth interval (plate) where both light and H_2S occur. Therefore they are unlikely to occur in aerobic lakes.

The aerobic, non-photosynthetic bacteria have much the same problem of restriction. Because they require both O_2 and H_2S, their appearance is restricted to the interface of declining O_2 and H_2S levels, often at the bottom of the metalimnion.

SO_4^{2-} enters aquatic ecosystems through atmospheric depositions of sea salt and as a combustion product of fossil fuels, as well as through natural weathering processes in the watershed. The acidity of precipitation downwind of industrialized areas is caused primarily by H_2SO_4. While lakes are acidified by deposition of such anthropogenic SO_4^{2-}, a significant fraction of SO_4^{2-} may be consumed through step 1, resulting in a pro-duction of alkalinity, even in unproductive oxic lakes (Schindler, 1986). Apparently there is sufficient anoxia in sediment-pore water for the process to proceed. The process and effects of acidification will be dis-cussed further in Chapter 13.

4.4   CARBON

Carbon is the element used most frequently to study productivity and trophic transfer of energy in aquatic ecosystems. This is logical because C comprises nearly 50% of the dry organic matter in living organisms. Also, the C content tends to be more stable, per unit dry matter, than N or P, because C usually is not limiting to growth in the long term. The pertinent pools, sources and transfers for C are shown in Figure 4.13.

The sediment pool, although omitted for clarity, is a sink as in the P

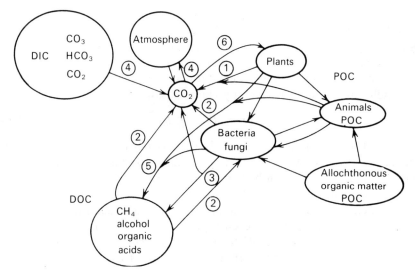

Figure 4.13. The carbon cycle with some important biomediated process.

1 This process is normal aerobic respiration: $CH_2O + O_2 \rightarrow CO_2 + H_2O$.

2 The same respiratory process is performed by microorganisms through decomposition.

3 Bacterial decomposition is also performed by anaerobic respiration. Since anaerobic metabolism is incomplete, only a portion of the carbon goes to CO_2. The remainder is contributed to the dissolved organic pool (DOC), which is slowly metabolized to CO_2. Some anaerobic processes in which a carbon by-product other than CO_2 is produced are performed by methane bacteria, which can carry on the following overall reaction (Brock, 1970):

$$C_6H_{12}O_6 + X H_2O \rightarrow Y CO_2 + 2CH_4$$

4 The principal source of free CO_2 is dissociated from the dissolved inorganic carbon (DIC) pool. This supply depends on the total inorganic carbon content, temperature, and pH.

5 Organism excretion contributes to the DOC pool

6 CO_2 is assimilated through photosynthesis into particulate organic carbon (POC) to initiate the cycle

and N cycles. In contrast to P, the atmosphere is a source as in the case of N. However, the rate of input of CO_2 from the atmosphere into the cycle is dependent solely on a physical process: diffusion across the air–water interface as defined by Henry's Law. CO_2 is the principal form of C that is utilized by plants, although bicarbonate (HCO_3^-) is used by some species (Goldman *et al.*, 1971). Another difference in the C cycle, compared to the N and P cycles, is that the concentration of free CO_2 is controlled predictably by pH, temperature and the total C content ($C_T = CO_2 + HCO_3^- + CO_3^{2-}$) and, because it is a gas, the diffusion rate from the atmosphere where it comprises 0.03%.

The reactions and equilibria of the CO_2 system will be described briefly in order to clarify the availability of C for plant uptake and the effect of plant photosynthesis on water quality. CO_2 diffuses into water from the atmosphere when the water is undersaturated relative to the air and from water to air when supersaturated. The amount of CO_2 in water at equilibrium with air is given by Henry's Law:

$$CO_2 = P_{CO_2} \times K_H \tag{4.1}$$

where CO_2 is expressed as $mol\,l^{-1}$, P_{CO_2} is partial pressure in atmospheres (atm) and K_H is the Henry's Law constant in $mol\,l^{-1}\,atm^{-1}$ ($10^{-1.37}$ at 25°C). CO_2 can be highly supersaturated in lakes following ice cover, especially in oligotrophic lakes where photosynthetic activity is low, reaching values of several hundred per cent saturation or more (Wright, 1983). In productive eutrophic lakes, CO_2 can by depleted to very low levels causing the pH to rise to 10 or more (Andersen, 1975).

CO_2 reacts with water to form carbonic acid, H_2CO_3, as follows:

$$CO_2 + H_2O \rightarrow H_2CO_3^* \tag{4.2}$$

where $H_2CO_3^*$ indicates the sum of aqueous CO_2 and H_2CO_3, where the latter is only $\approx 1\%$ of the total. Free aqueous CO_2 is actually calculated as $H_2CO_3^*$.

H_2CO_3 dissociates in water according to the following reactions:

$$H_2CO_3^* \rightarrow H^+ + HCO_3^- \tag{4.3}$$

$$HCO_3^- \rightarrow H^+ + CO_3^{2-} \tag{4.4}$$

These reactions attain an equilibrium that is dependent on pH and temperature as follows:

$$K_1 = [H^+][HCO_3^-]/[H_2CO_3^*] \tag{4.5}$$

$$K_2 = [H^+][CO_3^{2-}]/[HCO_3^-] \tag{4.6}$$

where K_1 and K_2 are, respectively, the first and second dissociation constants for the system and $= 10^{-6.3}\,mol\,l^{-1}$ and $10^{-10.3}\,mol\,l^{-1}$ at 25°C. These equilibria are shown graphically in Figure 4.14 as a function of pH. From the equilibrium expressions above it is clear that at a pH of 6.3, half of the C_T is $H_2CO_3^*$ and half is HCO_3^- and at a pH of 10.3 the same is true for HCO_3^- and CO_3^{2-}. This is also evident in Figure 4.14 where C_T is constant and the system is closed to the atmosphere. If that were not so, $H_2CO_3^*$ would remain constant at any pH if in equilibrium with the atmosphere.

Photosynthesis and respiration are two major factors that cause a significant departure from equilibrium of the system with the atmosphere. As algae photosynthesize, depleted CO_2 can be replaced in water by the following two reactions:

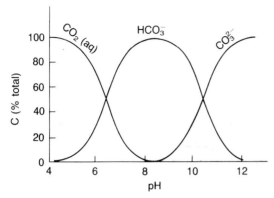

Figure 4.14. The distribution of the three forms of total C, when C_T is constant, as a function of pH (modified from Golterman, 1972)

$$H_2CO_3 \rightarrow CO_2 + H_2O \qquad (4.7)$$

$$HCO_3^- \rightarrow CO_2 + OH^- \qquad (4.8)$$

If CO_2 is replenished from the atmosphere as fast as it is removed by algae, the pH will not change, because H_2CO_3 and H^+ in the first equilibrium expression remain constant. However, pH will decrease if CO_2 production through respiration is in excess of CO_2 loss through diffusion to the atmosphere. That is, H_2CO_3 would increase above saturation and according to the first equilibrium expression, so would H^+.

From the first equilibrium expression, it is also clear that as algae consume CO_2 faster than it can be replaced by diffusion from the atmosphere, H^+ will decrease and, thus, pH will rise. That is because HCO_3^- can be considered to represent alkalinity at pH < 8.3, and would tend to remain constant. HCO_3^- can represent alkalinity at pH < 8.3 because as Figure 4.14 shows, CO_3^{2-} does not exist to any appreciable extent below that pH.

In reality, however, HCO_3^- is not constant as CO_2 is utilized but is dissociated to H_2CO_3, CO_2 and H_2O, as shown earlier. This, as well as the second dissociation reaction, are several times faster than necessary to match any conceivable algal demand (Goldman *et al.*, 1971). Considering the change in HCO_3^- as pH raises, alkalinity at pH < 8.3 can be defined as:

$$\text{Alkalinity} = HCO_3^- - H^+ + OH^- \qquad (4.9)$$

Alkalinity, however, will remain constant as CO_2 (and HCO_3^-) is consumed, because as HCO_3^- and H^+ decrease, OH^- will increase in order to maintain electrical neutrality.

The CO_2 system provides a buffer against the addition of H^+ and OH^- through the first and second dissociation reactions. Thus, pH remains

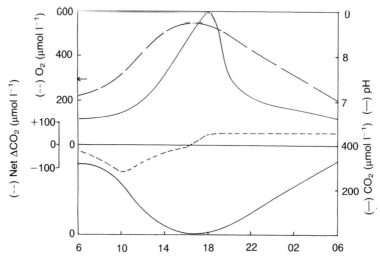

Figure 4.15. Diurnal changes in net CO_2 uptake/release, through photosynthesis/ respiration, and resulting concentrations of CO_2 (aq) and O_2 and pH in a system with $30\,mg\,l^{-1}$ $CaCO_3$ alkalinity (e.g. Lake Washington). No exchange with the atmosphere is assumed, as well as a $2:1$ ratio of daily photosynthesis ($3780\,mg$ $C\,m^{-3}day^{-1}$) to respiration, which is not unrealistic. Photosynthetic rate is reduced when CO_2 becomes depleted. pH was calculated using $K_1 = 10^{-6.3}$, $25°C$ and alkalinity remains constant. Arrow shows 100% saturation

relatively constant so long as C_T remains constant and in equilibrium with the atmosphere. However, the effect of photosynthesis is to remove CO_2 and, thus, C_T. If CO_2 were replaced from the atmosphere as fast as algae removed it, the pH would not rise. But the mere fact that C_T is observed to decrease and pH rise is evidence that algal uptake does exceed atmospheric resupply (Schindler, 1971b; King, 1972). In this way, photosynthesis tends to be a self-limiting process; by effectively reducing the CO_2 concentration, HCO_3^- and C_T are depleted and the pH rises. As the pH rises, the free CO_2 concentration decreases and both effects can contribute to inhibition of photosynthesis.

A hypothetical diurnal pattern of the effect of photosynthesis and respiration on DO and pH, and the self-limiting effect of pH, is shown in Figure 4.15. In this example, a very high but realistic photosynthetic rate of C uptake of $3.7\,g\,m^{-3}day^{-1}$ is assumed. A rate of about this magnitude was measured in Lake Trummen, Sweden before restoration (Cronberg et al., 1975). The ratio of photosynthesis to respiration was assumed to be 2/1, also reasonable. The uptake of CO_2 (photosynthetic rate) progressively decreases as the residual CO_2 concentration declines, completely ceasing as light and CO_2 extinguish at 6 p.m. At that point DO and pH reach their maximums at $\approx$175% saturation and 9.0, respectively. The first dissociation constant at 25°C was used to calculate pH.

DO and pH frequently exceed such values even if photosynthetic rates are lower. This is because reaeration and respiration are not sufficient to completely replenish CO_2 and deplete DO during the night. Thus, the initial concentrations of CO_2 and DO are lower and higher, respectively, with each successive day of calm, sunny weather. As a result high pH and DO values persist for increasingly longer periods.

5
CHARACTERISTICS OF WASTEWATER

Much of the waste of civilization enters water bodies through the discharge of water-borne wastes, termed wastewater. That is, the water used by the urban population for drinking/washing or by industry for cooling/washing/ processing is discharged carrying the unwanted and unrecovered substances. Wastewater can be treated or untreated and has conventionally referred primarily to that which enters water bodies at points of concentrated flow (through pipes), or point sources. However, the waste products of civilization can enter water bodies indirectly and through diffuse sources, i.e. non-point sources. Therefore, runoff water from parking lots and streets, from agricultural land or remote watersheds receiving waste products from the atmosphere must also be included under the general heading wastewater, because those inputs also contain unwanted and unrecovered substances. The only difference is that those substances were first distributed over the landscape.

Pollutants or pollution load are terms often used to describe wastes from civilization. Those terms imply some adverse effect and would be inappropriate in cases where wastes have no adverse, and may even have a beneficial effect. Wastewater was chosen here in order to remain more objective is discussing the effects of civilization's discarded products.

5.1 DOMESTIC SOURCES

Untreated domestic waste, or raw sewage, is usually greyish-brown, odiferous and relatively dilute (99% water). Those characteristics are often apparent in grossly polluted streams. There are four important constituents of domestic wastewater that are targeted for removal through treatment: total suspended solids (TSS), biochemical oxygen demand (BOD), nutrients nitrogen (N) and phosphorus (P), and pathogenic bacteria. The first four of these constituents have average concentrations ($mg l^{-1}$) in raw sewage of 200, 200, 40 and 10, respectively (Lager and Smith, 1975). Pathogenic bacteria are determined indirectly using the most probable

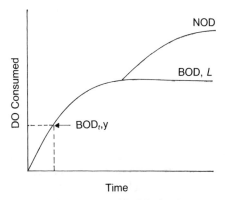

Figure 5.1. Relationship of BOD (carbonaceous DO demand) and NOD (nitrogenous DO demand) with time

number of bacteria from the coliform group as an index (5×10^7 most probable number (MPN)/100 ml in raw sewage).

BOD is an indirect determination of the readily oxidizable organic matter present. The reaction, by which organic matter is oxidized, can be described by first-order kinetics:

$$dL/dt = -kL \tag{5.1}$$

where L is BOD in $mg\,l^{-1}$, k is the reaction rate constant, and t is time. The integrated expression is:

$$L_t = L\,e^{-kt} \tag{5.2}$$

and if $y = L - L_t$, then:

$$y = L(1 - e^{-kt}) \tag{5.3}$$

so that y = BOD at time t and L is the ultimate BOD (Figure 5.1). The 5-day BOD (BOD_5) is a common measurement of effluent strength.

Domestic wastewater usually contains high concentrations of ammonium N ($15\,mg\,N\,l^{-1}$), because insufficient time and oxygen availability have prevented nitrification. Although carbonaceous BOD may be satisfied through biological treatment, oxidation of the ammonium represents a nitrogenous oxygen demand (NOD) to be satisfied in the receiving water through nitrification.

Domestic wastewater can be treated by settling the heavy solids and skimming off the floatable materials. This is referred to as primary treatment and normally removes $\approx 35\%$ of the BOD, 30–50% of the TSS, but only ≈ 10–20% of the N and P (Viessman and Welty, 1985). Secondary treatment, which was mandated by the Clean Water Act of 1972, should remove at least 85% of TSS and BOD to concentrations of $30\,mg\,l^{-1}$ for each (USEPA, 1982). This is usually accomplished by the activated sludge process, which is a system providing large amounts of oxygen to actively growing bacteria ('activated sludge'), trickling filters, biological discs or aerated or unaerated lagoons (Viessman and Welty, 1985). What is accomplished with secondary treatment is the removal of dissolved and non-settleable BOD by bacterial decomposition and the finer particulate matter through chemical coagulation and clarification.

Advanced wastewater treatment is usually employed for the purpose of P removal, and sometimes N, because only ≈ 50 and 25% of these two nutrients, respectively, are removed through the primary and secondary processes (USEPA, 1982). Also, a large fraction of TN and TP is soluble. P removal can be efficiently accomplished by chemical precipitation with alum (aluminium sulphate), iron salts (e.g. $FeCl_3$), or lime. These processes are often >90% efficient, producing residual P concentrations of $\approx 0.1\,mg\,l^{-1}$ ($100\,\mu g\,l^{-1}$), which, unless sufficiently diluted, is still a rather high inflow concentration to a lake (see Chapter 7). N can be removed by biological denitrification, ammonia stripping, breakpoint chlorination or ion exchange (Viessman and Welty, 1985).

Coliform organisms are usually removed by chlorination, which can be employed after either primary or secondary treatment. This treatment is referred to as disinfection, which could be accomplished with other oxidants, such as ozone. The addition of free chlorine dissociates into molecular hypochlorite (HOCl) and the hypochlorite ion (OCl^-), which represent available free chlorine. Chlorine is consumed by organic matter, so dose is adjusted to maintain a free chlorine residual of $0.2\,mg\,l^{-1}$ and should remove coliforms to less than 200 MPN/100 ml (Viessman and Welty, 1985). Free chlorine reacts with ammonia to produce chloramine, which is also toxic, and together are referred to as total residual chlorine ($TRCl_2$). The problems caused by $TRCl_2$, if not removed by dechlorinating domestic wastewater effluents, will be discussed under toxicants and fish.

Domestic wastewater is also treated by septic-tank drainfields, or 'on-site' systems, in suburban and rural areas where sewer collection systems are not available. If the drainfield is not overloaded with wastewater, so that it clogs with organic matter, and if composed of soils that allow sufficient percolation, treatment (including P removal) can be highly effective (see Chapter 7.3.7).

5.2 URBAN RUNOFF

Runoff from urban watersheds is usually transported in what are called 'separated systems' (separated from sewage) or 'combined systems' (combined with sewage). During and following storms, stormwater flows and combined sewer overflows (CSOs) can contribute the same constituents as in sewage and in relatively high concentrations as well. Table 5.1 shows a

TABLE 5.1.   Comparison of the contributions of common constituents in sewage effluent between a natural pristine area and urban area of 1 km² with 2500 people (from Weibel, 1969; metal data from METRO, 1987)

| Constituent | Concentration $(mg\,l^{-1})$ | | | Yield $(10^6\ tons\,km^{-2}\,yr^{-1})$ | |
	Natural	Sewage	Runoff	Sewage	Runoff
TSS	0.8	200	227	68	64
BOD	1.0	200	17	68	5
TN	0.5	40	3.1	14	1
TP	0.02	10	0.4	3.5	0.1
Zn	0.002	0.16	0.3	0.06	0.3
Cu	0.002	0.11	0.2	0.04	0.2
Pb	0.002	0.04	0.7	0.01	0.7

comparison of the concentration and yield of the important constituents in raw sewage effluent and urban runoff from a typical urban area.

The interesting points about these average values are: the concentration and yield for TSS from stormwater runoff and sewage are similar. Streets and parking lots, etc. collect large amounts of solid matter, which washes off to streams and lakes; and BOD in urban runoff is much higher than in runoff from pristine areas, but much less than in raw sewage. Higher average concentrations for stormwater have also been given (TSS = 630 and BOD = $30\,mg\,l^{-1}$; USEPA, 1982). The important point is that streams can be composed entirely of stormwater following rains with BOD concentrations that would be the same as if a raw sewage effluent entered the stream with only a 6- to 12-fold dilution (200 : 30, 200 : 17). A stream BOD of $20\,mg\,l^{-1}$ was observed during intermediate flow conditions in Lytle Creek, Ohio, in which gross biological effects were described (Chapter 12).

While TN and TP concentrations are much higher in runoff water than in runoff from pristine watersheds, they are less than in raw sewage. Metal (Zn, Cu and Pb) concentrations and areal yields (1 m runoff assumed), on the other hand, are often higher in runoff than in sewage effluent. That could vary, however, depending on how much industrial waste enters the municipal system. Much of the nutrients (N and P) come from dirt washed off vehicles or moved by wind, animal faeces and fertilizers applied to lawns. Sources of nutrients will be discussed further in Chapter 10.

The metals in urban runoff are sorbed to particles and come from corrosion of vehicles and piping systems, as well as from atmospheric emissions. Lead in runoff comes from gasoline and has declined since the advent of lead-free gasoline in the early 1970s. A large fraction of these metals are complexed (tied up) with particulate matter and are not in a form normally considered toxic.

Stormwater runoff can have rather high concentrations of coliforms (4×10^5 MPN/100 ml), although not as high as CSOs (5×10^6 MPN/100 ml; USEPA, 1982).

5.3 INDUSTRIAL WASTES

Wastewater from industrial processes are varied and complex often containing compounds not found in nature. Industrial wastewaters are often highly discoloured, turbid, alkaline or acid and unique to the industry in question. Wastewaters from some of the more common industries will be briefly described.

5.3.1 Pulping wastes

There are three pulping processes; Kraft, sulphite and ground wood. The magnitude of the waste problem from pulping is obvious from the fact that cellulose is the desirable commodity and cellulose comprises only about half of the tree. The structural portion of the tree, or the lignin, is separated from the cellulose by digestion with chemical additives. Alkaline solutions of sodium sulphate are used in the Kraft process, while acid solutions of sodium bisulphite are used in the sulphite process. After cooking, the pulp is washed, bleached and pressed. Each of those steps has a waste stream.

Sulphite plants produce very high BODs; effluents with 10 to 14% solids can have a BOD_5 of 30 000 mg l^{-1}. Kraft effluents have lower BODs, but yield more odiferous gases to the atmosphere, such as sulphides and mercaptans, and higher suspended solids to wastewater. Toxicity of pulping wastes has been associated with the resin acid component, but more recently dioxin has been found in bleach plant wastewater.

5.3.2 Petroleum refinery wastes

There are several operations at refineries, such as oil storage, desalting and fractionation of crude oil, thermal and catalytic cracking, solvent refining, dewaxing, deasphalting, and wax and hydrogen manufacture. Each of these can produce wastes that have BOD, toxicity and/or produce taste/odour problems, due to the dissolved organics. Phenols are an important constituent in refinery wastewater. In addition, spills of fuel and crude oil from tankers at sea represent a substantial waste source from the petroleum industry.

5.3.3 Food processing wastes

This industry is also varied, but generally the wastewater consists primarily of TSS and BOD. Such industries as fruit and vegetable canneries, meat packing plants, fish canneries, dairies and breweries produce wastewaters with very high BODs, often ten times that of domestic sewage.

5.3.4 Mining wastes

Acid mine drainage and leaching from abandoned dredge spoils often produce 'wastewater' with high concentrations of toxic metal ions, such as Cu, Zn and Fe. The problem of acid mine drainage develops because the mined-out portions of coal veins fill with water, which becomes anoxic producing reduced iron and sulphur compounds (FeS). When the water drains from the veins it is aerated, promoting the biological oxidation of iron and sulphur, yielding sulphuric acid. The metal ions are more soluble in the acid water and hence become more toxic. There are many abandoned, uncontrolled dredge-spoil deposits that continually leach acidity and toxic metals to streams. One of these cases is discussed in Chapter 12.

5.3.5 Acid precipitation

In a sense, acid precipitation is analogous to acid mine drainage and urban stormwater runoff, except that this acid runoff creates the most serious effects in high elevation, remote areas, rather than in urban settings. That is because those areas have less capacity to neutralize the acidity. The acid precipitation is caused by strong acids (nitric and sulphuric) that originate as sulphur dioxide from primarily fossil-fuel burning power plants and nitrogen oxides from largely auto emissions that are later oxidized in atmospheric clouds. Although NO_3^- may be nearly as important as SO_4^{2-} in causing a low pH in rain, SO_4^{2-} is considered more significant at acidifying surface waters because NO_3^- is usually a limiting nutrient in forested areas and is thus neutralized through biological uptake and growth. Although the equilibrium pH of pure water in contact with the atmosphere is about 5.6, precipitation is not considered acidified unless the pH is <5, because of the effect of natural background SO_4^{2-} (Charlson and Rodhe, 1982).

5.3.6 Toxic wastes

There are highly toxic, refractory organic compounds that have entered the environment via leachate from solid-waste landfills, direct discharge to surface water, atmospheric emissions that are deposited over broad areas, direct applications to control pests, and accidental spills. Some representatives of this group include polychlorinated biphenyls (PCBs), polybrominated biphenyls (PBBs), chlorinated hydrocarbon insecticides, and dioxins, which are chlorinated organic compounds. Many of these compounds were used because they continued to be effective over a long period of time, thereby making them economical. For the same reason, they are especially harmful if released into the environment. Use of PCBs and some chlorinated organic insecticides is banned in the US, but because of their long life there are high concentrations still remaining in landfills, sediments of lakes and rivers and in the flesh of animals.

PCBs were used extensively as an insulating material in transformers and capacitors, primarily in the electrical industry. They have also been used in paint solvents and plastics. Concentrated residues still remain in landfills and sediments (see Chapter 13.5). PBBs were used in a fire retardant. Chlorinated hydrocarbon insecticides, of which DDT is the well-known example (including endrin, aldrin, heptachlor and dieldrin), were used extensively in the US until the 1970s, when sale was banned. However, some use of residual stocks continued thereafter. They are still used in developing countries and often returned to the US in imported food (Revelle and Revelle, 1988). Many insecticide substitutes of shorter longevity have been developed, such as the pyrethrins. Dioxins are a group of chlorinated hydrocarbons that can be formed through combustion and were a byproduct in the manufacture of 2,4,5-T, a brush control herbicide used in the Vietnam war.

5.4   AGRICULTURAL WASTES

The wastewater from agriculture is more varied and extensive than any of the above types. Cultivation results in increased erosion and high TSS loads in receiving waters. Along with the TSS come nutrients and pesticides that were applied for fertilization and insect control. The increased waste loads can also occur in forested areas through timber harvesting, which is in a sense agriculture. BOD/TSS wastes from livestock enclosures can be highly concentrated. Return flows from rill (or flood) irrigation can carry high concentrations of TSS, fertilizers and pesticides.

There are many attempts underway to control the wastewater load from agriculture through best management practices. These involve minimizing erosion through improved cultivation and harvesting practices, treatment of wastewater from livestock enclosures, minimizing fertilizer applications through soil-testing procedures and conversion to spray irrigation, which yields much less TSS in the runoff water because return flow ditches are eliminated.

Part Two

Effects of Wastes in Standing Water

Standing bodies of freshwater, namely lakes, ponds and reservoirs, are known as lentic environments. Although standing or lentic in nature, there is movement of water within their basins that is determined by morphometry (depth, shape, etc.), solar heat input and wind. Plankton algae, or phytoplankton, are at the mercy of water movement and are usually the most important producers in lentic water bodies. The extent to which water moves (is mixed) vertically and horizontally determines the distribution of plankton and the amount of light available to mixed plankton cells, so water movement indirectly determines the productivity of lentic water bodies.

The gradation among lakes, reservoirs, rivers and streams is large and the distinction is not always clear-cut physically. The principal factor separating these environments is water residence time. Algal abundance per unit phosphorus in the water was found to increase along a residence time gradient from rivers to impoundments to natural lakes (Soballe and Kimmel, 1987). However, large, slow-moving rivers and the heads of reservoirs and estuaries may have significant plankton populations even though their residence times are relatively short. The effect of residence time on the dominant producer organisms present is illustrated in Chapter 11.

While the emphasis here is on the ecological principles and effects of nutrients in lakes, those principles and effects apply to reservoirs as well. However, reservoirs generally have higher flushing rates (shorter residence times) and consequently higher nutrient loading than lakes (Walker, 1981). Main stem reservoirs usually show a zonal transition from a riverine to a lacustrine type environment (Kimmel and Groeger, 1984). Estuaries show similar transitions from freshwater to marine, but the boundaries can change diurnally as a result of tidal effects.

Lakes/reservoirs are affected by sediment, toxicant and BOD wastes, but dilution capacity is usually sufficient to avoid violations of standards for DO, pH and temperature, except in enclosed embayments near wastewater sources. Chlorinated organic and/or metal contamination of fish and sediments are serious problems in such water bodies as the Great Lakes (see Chapter 13), Puget Sound (Washington state) and the Baltic Sea (see Figure 3.13). However, effects of and controls for these types of waste have typically occurred in rivers. The degradation of whole lakes/reservoirs is more likely to occur from eutrophication, the causes, effects and controls for which are emphasized in Chapters 6 to 10, or from acidification, which is treated with toxicity in Chapter 13. Global warming may affect whole watersheds, through both thermal restriction of population distribution and hydrologic modification.

6
PHYTOPLANKTON

The phytoplankton of lakes, estuaries and oceans is composed of single-celled algae. Cells of the plankton algae are largely microscopic in size ranging from a few microns to a few hundred microns in the longest dimension. Species may occur as single cells, or as colonies or filaments composed of many cells. The colonies/filaments of some species are often visible to the naked eye. For example, spherical colonies of the blue-green alga *Gloeotrichia* are ≈ 1 mm in diameter and clearly visible. The filamentous blue-green alga *Aphanizomenon* forms bundles of thichomes (filaments) that appear as grass clippings in the water. The phytoplankton are sometimes grouped by size into ultra-, nano- and net plankton (those caught with a net), with respective separations at about 10 and $50\,\mu$m (Wetzel, 1983). Phytoplankton are usually quantified by microscopic analysis of preserved water samples, rather than net samples, because of the significance of nanoplankton.

The taxonomic orders of algae representing the major portion of the phytoplankton are the Chlorophyta (green algae), Chrysophyta (diatoms and yellow-green algae), Cyanophyta (blue-green algae, also Cyanobacteria), Pyrrhophyta (dinoflagellates), Euglenophyta (euglenoids) and Cryptophyta (cryptomonads), although groupings may vary among authors. For general and taxonomic characteristics see Prescott (1954), Hutchinson (1967) or Wetzel (1983).

Most phytoplankton are maintained in the water column by turbulence caused by the wind, hence the term plankton. Many have irregular shapes that increase their surface to volume ratio and accordingly decrease density, so as to resist sinking. However, except for blue-green algae, which contain gas vesicles that render them buoyant, and some flagellated motile species, most will readily sink to the sediments under quiescent conditions. For example, *in situ* experiments conducted in plastic water columns have resulted in increased sinking loss rates because they are isolated from the normal water column turbulence. As the cells senesce they become denser, which increases their loss through sinking.

The three determinations usually made in characterizing the phytoplankton in standing waters are productivity, biomass and species composition. How much is being produced, how much is there and what is it? Productivity and biomass can be determined by chemical or physical procedures (Vollenweider, 1969b) and related to environmental factors without necessarily knowing the major organisms responsible. Although productivity and/or biomass may be sufficient to characterize the phytoplankton in certain situations, as will be seen later in this chapter, the taxa (species, genera, etc.) comprising most of the biomass may be equally important to water quality in many cases. In the same sense, the trophic state of a lake can be defined quantitatively, based on measurements of productivity and biomass (see Chapter 7). However, quantitative criteria for species composition have not been defined for that purpose.

Although detailed descriptions of seasonal changes in phytoplankton taxa, biomass and productivity in a particular lake are interesting and useful, it is usually more appropriate for water-quality management to know how the average state of lakes in general changes with an increasing or decreasing load of nutrient, acidity or sediment. That is, effort expended on developing relationships between substance input and average response, based on studies of large populations of lakes (Vollenweider, 1969a, 1976; Dillon and Rigler, 1974a; Jones and Bachmann, 1976; Chapra and Reckhow, 1979; Smith, 1982, 1986, 1990a,b; Prairie *et al.*, 1989), has proven more valuable to lake management than if that effort had been expended on understanding the more detailed interactions within a given lake. Also, the value of whole-lake manipulations to limnology and lake management is apparent. Although small-container experiments (bottles, bags, etc.) give useful indications of lake response, they are largely inadequate because of the many factors operating in a whole-lake ecosystem that interact to buffer or magnify the response. Therefore, deliberate or accidental manipulations of whole-lake systems (Edmondson, 1972; Bjork, 1974; Schindler, 1974; Edmondson and Litt, 1982) have presented more reliable information in that regard. Nevertheless, small-scale experiments are usually necessary to understand the underlying mechanisms that combine to produce the overall response.

6.1   SEASONAL PATTERN

The general seasonal pattern shown for a northern temperate sea (Figure 6.1) is quite similar to that for a large temperate monomictic or dimictic lake. The nutrient content is normally high following autumn overturn and may remain high during winter in a monomictic lake; it may increase

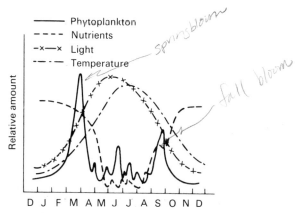

Figure 6.1. General seasonal cycle in north temperate oceanic phytoplankton and ecological factors (modified from Raymont, 1963)

during spring overturn in a dimictic lake having declined some during winter ice cover. Both situations provide a large available nutrient supply for phytoplankton at a time when light is increasing. The large nutrient supply exists principally because there is little or no growth to remove nutrients during winter when light and temperature are low. A spring diatom 'outburst', or 'bloom', usually occurs when light intensity reaches a level so that gross photosynthesis exceeds respiration. Species with low temperature requirements are usually responsible for the spring outburst.

The spring bloom is followed by a mid-summer minimum in algal biomass and productivity, largely because nutrients have reached a low and production-limiting level. The controlling effect of nutrients is particularly evident because light and temperature are at their maximums, thus the production potential is high. Grazing by herbivorous zooplankton is also occurring and may represent a significant loss of cells in addition to sedimentation.

In autumn, the significance of nutrients is again readily apparent because increased mixing, which results from lowered surface temperatures, entrains regenerated nutrients into the trophogenic (productive) zone. More nutrients from depth may become available during late summer in lakes because of the continual deepening of the mixed layer. A fall outburst of diatoms often results in the temperate ocean, as is also the case for poorly enriched freshwater lakes. In highly enriched temperate lakes, however, blooms of blue-green algae (or cyanobacteria) tend to develop in late summer and autumn or may occur throughout the summer, especially in shallow unstratified lakes. Seasonal cycles of stratification and algae also occur in tropical lakes, although less pronounced than in their temperate counterparts (Thornton, 1987a).

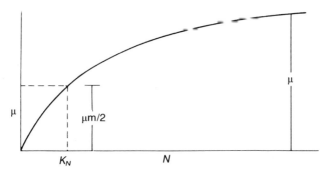

Figure 6.2. A Michaelis–Menton type relationship between microorganism growth rate and the concentration of limiting nutrient

6.2 POPULATION GROWTH KINETICS AND CONCEPT OF LIMITATION

Growth of a population of microorganisms can be described as exponential if resources are unlimited as shown in Chapter 2. This unlimited growth can be represented by a first-order expression where the rate of increase in the population, dX/dt, which is also net productivity, is dependent on the size of the population, X, and the growth rate constant, μ, is symbolized by:

$$dX/dt = \mu X \tag{6.1}$$

The same relationship of population size with time shown in Figure 2.3 holds here; only the symbols for biomass and growth rate have changed to conform with those commonly used to describe growth of microorganisms.

The growth rate, μ, is controlled by the concentration of most limiting nutrient at any time, as well as other environmental factors. The control exerted on growth by a limiting nutrient can be described by the Michaelis–Menten relationship developed from enzyme kinetics; bacterial growth being dependent of the concentration of an organic substrate and algal growth on that of a limiting inorganic nutrient (Herbert *et al.*, 1956; Droop, 1973). The equation for growth rate of a microorganism limited by a single nutrient is:

$$\mu = \mu_{max}N/(K_N + N) \tag{6.2}$$

where μ_{max} is the maximum rate attainable for that population under optimal environmental conditions and is determined by its genetic character, N is the concentration of limiting nutrient, or in the case of bacterial growth it is indicated by S for substrate, and K_N (or K_S) is the half-saturation constant or the nutrient (or substrate) concentration at one half the maximum rate of growth. This relationship, and a definition of the parameters, is shown in Figure 6.2. The expression shows that at very high concentrations of nutrient and a relatively low K_N, growth rate approaches the maximum. As N decreases, especially when it drops below K_N, μ becomes increasingly dependent on N.

Substituting Equation 6.2 into the first-order population growth Equation 6.1 gives the following expression for the change in population size controlled by a limiting nutrient:

$$dX/dt = \mu_{max} \times N/(K_N + N) \qquad (6.3)$$

6.2.1 Mixed reactor

To understand better how population growth is controlled and the relationship of biomass to nutrient concentration, growth in a continuously stirred tank reactor (CSTR) or chemostat will be considered. If algae are grown in such a mixed reactor, there will be an inflow of water and dissolved nutrient (N_i) and an outflow of nutrient (N), water and algal cells (X). The population growth in the reactor is given by:

$$dX/dt = \mu_{max} \times N(K_N + N) - DX \qquad (6.4)$$

where D is the dilution rate determined by the inflow rate of water/reactor volume. That parameter is also known as the water exchange rate or flushing rate when referring to lakes. When nutrient rich water is added to a algal-seeded chemostat, the population growth and loss begin, the loss being the fraction of reactor volume replaced per unit time (D) multiplied by the concentration of algal biomass in the reactor (X). The population will continue to increase faster than cells are lost through the outlet until the concentration of cells (X) becomes large enough that the loss equals the growth. At that point, $dX/dt = 0$ and a steady-state condition is reached. The population will continue to grow and maintain a steady-state biomass so long as the inflow rate of water and inflow nutrient concentration (N_i) remain constant. The dynamics of phytoplankton growth in lakes can begin to be understood by considering further the steady-state growth in a chemostat.

The change in nutrient concentration (N) in the reactor with time is given by:

$$dN/dt = DN_i - DN - [\mu_{max}X/Y \ N/(K_N + N)] \qquad (6.5)$$

where Y is the yield coefficient or the mass of cells per unit mass of nutrient. The equation simply says that the rate of change of N in the reactor is equal to the difference between the inflow rate of nutrient and the uptake rate by algae. To illustrate the steady-state phenomenon, the calculated response of algal biomass and nutrient concentration in a reactor, starting from initial conditions of $N_i = 5.0\,\mathrm{mg\,l^{-1}}$, $X = 5.0\,\mathrm{mg\,l^{-1}}$, $D = 0.5\,\mathrm{day^{-1}}$, $\mu_{max} = 1.0\,\mathrm{day^{-1}}$, $Y = 64$ and $K_N = 0.3\,\mathrm{mg\,l^{-1}}$,

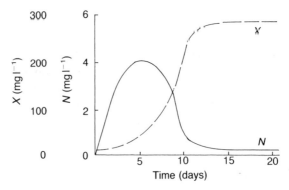

Figure 6.3. Calculated response of microorganism biomass (X) to nutrient content (N) in a mixed reactor over time

is shown in Figure 6.3. N is observed to increase initially as the inflow of nutrient exceeds the removal rate by the algae. But as the algal biomass increases the removal rate of N increases and the concentration falls. The system reaches a steady state for biomass and nutrient after about 12 days.

The steady-state result of Equation 6.5 for N in a reactor is:

$$N = K_N D/(\mu_{max} - D) \tag{6.6}$$

This equation shows that at a very low dilution rate there is ample time for algal uptake to reduce N to low levels. As D increases, however, there is less and less time for utilization of the incoming nutrient and the residual N in the reactor increases. At the point where $D = 0.5\mu_{max}$, $N = K_N$. When $D > \mu_{max}$, the loss exceeds the growth, washout occurs and $N = N_i$.

The steady-state biomass is determined by the difference between the inflow (N_i) and reactor (N) concentrations of nutrient:

$$X = Y[N_i - K_N D/(\mu_{max} - D)] \tag{6.7}$$

or, substituting from Equation 6.6:

$$X = Y(N_i - N) \tag{6.8}$$

Multiplying both sides by the dilution rate one has steady-state productivity as follows:

$$DX = DY(N_i - N) \tag{6.9}$$

The expected response of steady-state biomass, residual nutrient and productivity to changing dilution rate is shown in Figure 6.4. The results in Figure 6.4 are easily explained with the aid of Equations 6.6, 6.8 and

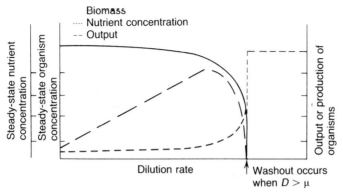

Figure 6.4. Dilution rate versus biomass, production and nutrient concentration in a chemostat culture (Herbert *et al.*, 1956)

6.9. The residual nutrient concentration (N) is independent of the inflow concentration, but is controlled by the dilution rate. This is because algal cells rapidly take up any increase in nutrient that results from continually adding the nutrient-rich inflow and turn it into biomass. N begins to increase only when D increases to a point where there is insufficient time for the algae to absorb the difference between N_i and N. According to Equation 6, this increasingly happens when D exceeds $0.5\mu_{max}$. As washout is approached, N approaches N_i. This illustrates why the soluble form of the limiting nutrient is usually in very low concentration in lakes.

Biomass in Figure 6.4 remains rather constant over the lower range of dilution rate; again, dilution rate is slow enough to permit near complete removal of nutrient and biomass is simply proportional to the difference between inflow and in-reactor concentrations of nutrient (Equation 6.8). As D increases and μ_{max} is approached, the utilization of N is less complete and biomass declines with washout occurring when $D > \mu_{max}$. If the dilution rate is held constant and the inflow nutrient concentration is increased, the time for uptake is the same so the added nutrient is simply converted to biomass and the latter increases proportionately. This process has great significance for managing the biomass of algae in lakes; all other loss rates being constant, biomass should be a function of the inflow concentration of available (soluble) nutrient.

As Equation 6.9 shows, productivity or output of algal cells is simply the product of dilution rate and biomass in the reactor. Hence, productivity increases in proportion to dilution rate until the loss rate through dilution begins to approach washout. Productivity, or efficiency of the system to convert nutrients or substrate, is more efficient at high dilution rates than the simple product of X and D may indicate.

Example problem

Calculate the steady-state biomass in a continuous flow culture of algae for each of three dilution rates ($D = 0.3$, 0.6 and $0.9\,\text{day}^{-1}$) at each of three inflow concentrations of limiting nutrient ($N_i = 2.0$, 4.0 and $6.0\,\text{mg}\,\text{l}^{-1}$). Assume $\mu_{max} = 1.0\,\text{day}^{-1}$, $K_N = 0.3\,\text{day}^{-1}$ and a yield ratio = 64. When is biomass most dependent on N_i and when on D?

Using Equation 6.7, because N varies with D and has some impact on X:

$$X = Y[N_i - K_N D/(\mu_{max} - D)]$$

gives the following X values:

		D	
N_i	*0.3*	*0.6*	*0.9*
2.0	120	99	−45
4.0	248	227	83
6.0	376	355	211

At D values substantially below μ_{max}, when washout is not occurring, X is linearly dependent on N_i (at low D, N is reduced to very low levels). As D approaches μ_{max} (near washout), this linear relationship ceases to exist and D has a larger control on X (and residual N increases, since time is insufficient for complete uptake).

Changes in other variables as dilution rate increases are also of interest. The growth rate, μ, also increases as D increases. This is because residual N increases allowing a higher growth rate (Equation 6.2). Growth rate (μ) must increase as D increases because at steady state ($dX/dt = 0$), $D = \mu$. This is obvious from the following expression for constant biomass (X):

$$dX/dt = \mu X - DX \tag{6.10}$$

As indicated dependence of μ on D is apparent from the Michaelis–Menten relationship. As D increases, N increases, especially after $D > 0.5\,\mu_{max}$ (see Equation 6.6), and the $\Delta\mu/\Delta N$ decreases. At low D ($D < 0.5\,\mu_{max}$), $\Delta\mu/\Delta N$ is large, representing the steeper slope in the Michaelis–Menten relation.

In summary, an increase in limiting nutrient concentration in the inflow increases biomass and production, whereas an increase in the dilution rate only increases production. Environments where dilution or flushing rate can have a large influence on biomass and soluble nutrient control are small lakes or bays with large watersheds or large volume tributary streams, constricted estuaries and sewage lagoons. These are ecosystems with detention times of the order of 10 days or less, or expressed as dilution rate, $0.1\,\mathrm{day}^{-1}$ or more. For example, Sobelle and Threlkeld (1985) observed that plankton algal biomass varied directly with detention time in a small reservoir with a mean detention time of 2.4 days ($0.4\,\mathrm{day}^{-1} = \rho$) over a range up to 12 days detention time ($0.08\,\mathrm{day}^{-1}$). Advection controlled algal biomass, because nutrient content was high, supporting a relatively high growth rate.

The concept of continuous-culture kinetics has been employed to predict the algal biomass leaving reservoirs with detention times less than two weeks (Pridmore and McBride, 1984; McBride and Pridmore, 1988). A continuous culture, logistic model, in which growth rate was restricted by a chlorophyll a maximum based on total P concentration, gave very reliable predictions of steady state, growing season chlorophyll a concentrations at the outlet of reservoirs with detention times of 1.3 and 5.6 days. In the short detention time reservoir, outflow chlorophyll a was only $\approx 14\,\mu\mathrm{g}\,\mathrm{l}^{-1}$, while the TP content of $48\,\mu\mathrm{g}\,\mathrm{l}^{-1}$ would have predicted a chlorophyll a of $39\,\mu\mathrm{g}\,\mathrm{l}^{-1}$, clearly a situation where detention time was most limiting. Their model predicted a chlorophyll a of $\approx 15\,\mu\mathrm{g}\,\mathrm{l}^{-1}$.

6.2.2 Batch reactor

Population growth in ecosystems with longer detention times is more typical of growth in batch cultures or closed containers. For this system, the dynamic equations for biomass (Equations 6.1–6.3) still apply, but because there is not a constant loss of biomass, a steady state cannot be reached. The batch culture is more typical of a natural system in which biomass and nutrient concentration continually change. Nutrient concentration declines as biomass increases as shown in Figure 6.5. As N is depleted, the resulting growth rate declines causing a slowing in the biomass increase, which ultimately ceases as N reaches the minimum concentration. There is some cell loss through death and decay, which increases as growth slows. If conducted at optimum light and temperature for the species involved, nutrient utilization and growth will be rapid so that maximum biomass or yield will be consistently proportional to the initial nutrient concentration. Figure 6.5 is a typical result of an algal growth potential (AGP) test with *Selenastrum* or *Scenedesmus*.

6.2.3 Other losses

There are other losses of phytoplankton in lakes, such as grazing by zooplankton and sedimentation through sinking, which are usually more important than losses through dilution. The following equation shows a simplified version of how these losses are frequently considered:

$$dX/dt = \mu X - DX - GX - SX \qquad (6.11)$$

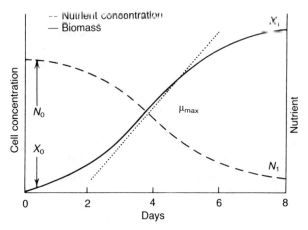

Figure 6.5. Nutrient concentration and biomass in an algal batch culture showing initial and final levels (after McGauhey *et al.*, 1968, p. 8)

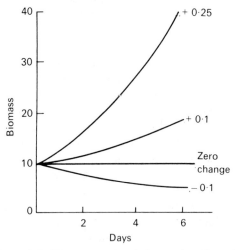

Figure 6.6. Biomass rates of change in a hypothetical algal population with $\mu = 1.0\,\text{day}^{-1}$ and combined loss rates of 0.75, 0.9 and $1.1\,\text{day}^{-1}$

where G and S are, respectively, the grazing and sinking rates. These rates can be represented by functional submodels, such as described for μ (Equation 6.2). To illustrate the significance of loss rates in controlling phytoplankton biomass, biomass change in response to three different rates of combined loss have been calculated and are shown in Figure 6.6. For the example, μ was held constant at $1.0\,\text{day}^{-1}$ as were G and S at $0.5\,\text{day}^{-1}$ and $0.2\,\text{day}^{-1}$, respectively. The effect of D was tested at 0.5,

Figure 6.7. Idealized steady-state system for plankton limited by nitrogen (after Dugdale, 1967)

0.2 and $0.4\,day^{-1}$. The resulting biomass change rates were $+0.25$, $+0.1$ and $-0.1\,day^{-1}$ and the biomass change in Figure 6.6 was calculated with the integrated equation below:

$$X_t = X_0\,e^{(\mu - D - G - S)t} \qquad (6.12)$$

Although grossly oversimplified, this example, with reasonable values for the rates involved, shows how sensitive phytoplankton biomass is to loss rates. Changes in loss rates can actually have as much or more influence on phytoplankton biomass than changes in limiting nutrient concentration. For example, consider perturbing the simplified, steady-state system shown in Figure 6.7. An increase in nutrient supply rate (N/t) to an actively growing population should result in increased productivity (nutrient uptake rate ($N_B N_{up}/N_b t$)) and possibly biomass, if loss rates remain low, without a noticeable increase in the nutrient pool. Supply rate could increase by either increasing the recycle rate or the inflow rate (e.g. dilution rate in the chemostat). The nutrient pool size could remain relatively low because of increased uptake rate. Biomass may not increase appreciably if loss rates, especially grazing, would also increase. However, some increase in nutrient pool size and biomass are necessary if uptake rate and grazing loss rate are to increase, because such rates are concentration dependent.

6.3 EFFECTS OF LIGHT AND MIXING

6.3.1 Light quality

Lakes may appear blue, green, red-brown or yellow. This is due to the selective transmission of different wavelengths of light. Dissolved and particulate matter scatters, absorbs and reflects light at some wavelengths more than at others. Pure water transmits blue light better than green and ultraviolet (UV) better than red and infrared (IR). In water with dissolved material, green light transmits farthest, then blue, red, UV and IR (Goldman and Horne, 1983). This is why pristine lakes, with very little particulate matter, appear blue or blue-green. Other wavelengths are

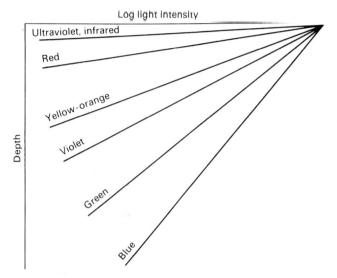

Figure 6.8. Relative peneltation of different wavelengths of light through lake water

absorbed by water first allowing blue light to penetrate farthest (Figure 6.8). Also, short wavelengths are reflected most and thus blue light is reflected back to the surface, further enhancing the blueness of pristine lakes. However, with increasing dissolved matter, lake water will appear more green than blue. Lake water that is brown or reddish-brown has high concentrations of particulate matter that absorb the shorter blue and green wavelengths first.

Chlorophyll *a* is the active pigment in photosynthesis, absorbing energy in the short and long wavelengths within the visible spectrum, which represents roughly one half the incident light energy over the range of 400–700 nm (Figure 6.8). Although chlorophyll *a* absorbs most in the red and violet areas, those wavelengths penetrate least (Figures 6.8 and 6.9). The energy from blue and green wavelengths, however, is available for photosynthesis through its absorbance by accessory pigments and transport to chlorophyll *a* (Schiff, 1964):

Cartonoids → Phycocyanins → Phycoerythrins → Chlorophyll

This is consistent with the observation that water column photosynthesis is predictable from the available quantity of the most penetrating wavelength, which is most often green light (Rodhe, 1963). In a comparison of photosynthetic rates in twelve European lakes with a range of light transmission characteristics, Rodhe found that relationships between photosynthetic rate and the intensity of the most penetrating wavelength

Figure 6.9. Absorption spectrum for chlorophyll compared with the penetration in water and emission from sunlight. Letters indicate colours (Schiff, 1964)

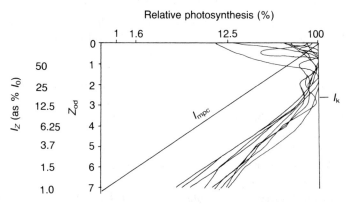

Figure 6.10. Relative photosynthesis versus optical depth (Z_{od}) and light intensity I_z ($= 100\% \, Z^{-Z_{od}}$) in a group of 12 European lakes compared to the most penetrating component (I_{mpc}, straight line) (modified from Rodhe, 1966) $I_k =$ saturation

were very similar for all the lakes for depths well below saturation where light is the limiting factor (Figure 6.10). Thus, the total incident visible light is used reliably as photosynthetically available radiation (PAR) and is generally related to photosynthetic rate as shown in Figure 6.11. Three regions can be depicted in the relationship as a light-limited or direct-response portion, a light-saturated portion and a light-inhibited portion where photooxidative destruction of enzymes occurs (Wetzel, 1983). The

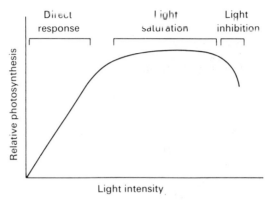

Figure 6.11. Hypothetical relationship between photosynthesis and light intensity

relationship could be expected to vary with species, temperature and adaptation.

6.3.2 Light quantity and limitation to photosynthesis

Light of any wavelength (colour) is attenuated with depth, by water and dissolved and particulate matter, according to the following equation:

$$I_Z = I_0 \, e^{-K_t Z} \tag{6.13}$$

where I_0 is the incident light intensity, I_Z is the remaining light intensity at depth, K_t is the attenuation coefficient for water and dissolved and particulate matter combined, and Z is depth. According to this equation, light penetrating the lake surface is reduced very rapidly at first and then more slowly as depth increases. Also, the greater the dissolved and particulate matter, and hence the attenuation coefficient, the more quickly light will decrease with depth.

An important source of particulate matter in lakes is plankton algae, whose abundance can be estimated by measuring the chlorophyll a concentration. Light transmission can be determined with a Secchi disc, the depth of which is a function of particulate matter (chlorophyll a) and attenuation due to water itself (K_w) and dissolved matter (K_d). By including chlorophyll a (Chl$a$) as the measure of particulate matter and adding attenuation by water and dissolved matter, i.e. K_{wd}, the general attenuation equation can be written as follows:

Secchi disk depth $$Z_{SD} = (\ln I_0/I_Z)/(K_{wd} + K_p \, \text{Chl}a) \tag{6.14}$$

If the Secchi disc is assumed to disappear at about 10% of surface intensity (100%), then the numerator on the right side becomes 2.3. Chl$a$ is in mg m^{-3} and K_p is expressed as $\text{m}^2 \, \text{mg}^{-1}$ Chl$a$, which can be taken as

Figure 6.12. The relationship between surface chlorophyll a and Secchi depth (SD). Data from Green Lake during June–September 1959, 1965 and 1981 (SD = $10.3/\text{Chl}a^{0.56}$; $r = -0.87$) The other lines are based on data from several lakes (Carlson, 1977; Reckhow and Chapra, 1983)

0.025. There are several empirical relationships in the literature that were derived from lake data sets and that approximate the above equation. Reckhow and Chapra (1983) approximated the effect of K_{wd} in their equation by estimating maximum Secchi depth (SD) without Chl$a$ (8.7 m), while Carlson assumed non-algal attenuation was negligible:

$$\text{SD} = 8.7/(1 + 0.47 \text{ Chl}a) \quad \text{Reckhow and Chapra (1983)} \quad (6.15)$$

$$\text{SD} = 7.7/\text{Chl}a^{0.68} \quad \text{Carlson (1977)} \quad (6.16)$$

where Chl$a$ is expressed as $\mu\text{g}\,\text{l}^{-1}$ and SD is expressed as metres.

These relationships, and one derived to fit data from Green Lake, are shown in Figure 6.12. Equations 6.15 and 6.16 produce very similar relationships, but substantially underestimate transparency in Green Lake. Developing such specific relationships of SD-Chl$a$ is frequently possible to do for a given lake, even with one or two years' data, while it is not possible for other trophic state variables with so little data (see Chapter 7.3).

Light quantity can be expressed in several units. Flux of visible (PAR) light is expressed as $\text{g}\,\text{cal}\,\text{cm}^{-2}\,\text{min}^{-1}$ or $\text{Ly}\,\text{min}^{-1}$ or $\mu\text{E}\,\text{m}^{-2}\,\text{s}^{-1}$ (Ly =

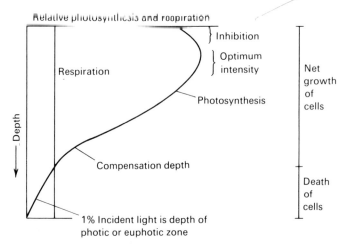

Figure 6.13. Relative photosynthesis and respiration versus lake depth, showing various characteristics of the water column

Langley, E = Einstein). Light saturation has been suggested to occur as low as $\approx 1.7\,\mathrm{Ly\,h^{-1}}$, or $100\,\mu\mathrm{E\,m^{-2}\,s^{-1}}$ or 4800 lux as a measure of intensity (Talling, 1957a, 1965). Inhibition should begin to occur at ≈ 8.6–$12.9\,\mathrm{Ly\,h^{-1}}$ or 510–$770\,\mu\mathrm{E\,m^{-2}\,s^{-1}}$ and $25\,000$–$37\,000$ lux. The bounds for light saturation are thus 7% and 35–50% of the full visible range of average daily June sunlight ($25\,\mathrm{Ly\,h^{-1}}$, $1500\,\mu\mathrm{E\,m^{-2}\,s^{-1}}$ and 71 000 lux).

The photosynthesis profile shown in Figure 6.13 is a summation of the curves of light attenuation with depth and the photosynthesis–light relationship.

Light intensity may be so great at the surface that photosynthetic rate is inhibited. At some greater depth the optimum occurs or the photosynthetic rate is saturated at that temperature. Photosynthesis decreases exponentially from that point in proportion to light and usually has decreased to insignificant levels at $\approx 1\%$ of surface intensity, which defines the photic-zone depth. In many clear lakes significant photosynthesis extends below 1% intensity. The compensation depth occurs where respiration equals photosynthesis and no growth results below that depth. This model applies only to a well-mixed photic zone, which unfortunately is not a typical situation in lakes, although it is more so in the ocean. Complete mixing is an acceptable assumption for explaining the process, but it must be remembered that the stratification of nutrients and algal biomass, which occurs in most lakes, will affect the photosynthetic profile. This process, however, does occur to the extent that light is the controlling factor and, consequently, these kinds of profiles will usually be observed. This pattern was evident in Chester Morse Lake, an oligotrophic lake in western Washington (Figure 6.14).

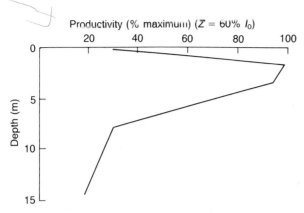

Figure 6.14. Distribution of productivity (relative) with depth in Chester Morse Lake, Washington. Values are based on means over a 1.5-year period (after Hendrey, 1973)

6.3.3 Efficiency of photosynthesis

Only a small percentage of available light is actually used because there are too few algae and too much light. About 1% of total incident light is fixed in photosynthesis as a biosphere mean, or $\approx 2\%$ of the visible light because the visible fraction represents about one-half of the total light energy. In water this efficiency drops below 1%; Riley reported 0.18% for an oceanic mean (Odum, 1959).

Efficiency can be increased greatly by decreasing the light intensity and maximizing the cells' exposure to it. Cultures of *Scenedesmus* and *Chlorella* have utilized as much as 50% of the incident light under these conditions (Brock, 1970). The cell itself can increase efficiency by adjusting its chlorophyll content – shade-adapted cells tend to contain more chlorophyll than light-adapted cells. For example, in Chester Morse Lake a photosynthetic efficiency index ($P/B-I$) was shown to increase greatly with depth, whereas chlorophyll *a* increased only slightly (Figure 6.15).

6.3.4 Predicting light-limited photosynthesis

The photosynthesis–depth profile can be predicted reasonably well by knowing P_{max}, the maximum photosynthetic rate, and the depth at one-half the light intensity at which P_{max} would occur without saturation (I_k). I_k and $0.5 I_k$ are defined in the light–photosynthesis relationship shown in Figure 6.16 (after Talling, 1957b). In this case, intensity is indicated by the percentage of incident light. As photosynthesis increases with increased light intensity, saturation begins to occur. That is, light becomes more abundant than the photosynthetic mechanism is able to utilize and the relationship departs from linearity, i.e. one unit of

Figure 6.15. Distribution of chlorophyll *a* and normalized productivity in Chester Morse Lake. Values are means over a 1.5-year period (after Hendrey, 1973)

photosynthesis per unit of light. I_k is defined as the intensity at which P_{max} would occur if this saturation phenomenon did not exist. Rather than saturation occurring abruptly with increased light, the approach to saturation is gradual, as shown in Figure 6.16(a), ultimately reaching a maximum at 50% I_0. Beyond 50%, P declines due to inhibition, reaching $0.5 P_{max}$ at 100% I_0. I_k, according to this model, occurs at 16% I_0 and, therefore, $0.5 I_k$ at 8% I_0.

The profile of photosynthesis with depth (Figure 6.16(b)) is simply the light response relationship that is tipped on end and reversed. The depth profile has a more constricted shape compared to the curve in Figure 6.16(a) because light intensity decreases exponentially with depth. The depths of I_k and $0.5 I_k$ can be located by knowing the extinction coefficients for water and chlorophyll ($K_w + K_c$) assuming K_w includes all non-chlorophyll attenuating material. Areal gross photosynthetic rate can be estimated by:

$$\text{Areal } P_{gross} = Z_{0.5 I_k} \times P_{max} \qquad (6.17)$$

The product of $Z_{0.5 I_k}$ and P_{max} produces that area of a rectangle in the profile that approximates the area under the photosynthetic curve including the tail below $Z_{0.5 I_k}$ (Figure 6.16b). For example, if $P_{max} = 50 \, mg \, m^{-3} day^{-1}$ and $K_w + K_c = 0.5 \, m^{-1}$, $Z_{0.5 I_k}$ would occur at 5.1 m $[\ln(I_0/I_{0.5 I_k})/0.5 \, m^{-1}]$. Thus, gross productivity would be, 5.1 m $\times$ 50 mg m^{-3}day^{-1} = 255 mg m^{-2}day^{-1}.

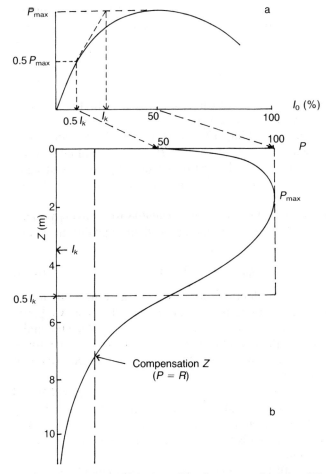

Figure 6.16. A hypothetical profile of gross productivity (P) with depth (Z) and light intensity (I) showing points of I_k, $0.5\ I_k$ and compensation (after Talling, 1957b)

This procedure resulted in predicted rates of productivity within 5% of observed for a group of 12 European lakes (Rodhe, 1963). Relative photosynthetic rate versus percentage surface light intensity, compared on log scales, for these lakes shows that at light intensities less than an average I_k, photosynthesis tends to be linearly related to the most penetrating wavelength (Figure 6.10). The success of this method is illustrated by the rather consistent shape of the photosynthetic curves, in spite of widely varying areal rates of productivity and P_{max}. It also says that 50% and 16% are reasonably good estimates for P_{max} and I_k, respectively.

To predict photosynthetic rate at depths throughout the water column, a model that represents the relationships in Figure 6.16 is needed. Steele's (1962) model, which relates photosynthetic rate to relative light intensity, with P_{max} occurring at 50% I_0 is often used for that purpose:

$$P = P_{max} 2I/I_0 \, e^{1-2I/I_0} \qquad (6.18)$$

Where $I = 50\% I_0$, $P = P_{max}$, which is consistent with the previous model. Where $P = 0.5 P_{max}$, $I = 12\% I_0$, which is analogous to $0.5 I_k = 8\% I_0$ in the formulation by Talling (1957b). However, this model reflects inhibition such that where $I = I_0$, $P = 74\% P_{max}$. An absolute rate for P_{max} is also required for this model. Also because P_{max} occurs at light-saturated conditions, its magnitude is dependent upon temperature and nutrient content.

The depth of mixing may limit the light available to a mixing population of plankton cells. The average amount of light available to mixing cells is given by:

$$I = I_0(l - e^{-kz})/(kz) \qquad (6.19)$$

where z is the mixed layer depth. Thus, by incorporating both mixed layer depth and light extinction, this equation provides a much better indication of light available in the water column over 24 h and under various mixing scenarios than would exist from incident light measurements alone. For example, I would be 20 and 43% I_0 for K_t values of 0.5 and $0.2 \, \text{m}^{-1}$, respectively.

Most plankton algae have a cell density greater than that of water and will sink from the water column unless continually mixed upward. If they are not returned to the photic zone frequently enough, growth will cease, which may also occur if mixing is too strong and/or at too great a depth, relative to light availability.

The settling velocity (W_s) of an algal cell can be estimated by assuming conformity to <u>Stokes' Law</u>:

$$W_s = 2gr^2(\rho' - \rho)/(9\eta\phi) \qquad (6.20)$$

where g is gravitational acceleration, ρ is the density of water, ρ' is density of the plankton cell, r is the radius of a sphere of equal volume as the cell, η is viscosity and ϕ is the 'coefficient of form resistance' (Reynolds, 1984). Thus, one can see that species-to-species differences in cell size (r), shape (ϕ) and density difference ($\rho' - \rho$) can account for the relative success and dominance of one species over another depending on strength and depth of mixing, which can determine the amount of light available. Some algae can resist settling through motility (flagellated green algae and dinoflagellates) and gas vacuoles (some blue-green algae), while diatoms, with cell densities on the order of $1.003 \, \text{g cm}^{-3}$, have little resistance and will sink on the order of a metre per day.

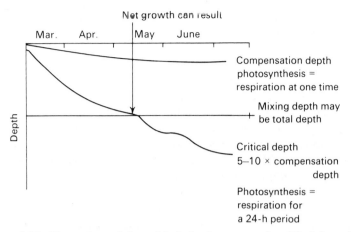

Figure 6.17. Illustration of the critical depth concept (modified from Marshall, 1958)

Under non-limiting nutrient conditions, i.e. during the well-mixed, pre-growth period in the spring in both deep temperate lakes and the northern ocean, the onset of the spring diatom bloom can be predicted relatively well. The approach involves the prediction of the critical depth, which is that depth below which the water column total of algal cells will not produce net growth over the duration of a day. To describe the critical-depth concept, it is appropriate to begin with the model first developed by Sverdrup (1953) as follows (see also Murphy, 1962):

$$P = m/k\, I_0(1 - e^{-kz})t - ntz \tag{6.21}$$

where P is a factor proportional to daily net production, m is a factor related to O_2 production, n is a factor related to O_2 consumption, k is the vertical extinction coefficient, z is the mixing depth in metres, I_0 is the effective radiation passing the sea surface on a daily basis and t is time. At the compensation light intensity (I_c), $n = mI_c$, then:

$$P = I_0(1 - e^{-kz})/k - I_cZ \tag{6.22}$$

At the critical depth ($D_{cr} = z$), net photosynthesis is zero so $P = 0$, thus:

$$D_{cr} = I_0/I_c(1 - e^{-kD_{cr}})/k \tag{6.23}$$

And if $I_c = 4.3\,\mathrm{Ly\,day^{-1}}$, then:

$$D_{cr} = I_0/4.3(1 - e^{-kD_{cr}})/k \tag{6.24}$$

Thus, the necessary information to know in order to predict D_{cr} for a lake, estuary or ocean area is k, I_0 and I_c. From the last equation it is clear that D_{cr} is primarily a function of I_0. As incident light increases, the compensation depth, and hence D_{cr}, increases as shown in Figure 6.17. At the time when D_{cr} exceeds the mixing depth, net productivity is possible and a plankton bloom can occur.

The critical depth may be confused with compensation depth. Compensation depth represents a light intensity where photosynthesis equals respiration, i.e. if a plankton cell remained stationary at that depth it would receive just enough light to exist, but no growth could occur. However, that is an unrealistic condition, because cells sink and are mixed vertically in the water column, such that some of the time they are below the compensation depth, and essentially dying, while some of the time they are above the compensation depth where the population is growing. The question is, what is happening in the water column overall on a daily basis? How does the amount of light available above the compensation depth compare with the depth of the mixed layer, most of which may be below the compensation depth, through which the respiration demand must be met? The first term on the right-hand side in Equation 6.21 represents gross productivity, or total available light converted into O_2, while the most right-hand term is respiration throughout the depth of the mixed layer or O_2 consumed. Thus, considered from another perspective, critical depth is clearly the depth that would be necessary to generate enough respiration to exactly use up the photosynthate produced from the amount of light available above the mixing depth.

The net effect is that if the average light intensity received by a mixing plankton population is greater than the compensation depth light intensity, net production will result. In many clear-water bodies, such as the ocean, that sustainable average light intensity can be received if cells are mixed to depths not greater than 5 to 10 times the compensation depth. In very clear lakes, the compensation depth may be from 50 to 100 m. Few lakes are that deep over much of their area, so in spring the mixing depth may be, in effect, the lake bottom and as light intensity increases, the critical depth may exceed the maximum lake depth, allowing net growth and a bloom, as early as February. In other cases, such as when turbidity is high, the decreasing mixing depth as thermal stratification increases can determine the bloom timing. Examples of the effect of mixing depth on productivity will follow.

Although the critical depth concept developed for oceanic phytoplankton may not be entirely appropriate for some lakes, because they are usually shallower than Z_{cr} (D_{cr}), the ratio of $Z_{eu} : Z_m$ (euphotic : mixing) is nonetheless an important concept for lake productivity (Talling, 1971). This can be demonstrated with the following hypothetical example. Given two morphologically different lakes with the same mixing and critical depths, the lake with the greatest ratio of $Z_{eu} : Z_m$ will have the greatest productivity, all other things being equal (Figure 6.18). To illustrate better the morphometric differences, use of mean depths of the euphotic

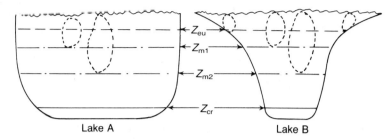

Figure 6.18. Two hypothetical lakes with different morphometry but similar euphotic, mixing and critical depths. The dashed ovals represent mixing patterns for plankton under the two mixing depths

and mixing depths would be more appropriate. Thus, for mixing depth 1 the potential productivity ratio would be about 1.0:

$$\frac{\text{Productivity lake A}}{\text{Productivity lake B}} = \frac{\bar{Z}_{eu}/\bar{Z}_{m1}}{\bar{Z}_{eu}/\bar{Z}_{m1}} = \frac{0.5/1.0}{0.4/0.8} \approx 1.0$$

However, with a doubling of the mixing depth, productivity in lake B should exceed that in lake A because of the increased shallow area in lake B.

$$\frac{\text{Productivity lake A}}{\text{Productivity lake B}} = \frac{\bar{Z}_{eu}/\bar{Z}_{m2}}{\bar{Z}_{eu}/\bar{Z}_{m2}} = \frac{0.5/2.0}{0.4/1.3} \approx 0.8$$

and therefore as mixing depth increases in the two lakes, the time spent by plankton in the photic zone will be relatively longer in lake B than in lake A. Thus, although the concept of Z_{cr} may not strictly apply it can help to explain why some lakes are more productive than others, notwithstanding similar enrichment levels.

The effect of mixing depth may also be examined using the model of Oskam (1978):

$$P_{net} = P_{max}C[F_i\lambda/(K_w + K_cC) - 24rZ_m] \tag{6.25}$$

where P_{net} is expressed as $mg\,C\,m^{-2}\,day^{-1}$, P_{max} as $mg\,C\,mg^{-1}\,Chla\,h^{-1}$, C as $mg\,m^{-3}\,Chla$, K_w is the extinction coefficient for water and non-algal matter (actually K_{wd}) in m^{-1} and K_c for algae in $m^2\,mg^{-1}\,Chla$ (nominally 0.025), F_i is a dimensionless function of light intensity to expand P_{max} to a real productivity (nominally 2.7), λ is photosynthetic hours, r is the respiratory fraction of photosynthesis, Z_m is the depth of mixing and 24 is the number of hours for respiration. As an example, a Z_m of 10, 20 and 30 m would give potential P_{net} productivities of about 139, 19 and $-101\,mg\,m^{-2}\,day^{-1}$, assuming a K_w of $0.5\,m^{-1}$, a P_{max} of $1\,mg\,C\,mg^{-1}$ $Chla\,h^{-1}$, a C (Chla) of $5\,mg\,m^{-3}$ and $12\,h\,day^{-1}$ of photosynthesis. In addition to illustrating the balance between respiration as a function of mixing depth and photosynthesis in the water column, this model also shows the effect of self absorption by plankton algae themselves.

By setting $P_{net} = 0$ and rearranging, Equation 6.25 can also predict the critical depth ($Z_m = Z_{cr}$):

$$Z_{cr} = F_i \lambda / (K_w + K_c C) 24r \qquad (6.26)$$

In contrast to Equation 6.23 for critical depth, this equation does not contain light intensity. Therefore, it is not useful for predicting the timing of the spring bloom, but rather to estimate the potential of mixing depth to control productivity given the clarity of the water. Thus, critical depth is a function of the extinction coefficients. For the above example, the Z_{cr} is 43 m. Thus, if the lake approaches that depth over much of its area, then mixing the lake artificially may control productivity and biomass as well.

Example problem
Given a lake with a mean depth of 10 m and a non-algal attenuation coefficient of $0.5 \, m^{-1}$, would it be possible to control algal biomass to a level less than $10 \, mg \, m^{-3}$ Chl$a$ by complete mixing? Why? Assume an absorption coefficient for Chl$a$ of 0.025, a 10 h photosynthetic day and a respiration:P_{max} quotient of 0.1.
From Equation 6.25:

$$Z_{cr} = \frac{2.7 \times \lambda}{24r(K_w + K_c C)} = \frac{2.7 \times 10}{24 \times 0.1(0.5 + 0.25)}$$
$$= 15 \, m$$

Biomass would probably not be controlled below $10 \, mg \, m^{-3}$ Chl$a$, because most of the lake is shallower than the critical depth that would allow a maximum of $10 \, mg \, m^{-3}$. Therefore, the potential maximum biomass possible with available light could be $>10 \, mg \, m^{-3}$, assuming nutrients were not limiting.

The model can be rearranged to predict maximum biomass possible under non-nutrient-limited conditions before being limited by self absorption:

$$C_{max} = (1/K_c)[F_i \lambda / (24r Z_m) - K_w] \qquad (6.27)$$

This model tends to give values consistent with maximum possible Chl$a$ of $250 \, mg \, m^{-2}$ (also in $mg \, m^{-3}$ in a 1 m water column) (Wetzel, 1975; p. 337). Oskam (1978) showed that for a K_w of $1.0 \, m^{-1}$, the biomass of some highly enriched German reservoirs should be controlled if the mixing depth were maintained at 15 m. In reality there was less biomass realized than predicted by the model presumably due to zooplankton

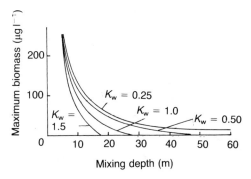

Figure 6.19. Relation of maximum phytoplankton biomass (Chl*a*) to mixing depth for different levels of non-algal attenuation of light (K_w) where $F_i = 2.7$, $\lambda = 12\,h$, and $r = 0.05$ (see text, after Oskam, 1978)

grazing. The potential maximum biomass as a function of mixing depth is shown in Figure 6.19.

6.3.5   Examples of light-mixing effects

Most temperate lakes greater than 6 to 8 m in depth will stratify thermally during the summer. Lakes shallower than that may stratify temporarily. The depth of the thermocline can be considered as the depth of mixing, and that depth results from a balance between the force of the wind mixing the water downward and the resistance to that mixing due to the buoyancy of the water, caused by heating, and density reduction. The position of the thermocline varies from lake to lake, depending upon wind velocity and fetch, light intensity and water transparency. A stable thermocline will initially favour productivity because it results in increased available light to mixing algal populations and may stimulate a bloom. However, if stratification persists, nutrients become depleted, because the thermocline usually represents a barrier to circulation.

A common way to indicate the degree of stratification is by the change in density per unit depth or between surface and bottom. Density change can be indicated by temperature in freshwater, because dissolved materials usually are insufficient to influence density. However, in seawater, σ_t is used and is defined as the density of water resulting from salinity, temperature and pressure minus 1 multiplied by 10^3.

Another important aspect of stratification in freshwater is that the change in density of water becomes greater as temperature increases. Thus, a few degrees difference between surface and bottom at high temperature will provide more stability (resistance to mixing) than the same degree change at low temperature. This is illustrated by the term

Figure 6.20. Seasonal relation among compensation depth, water column stability, and net primary production in a British Columbia fjord (modified from Gilmartin, 1964)

relative thermal resistance to mixing (RTRM), which compares the density gradient against the density gradient between 4 and 5°C:

$$RTRM = (D_s - D_b)/(D_4 - D_5) \qquad (6.28)$$

where D_s is surface density, D_b is bottom density, D_4 is density at 4°C and D_5 density at 5°C. RTRM would be greater for a high compared to a low temperature range if the surface-to-bottom gradient were the same numerically. This has implications for shallow polymictic lakes, which could develop significant temporary stability with differences of only a couple of degrees between surface and bottom.

In the Indian Arm, a fjord area in British Columbia, productivity was related to the degree of stability and to the compensation depth, which increased with time as incident light increased. When these two factors, which determine the amount of light received by a mixed plankton cell, were greatest, productivity was greatest (Figure 6.20). The periods of low stability were favourable to nutrient replenishment.

It is useful to compare the degree of stability in Indian Arm with that in Puget Sound. The unit $(10^5 \, m^{-1})$ is actually $(\rho - 1) \times 10^5 \, m^{-1}$. The values in Indian Arm ranged from 40 to 100 over the first 10 m, whereas in Puget Sound productivity was strongly affected by stability factors ranging from only 2 to 6 over 50 m depth (Winter *et al.*, 1975). Peaks in

Figure 6.21. Productivity, biomass, chlorophyll *a*, stability and tidal prism thickness in Puget Sound, 1966 (Winter *et al.*, 1975)

productivity and biomass usually occurred at minimum tidal-prism thickness (amplitude), which in turn allowed maximum stability and light availability during the spring months (Figure 6.21).

Water column productivity was related to available light in the mixed depth (Equation 6.19) within and outside the Frazier River plume in the Strait of Georgia, British Columbia (Parsons *et al.*, 1969). The strong stratification within the plume resulted in relatively shallow mixed depths and usually much greater available light to circulating plankton cells than was the case outside the plume where light was seldom sufficient for net productivity to occur.

In the Duwamish estuary at Seattle, there was no real phytoplankton activity until August when river flow was low and tidal flushing action was least. Such tidal conditions are characterized by low-high and high-low tides. The tidal conditions determine the turbulence and stability and are indicated, in addition to river flow, by tidal-prism thickness, which can be calculated as follows (units of metres):

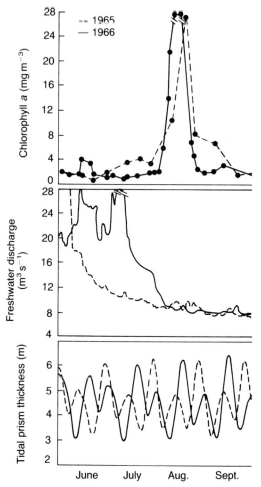

Figure 6.22. Tidal prism thickness, river discharge and chlorophyll *a* content in the Duwamish River estuary, Washington, 1965 and 1966 (Welch, 1969a)

Unfavourable turbulence: TPT $= (HH + LH) - (HL + LL)$
$$= (3.6 + 3.0) - [(1.5 + (-0.9))] \quad (6.28)$$
$$= 6.0$$

Favourable stability: TPT $= (3.0 + 2.7) - (1.8 + 0.9) = 3.0$

where TPT = tidal prism thickness, HH = high-high, LH = low-high, HL = high-low and LL = low-low tides.

Figure 6.22 shows the timing of phytoplankton blooms at one point in the Duwamish related to freshwater discharge and tidal-prism thickness in 1965 and 1966. The water column reached maximum stability and minimum turbulence and flushing when tidal-prism thickness (difference between sum of two daily high tides and sum of two daily low tides) and freshwater discharge were least. At that time tidal excursions were also minimal and the net result was that surface water moved slowly up and down the estuary and was allowed to retain heat and become even more stable because of the increased temperature gradient. Primary productivity correlated with degree of stratification (temperature difference from surface to bottom) and river discharge, and these two variables explained 65% of the variation in that parameter (Welch, 1969a).

Another way to examine the relationship between the physical limitation and productivity in the Duwamish is with a simple growth model. That is, the change in biomass X is a result of growth and loss (Equation 6.10):

$$\frac{dX}{dt} = \mu X - DX$$

where μX is growth and DX is loss through washout. Because nutrient content is very high and non-limiting (Welch, 1969a), the observed change in mixing depth can be hypothesized to affect the daily maximum growth rate based on ^{14}C productivity measurements made during the bloom. Thus, during neap-tide conditions (TPT = 3 m), the mixing depth was only 1 m and the average μ for a plankton cell mixed to that depth and average light exposure was about $0.6\,day^{-1}$. On the other hand, the spring-tide situation usually resulted in a mixing depth of 4 m and about one-half the μ ($0.3\,day^{-1}$) because of the much reduced light per cell. The extinction coefficient was not considered in this case, but higher values usually occur with greater river discharge, which would cause further limitation during the spring runoff period.

The loss rate through flushing can be approximated from a dye study conducted in the estuary, which showed an average flushing (dilution) rate of about $0.2\,day^{-1}$ over neap and spring tides alike. Knowing that spring tides allow more plankton loss than neap tides, the flushing rate was adjusted linearly so that a loss rate of $0.13\,day^{-1}$ was assigned to the neap tides and $0.27\,day^{-1}$ to the spring tides. Even so, it is clear from these values that mixing depth produced a much larger effect on population growth than flushing rate did on loss during the change from spring to neap tides. That is a greater growth of $0.3\,day^{-1}$, but a smaller loss of only $0.13\,day^{-1}$.

The net rates of population increase for the neap and spring tide situations, respectively, are:

$$\frac{dX}{dt} = 0.6X - 0.13X = 0.47X$$

and

Figure 6.23. Succession among diatoms, green algae and blue-green algae in a Thames River reservoir in years with and without destratification. Arrows indicate time of destratification (Taylor, 1966). Curves are idealized

$$\frac{dX}{dt} = 0.3X - 0.27X = 0.03X$$

Clearly a bloom would not be expected in the Duwamish when spring tidal conditions persist, even at low river flow, and this is precisely what has been observed. Moreover, if the μ of $0.47\,\mathrm{day}^{-1}$ and an initial biomass of $3\,\mu\mathrm{g}\,\mathrm{l}^{-1}$ Chl$a$ are used to simulate the bloom initiation of 1966 (Figure 6.22), the bloom seems to be explicable. The time required for the biomass to reach the observed maximum of $70\,\mu\mathrm{g}\,\mathrm{l}^{-1}$ Chl$a$ is 6.7 days from the first-order growth equation;

$$70\,\mu\mathrm{g}\,\mathrm{l}^{-1} = 3\,\mu\mathrm{g}\,\mathrm{l}^{-1}\mathrm{e}^{(0.47)(6.7)}$$

Although the dynamics of the phytoplankton in that estuary are much more complex, this analysis nevertheless indicates how strong the effect is from physical factors alone, particularly as they affect the amount of light received.

Stratification can limit production, however, if the condition is prolonged, because of nutrient depletion in the epilimnion. Partial destratification during midsummer in a Thames River reservoir illustrates the part played in production and species succession by stratification and nutrient depletion (Taylor, 1966).

The typical plankton succession occurs during spring and summer, starting with diatoms dominating in the spring, greens and yellow-greens in the summer, blue-greens in late summer, and a small diatom bloom again in late autumn (Figure 6.23).

In mid-July a partial destratification eliminated the typical midsummer growth of green algae and stimulated *Asterionella formosa*, a diatom that usually blooms from February to March. The apparent cause for that summer bloom was the transfer to water at $9-12.5\,\mathrm{m}$ with $2.8\,\mathrm{mg}\,\mathrm{l}^{-1}$ SiO$_2$ to the photic zone that contained only $0.5\,\mathrm{mg}\,\mathrm{l}^{-1}$ SiO$_2$. Concentrations below that have been shown to be limiting to *Asterionella* (see Chapter 6.5).

The *Asterionella* bloom subsequently declined after a second destratification and transfer of fertile water that further increased nutrient levels. Blue-greens continued to increase during destratification probably due to their buoyancy. Population increase was more rapid for *A. formosa* in summer than in spring, but its persistence was much less, possibly a result of temperature that was higher than required for best population maintenance even though growth rate was higher (see Chapter 6.4). Nevertheless, *A. formosa* actually showed greater growth in summer if nutrients were available, possibly because of higher light-saturated photosynthetic rates.

Thus the thermoclines in stratified lakes are usually a barrier to the recycling of hypolimnetic nutrients, and they can have a considerable range of negative effects on plankton algal production, depending on the permanency of the thermocline (see Chapter 4).

6.4 EFFECTS OF TEMPERATURE

6.4.1 Growth rate

The question is how do phytoplankton respond to temperature and what is an appropriate model to describe that response? The most common description of the growth response of phytoplankton, and other microorganisms as well, is the Q_{10} rule:

$$Q_{10} = (\mu_2/\mu_1)^{10/t_2-t_1} \tag{6.29}$$

where μ_2 and μ_1 are growth rates in response to a temperature increase from t_1 to t_2. This is a simplified form of the Van't Hoff (1884)–Arrhenius (1889) equations and is also known as the RGT rule (Reaktidnsgeschwindigkeit-Temperatur-Regel), as reviewed by Ahlgren (1987). The equation says that the Q_{10} will be 2 if the growth rate doubles for a 10°C rise in temperature.

As shown in Figure 6.24, the light-saturated rate of photosynthesis (flat portions of the curves) is limited by temperature as it affects the rate of enzymatic processes. At low light intensities, photosynthesis increases in direct proportion to increased light, but reaches a maximum that depends upon temperature. This has been interpreted as a temperature-dependent limit of the organism's enzyme systems, in contrast to the photochemical pigment response (Wetzel, 1975). Photosynthesis is seen to roughly double in the mid-part of the range, from about 25 to 35°C, but the Q_{10} tends to decrease as temperature increases and, therefore, is not a true constant (Ahlgren, 1987).

Figure 6.24. Relationship between growth rate and light intensity over a range in temperature for *Chlorella pyrenoldosa* (modified from Sorokin and Krauss, 1962)

Eppley (1972) summarized a large amount of data from experiments with a variety of phytoplankton species in batch culture and showed that the maximum growth rates observed increased exponentially with temperature (Figure 6.25). From inspection, it can be seen that μ_{max} roughly doubles with each 10°C increase in temperature. The calculated Q_{10} for this relationship is 1.88 and agrees well with a similar relationship developed by Goldman and Carpenter (1974) from continuous culture experiments with a variety of species. These relationships may be more useful in describing the growth rate response of a mixture of species which have optima that vary over a seasonal range in temperature. However, the Q_{10} rule may not be appropriate for a single species over a seasonal temperature range.

Ahlgren (1987) compared several equations describing temperature–growth rate relationships using data from unialgal continuous culture experiments. She found that the more linear model of Belehrádek (1926) fit the experimental data more closely, and the same was generally true for data from other investigators if the linear model were used. That equation using velocity (v) to indicate growth rate is:

$$v = a(t - \alpha)^b \tag{6.30}$$

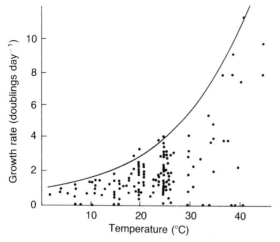

Figure 6.25. Relation of growth rate (μ_{max}) to temperature for a variety of culture experiments with different species (points) (Eppley, 1972)

where a and b are empirical constants, t is temperature and α is biological zero.

Belehrádak (1957) argued that the Q_{10} rule is not a true constant over each species' thermal tolerance range and that it is based on chemical reaction rate processes, which take an exponential form as shown in Figure 6.25. He observed artificial 'breaks' in the temperature–growth rate response over a range in temperature if the data were expressed in the form of the exponential equations, but such problems were overcome if expressed in his more linear form.

According to Ahlgren, Belehrádek developed the more linear equation (Equation 6.30) because biological processes are highly complex and are probably controlled by physical processes at the cell surface, such as diffusion and viscosity. Ahlgren obtained very good correlations using Equation 6.30, with exponent b ranging between about 0.3 and 2.5. She found also that the half-saturation constant, K_s, and the yield coefficient, Y, were generally independent of temperature with *Scenedesmus quadricauda*.

The equation for temperature correction used in the Streeter–Phelps DO deficit model is given by:

$$K_1 = K_2\theta^{(t_1-t_2)} \tag{6.31}$$

where K_1 and K_2 are reaction rates at temperatures of t_1 and t_2, respectively, and θ is an empirical coefficient. This equation can be derived from Arrhenius' equation and, therefore, simulates chemical reaction rates.

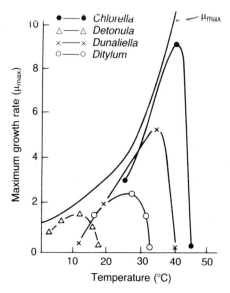

Figure 6.26. Maximum growth rate (μ_{max}) versus temperature with four species of phytoplankton (modified from Eppley, 1972)

Although of an exponential form, its success may be because prediction is based on a mixture of species over the seasonal temperature range.

6.4.2   Seasonal succession

Lake phytoplankton contain many species with widely different thermal optima for growth rate. Examples for four species are shown inserted under the maximum growth rate curve from Eppley (1972) (Figure 6.26). With such a distribution of thermal optima within a phytoplankton community, changes in temperature alone from season to season should force succession among species according to those optima. If optima for light are also considered, a further dimension is added to account for succession. For example, the following pattern might be expected (Hutchinson, 1967). Winter: low light and low temperature (no growth); spring: high light and low temperature (diatoms); summer: high light and high temperature (greens); autumn: low light and high temperature (blue-greens).

The observed patterns of plankton algae succession are to some extent explicable by light and temperature changes. For example, Cairns (1956) has shown that algae in stream water from Darby Creek in Pennsylvania demonstrated the following shifts in response to raised and lowered temperature from ambient after sufficient time was allowed for population equilibration at each temperature:

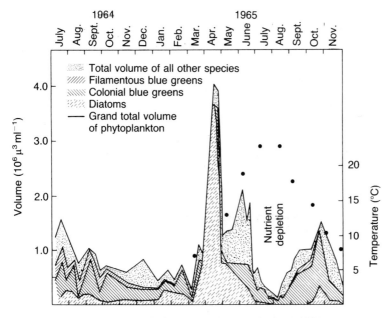

Figure 6.27. Surface phytoplankton composition at a centrally located station in Lake Sammamish. Temperature indicated by solid circles (after Isaac *et al.*, 1966)

$$20°C \quad \leftrightarrow \quad 30\ °C \quad \leftrightarrow \quad 35-40°C$$
diatoms   green algae   blue-green algae

Adequate evidence exists that plankton algae do have readily definable thermal optima as shown by Rodhe (1948). He also showed that if such thermal optima are designated by that temperature at which population growth is most prolonged, the temperature is lower than if the criterion of rapid growth were chosen. For *Melosira islandica*, Rodhe found that the most prolonged growth occurred at 5°C, growth was less prolonged at 10°C, and at 20°C growth was fast but death was rapid and the population did not persist. Typical optima are as follows: *Melosira islandica* (diatom) ≈5°C; *Synura uvella* (yellow-green) ≈5°C; *Asterionella formosa* (diatom) 10–20°C (≈15°C); *Fragilaria crotonensis* ≈15°C; *Ankistrodesmus falcatus* ≈25°C; *Scenedesmus, Chlorella, Pediastrum, Coelenastrum* 20–25°C.

A typical pattern of plankton algal succession is shown for Lake Sammamish (Figure 6.27). Note the spring diatom outburst and the summer and autumn blue-green dominance. The diatom increase was greatest at temperatures between 5 and 10°C, whereas blue-greens increased when temperature was between about 15 and 20°C. *Fragilaria* and *Asterionella* were important representatives in the diatom outburst.

An explanation for thermal optima being lower than the temperature

Figure 6.28. Observations of productivity and temperature, Clear Lake, California, on 15 October 1959 (a calm day). *Aphanizomenon* is the dominant species (modified from Goldman and Wetzel, 1963)

for maximum growth rate is that as μ increases exponentially (or linearly) with increasing temperature, the death rate should also increase, but probably lags behind μ. Because biomass at any time is the difference between growth and loss (death) (see Equation 6.11), its maximum should occur at a lower temperature than for μ.

There are many exceptions to light and temperature as an explanation for such successional patterns in phytoplankton. *Oscillatoria rubescens* has occurred in Lake Washington at temperatures >20°C and was the principal bloom species that first indicated the degradation in that lake, yet it prefers low light and low temperature, ≈10°C (Zevenboom and Mur, 1980). *Fragilaria crotonensis* should occur most prominently in spring and autumn because its optimum is from 12 to 15°C, yet often reaches a maximum at temperatures from 15–29°C. Moses Lake produces massive blooms of blue-green algae of the genera *Anabaena*, *Aphanizomenon* and *Microcystis* (often referred to as 'Anny', 'Fanny', and 'Mike') at an average summer temperature of 22°C (Bush *et al.*, 1972). Clear Lake, California shows the same situation of blue-green blooms occurring during moderate temperatures. Results are shown in Figure 6.28 for a calm autumn day during which *Aphanizomenon* was the dominant species (Goldman and Wetzel, 1963).

The occurrence of blue-green algae tends to be initially confined to late summer or autumn in moderately enriched lakes, but as enrichment proceeds they begin to dominate in spring, even if temperature is low (e.g. *Aphizomenon* in Pine Lake, Washington, in early spring at a temperature of 8°C).

Why are light and temperature and their respective optima not adequate in predicting successional patterns within and among lakes?

TABLE 6.1.   Effect of acclimation temperature on respiration in two strains of *Chlorella pyrenidosa*

Test temperature (°C)	Respiration for two acclimation temperatures (mm^3 O_2 mm^{-3} cells h^{-1})	
	25°C	38°C
25	4.5	8
39	1.6	18

Source: Sorokin (1959)

6.4.3   Species adaptation

Algae are capable of adaptation to other temperature regimes even though somewhat removed from their optima. For example, the respiration rates of *Chlorella* show that an adaptation of strains can occur (Table 6.1). Although one strain obviously grows best at 30–35°C, as was indicated in Figure 6.26 (same sp.), another strain nevertheless can be most active at lower temperatures.

Blooms of *Oscillatoria* can be observed below the ice in lakes during winter. Of course, their growth rate is low at that time, just as is the rate of *Chlorella* at less than optimum temperature (Table 6.1). However, if loss rates are also low, a bloom can result.

Thus, algae should be able to compete at temperatures slightly removed from their range and possibly dominate the community if some other factor(s) is(are) provided nearer their optimum and to some extent compensates the organism for the out-of-optimum temperature.

6.4.4   Interaction with other factors

If provided closer to a species optimum, nutrient supply may be one such factor that may allow that species to grow outside its temperature optimum. If the supply is great enough, even if the growth rate is relatively low, a bloom can result. This may, for example, explain blue-green algal blooms (*Aphanizomenon*) at 8°C in spring during periods of relatively high light intensities. Even if temperature is sub-optimum, blue-greens may be able to outcompete diatoms and greens if the nutrient supply is more to their liking. This may hold for *O. rubescens*, which adapted to the warm season in Lake Washington because the nutrient supply was increased. Such selection for higher P levels and low light is indicated in Table 6.2 for filamentous blue-greens.

TABLE 6.2. Variation in the relative abundance of greens, blue-greens and diatoms resulting from five levels of phosphorus enrichment at three light intensities in Lake Washington water after 2 weeks

Added P ($\mu g\,l^{-1}$)	Light intensity (lux)			Row means
	4000	2000	1000	
Greens (Number of cells)				
0	43	58	32	44
10	132	196	57	128
20	211	143	65	140
30	243	285	86	204
40	334	232	92	219 ↓
Column means	193	183	60	
	←			
Blue-greens (μm filament length)				
0	1220	1749	2680	1883
10	1860	4540	4160	3520
20	3120	5010	5620	4583
30	1840	3980	6000	3940
40	2840	6960	5440	5080 ↓
Column means	2176	4448	4780	
		→		
Diatoms (Number of cells)				
0	5	2	9	5
10	153	69	30	84 ↓
20	91	35	24	50
30	168	86	15	90
40	135	30	13	59
Column means	110	44	18	
	←			

Source: Hendrey (1973)

The buoyancy mechanism allows blue-greens to compete with non-buoyant diatoms for light in warm quiescent conditions (Figure 6.28). Diatoms depend on turbulence to remain suspended in the lighted zone. Thus, with the lower sinking rate, the blue-greens are able to accumulate large masses when turbulence is minimal, assuming the nutrient supply is large (see Chapter 6.5.6).

Although temperature and light contribute to species dominance and succession, they do not tell the whole story because the available nutrient supply and other factors are also elements in the adaptation of species to sub- or supra-optimum temperature and light. The blue-green species that usually prefer warm summer temperatures can apparently outcompete other algae for the available nutrients at that time. However, they can also adapt to other seasons when nutrient availability is high. In spite of sub-optimum temperature, blue-green species can dominate for most of the year in highly fertilized lakes. Species succession and blue-green dominance will be discussed further at the end of this chapter.

Although a given temperature is not entirely indicative of species dominance among different waters, succession from diatoms to greens to blue-greens is often reasonably predictable from a temperature increase imposed upon a given water (and nutrient supply) and temperature regime.

The interacting physiological effects of temperature and light on production could produce two results from a temperature increase.

1. A short-term effect of a temperature increase may result in some adaptation of existing species, with the possible occurrence of even a decrease in production if the original temperature optimum and the new temperature are too far apart.

2. A long-term species shift may occur with time, and species with optima at the new temperature will dominate and production will increase because the light-saturated rate of photosynthesis goes up with a temperature increase across the tolerable range for all species.

6.4.5 Heated waters from power plants

Not many examples exist to examine the effect of the addition of heated water, but results from some Polish lakes indicate that fairly extensive changes in productivity and species composition could develop from relatively small changes in temperature (Hawkes, 1969). The test lake was Lichen Lake which received heated waters from a power plant: annual temperature range 7.4–27.5°C; taxonomic components (species, varieties and forms) 285; dominant species *Melosira amgibua*, *Microcystis aeruginosa*; primary production $7.3\,g\,m^{-2}\,day^{-1}$. The control lake was Slesin Lake which had no heated water added: annual temperature range 0.8–20.7°C; taxonomic components 198; dominant species *Stephanodiscus astraea* (no blue-greens); primary production $3.75\,g\,m^{-2}\,day^{-1}$.

6.5 NUTRIENT LIMITATION

The concept of nutrient limitation has been defined previously in the section on growth kinetics. The role of specific nutrients, both micro and macro will be discussed here in terms of whether they limit growth rate or biomass, how limitation by a particular nutrient and its prediction can be determined, and how nutrient limitation affects species succession, especially to blue-greens.

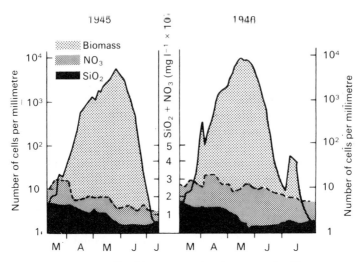

Figure 6.29. *Asterionella* biomass and nitrate and silica content in Lake Windermere in 1945 and 1946 (Lund, 1950 with permission of Blackwell Scientific Publications, Ltd)

6.5.1   Silica

Silica is considered a macronutrient (required in large amount by cells) for diatoms, which have cell frustules that are composed of silicious cell walls. It is available for uptake as an undissociated silicic acid or silicate ion (Hutchinson, 1957), but is also adsorbed on particulate matter. The rapid removal of silica from the photic zone during spring diatom blooms indicates that uptake and sedimentation occur much faster than regeneration in the water column.

Cessation of the spring bloom of the diatom *Asterionella* has been closely linked to ambient silica concentration and thus represents a rather clear case for nutrient control. Early laboratory experiments (batch cultures) showed that $0.5 \, mg \, l^{-1}$ of SiO_2 represented the minimum concentration that would sustain maximum growth of *Asterionella* (Lund, 1950). A consistent year-to-year pattern of bloom cessation was observed to occur once SiO_2 fell below $0.5 \, mg \, l^{-1}$ (Figure 6.29). Increased loss through sedimentation could also be an explanation for biomass decline, as stratification increases in the spring. N and/or P are also macronutrients that can limit growth and terminate blooms, but a clear link between ambient concentration and cessation is not as easily observed with those nutrients.

Later experiments with *Asterionella* in continuous culture have supported the earlier work of Lund. Kilham (1975) demonstrated that the uptake rate of SiO_2 could be described by Michaelis–Menten kinetics and that concentrations accounting for 90% of $\mu_{\max}$ were 0.82 and 0.39 mg l^{-1}, very close to Lund's 0.5 mg l^{-1}.

6.5.2 Nitrogen and phosphorus

N and P are the macronutrients (required in large amount by cells) that are usually most scarce relative to the needs of organisms and therefore most likely to limit growth and biomass. Except in very low alkalinity waters, C is usually in sufficient supply so as not to be limiting at all or to be limiting only in the short term if N and P are plentiful. The special conditions for C limitation will be discussed later. Trace elements have been shown to limit growth in only a few instances and usually in oligotrophic lakes (Goldman, 1960a, 1960b, 1962).

The deficiency in cells for a macronutrient, and therefore its growth and biomass limitation by that nutrient, can be determined by examining the ratios in cell material. The Redfield ratio, or the average ratio found in marine plankton, is often used to judge the requirements of phytoplankton (Redfield et al., 1963; Sverdrup et al., 1942):

$$O:C:N:P = 212:106:16:1 \text{ (by atoms)} = 109:41:7.2:1 \text{ (by weight)}$$

Applying this ratio to results of nutrient content in phytoplankton biomass from marine waters shows that N is usually in the shortest supply, or will be the nutrient to deplete first if all macronutrients are removed from water in the above ratios. Menzel and Ryther (1964) found the ratio of N:P in biomass to range from 5.4 to 17 by atoms and the C:P ratio by atoms to range from 67 to 91. Based on the Redfield ratio, these cells would range from deficient in N to just meeting the requirements for all nutrients.

In freshwater, on the other hand, P is usually the most limiting nutrient. Particulate matter, which normally in summer is predominately phytoplankton, in Lake Canadarago, New York showed a range in N:P (by atoms) from 17 to 24 during summer and a C:P (by atoms) of ≈ 200 (Fuhs et al., 1972). These results are in contrast to those from the marine environment above, and show that cells ranged from containing all nutrients in the appropriate ratio to being deficient in P.

The above comparisons only consider the cell content of nutrients and not the content in the ambient water. Vallentyne (1972) compiled data to evaluate the biospheric average of the demand ratio, which is the ratio of the average concentration of a given nutrient in plant tissue to the average concentration in the water. Thus, the higher the ratio for a given nutrient, the higher is the scarcity of that nutrient in the ambient environment, relative to the cellular needs. Vallentyne's data are summarized in Table 6.3 and show that P was the most limiting nutrient in both winter and summer.

TABLE 6.3.   Ratio of plant-to-water concentrations for important plant nutrients in a variety of freshwater habitats around the world

Nutrients	Winter	Summer
P	100 000	800 000
N	20–25 000	100–125 000
C	5–6 000	6–7 000

Source: after Vallentyne (1972)

The ratio of C:N:P in the ambient water is also often used to indicate which nutrient may be depleted or reach growth-rate-limiting concentrations first and thus could be identified as the potentially most limiting nutrient. Emphasis is on potential, because at the time of measurement concentrations of both N and P may be so high that neither is currently limiting. That process is evident in results of algal growth potential (AGP) experiments shown in Figure 6.30 (Skulberg, 1965). The initial N:P ratio in both experiments was 33:1 by atoms (15:1 by weight) and concentrations were so high as to not limit growth. Assuming that N and P were taken up by the algal cells in the required ratio of 16:1 (7.2:1 by weight), P should have been removed first and this is what happened; P reached undetectable amounts and ultimately stopped growth while N was still present at relatively high concentrations.

Under some conditions of low P and high N:P ratios in the ambient environment, cells may not actually be deficient in P, which is therefore technically not limiting. If the recycling rate of P is sufficiently high, its supply rate may be adequate to meet the growth rate demands of the cell in spite of apparent limitation based on the ambient N:P ratio. Also, the cells may have taken P up in luxury amounts (above growth needs) or are releasing alkaline phosphatase enzymes to break down organic forms of P (the latter is used as a test for P limitation). However, the ambient ratio would still indicate the potentially most limiting nutrient even though growth is not currently limited. In Figure 6.30, *Skeletonema* appears to have grown after P was depleted, which may indicate that the cells had stored P in luxury amounts and subsequent growth utilized those luxury stores. In addition to using ambient and cellular N:P ratios, one may want to employ a test for alkaline phosphatase activity, which, if present in relatively high concentrations, would indicate current limitation (Fitzgerald and Nelson, 1966; Goldman and Horne, 1983).

Assuming that changes in cellular nutrient content are proportional to those in the ambient environment, response will be related to changing N or P in a manner similar to that shown in Figure 6.31 (Sakamoto, 1971).

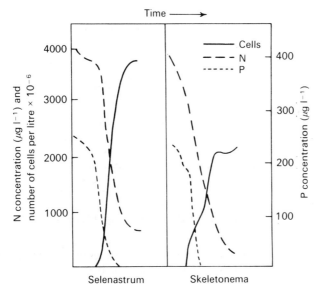

Figure 6.30. Growth of two species of algae related to N and P concentrations (soluble) in culture (Skulberg, 1965)

Figure 6.31. Response of phytoplankton ^{14}C uptake to P (a) and N (b) additions showing the significance of the ratio concept in determining which nutrient is limiting (generalized from Sakamoto, 1971)

^{14}C uptake in this case is proportional to biomass if experiment duration is several days. Limitation with P increase occurs when the ratio falls below the optimum level, nominally indicated by the Redfield ratio as $7:1$ by weight. Limitation occurs with increasing N when the ratio exceeds the optimum.

Other ratios in cells may also indicate nutrient limitation, such as the amount of chlorophyll a. Healy (1978) has suggested bounds in the chlorophyll a:biomass ratio in chemostat cultures that indicate moderate and severe nutrient limitation. Using a C:dry weight ratio of 0.2, the boundary values for chlorophyll a:C range from 1 to 2% for severe and moderate limitation, respectively. Figure 6.32 shows the response

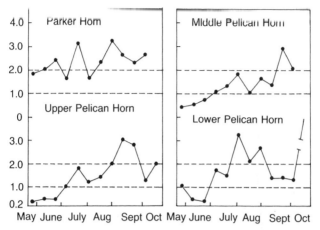

Figure 6.32. Ratios of Chl$a$:cell carbon for phytoplankton in Moses Lake, Washington during May to October 1982. Horizontal lines indicate Healy's (1978) zone of 'moderate' nutrient deficiency (from Carlson, 1983)

of severely N-limited (ambient NO_3-N:SRP = 0.1) phytoplankton in Pelican Horn of Moses Lake to the pumping of water high in NO_3-N, but low in SRP. Prior to pumping, the chlorophyll a:C ratio was <1%, but after pumping it increased to >1% and even rose above 2% on occasion.

The limitation of growth by either N or P may vary seasonally as changes in environmental factors, such as available light, temperature, stratification, internal loading, recycle rate, etc., affect the demand and availability of the two nutrients. There is a tendency for NO_3 to deplete more rapidly than SRP in the epilimnion during the growing season. This occurs in oligotrophic as well as eutrophic lakes and a large decrease in NO_3 is often accompanied by little or no change in SRP (Hendrey and Welch, 1974). From a consideration of ambient N:P ratios, limitation can change from P to N during the stratified period. In systems that have short detention times and that receive variable inputs (e.g. storm water) or are subject to other events (thermocline erosion), the limiting nutrient can change even more frequently. For example, Smayda (1974) has shown with bioassays using a marine diatom (*Thalassiosira pseudonana*) that frequent changes occur in the limiting nutrient throughout the growing season in Narragansett Bay and that more than one nutrient could limit at once. N was usually the limiting nutrient and to a lesser extent Si, but limitation also resulted from the exclusion of EDTA (ethylenediamine tetraacetic acid) as a chelator of algal excretory products and only occasionally by exclusion of P and trace metals.

The nutrient that limits growth also tends to change as eutrophication increases and the lake's trophic state changes. Oligotrophic lakes usually

TABLE 6.4.   Annual means of NO_3-N/PO_4-P (by weight) showing decreasing trends as eutrophication proceeds; unless specified, values are annual means

Kinds of lakes	Location	N/P ratios
Hypereutrophic	Hjälmaren (Sweden)	4/1
	Moses (WA)	1/1 (summer)
Eutrophic	Washington (WA)	10/1 (winter, 1963)
Mesotrophic	Washington	20/1 (winter, 1970)
	Mälaren (Sweden)	25/1
	Sammamish (WA)	27/1
Oligotrophic	Vättern (Sweden)	70/1
	Chester Morse (WA)	63/1

Source: Swedish lakes from National Environmental Protection Board

have higher N:P ratios than eutrophic lakes (Table 6.4). As these example ratios suggest, the process of eutrophication tends to change the nutrient that limits growth from P in oligotrophic and mesotrophic states to N in the mid and later stages of eutrophy. The winter N:P ratio in Lake Washington changed from $\approx 10:1$ during its most eutrophic condition, resulting from sewage effluent input, to values usually $>20:1$ during the years following diversion and recovery (Edmondson, 1978). The N:P ratio tended to be lower during the growing season, as indicated previously.

There are anomolies to the pattern of N or P limitation and trophic state. Bioassays using natural assemblages of phytoplankton species in oligotrophic lakes may often show no response, or even inhibition, to increased P in short-term exposures (Hamilton and Preslan, 1970; Hendrey, 1973), but with time for physiological adaptation and species selection, response to P in such waters is usually positive (Hendrey, 1973). There are also exceptions to the generalization that P will ultimately limit in oligotrophic lakes. N limits in Lakes Tahoe (Goldman and Carter, 1965; Goldman, 1981), where the spring–summer NO_3:SRP ratio is <3, and in Crater Lake where NO_3-N is $<1\,\mu g\,l^{-1}$ and SRP is $14\,\mu g\,l^{-1}$ during spring–summer (Larson *et al.*, 1987). These lakes are ultra-oligotrophic by the conventional criteria.

N limitation in eutrophic and hypereutrophic lakes is a common occurrence. A typical example is represented by the bioassay using water from two eutrophic Wisconsin lakes and the principal nuisance blue-green species, *Microcystis aeruginosa* (Gerloff and Skoog, 1954). Addition of the three elements together (N, P, Fe) resulted in growth about equal to that produced in the synthetic media which contained a surplus of everything this alga is known to require (Table 6.5). NO_3 addition produced the

TABLE 6.5.   Growth of *Microcystis aeruginosa* during 2 weeks in sterilized surface water from two lakes in Wisconsin containing various chemical additives

Additives	Lake Mendota water (cells mm^{-3})	Relative growth (%)	Lake Waubesa water (cells mm^{-3})	Relative growth (%)
Without additives	500	6	1 523	14
NO$_3$	5 780	64	7 737	73
PO$_4$	500	6	1 101	10
Fe	495	6	1 139	11
PO$_4$ + Fe	400	4	874	8
NO$_3$ + Fe	4 310	48	9 675	92
NO$_3$ + PO$_4$ + Fe	9 030	100	10 575	100
Synthetic nutrient solution	10 235	113	10 235	97

Source: Gerloff and Skoog (1954)

major portion of the biomass, as shown by results from those treatments in which each of the three elements was used alone. Although NO$_3$ was clearly the most limiting nutrient, there was added biomass produced from the three nutrients used in combination, with P being the second most important. *M. aeruginosa* is a non-N-fixing, blue-green species and, therefore, results could differ if an N fixer had been used.

Biomass increase of N-fixing blue greens should ultimately be controlled by P if sufficient time is allowed for N fixation. The depletion of NO$_3$, as a result of increasing SRP, would stimulate N fixation, which would allow biomass to ultimately increase to meet the P supply (Hutchinson, 1970a). Smith (1990a) has shown for lakes from temperate to tropical that N-fixation rate increases with TP concentration, except at very high TP. Schindler and Fee (1974) observed that P addition alone to eutrophic low-NO$_3$ lake water resulted in a biomass increase through N fixation requiring two to three weeks in plastic bag experiments *in situ*. Horne and Goldman (1972) found that N fixation contributed about one-half the N supply to Clear Lake, California. Although the growth rate of N-fixing blue-greens was relatively slow (growth rate 0.05 day^{-1}), fixation could supply the N to ultimately meet the available P supply.

N fixation does not always supply N to meet the ultimate supply of P, however. NO$_3$ addition to hypereutrophic Moses Lake water (plastic bottle bioassay *in situ*) in mid-summer, when NO$_3$ was low or undetectable, stimulated an increase in blue-green algal biomass and chlorophyll *a*. *Microcystis* was dominant and showed the greatest increase in biomass, as might be expected since it is a non-N fixer, although *Aphanizomenon* also increased (Welch *et al.*, 1972). However, the later (September) P-addition portion of the experiment had no effect on

Figure 6.33. Effect of treated sewage on biomass of natural phytoplankton populations in Moses Lake (Welch *et al.*, 1973)

blue-greens; diatoms had become dominant and blue-greens were much reduced, possibly due to much lower lake temperature. Additional bioassays *in situ*, using combinations of N and P additions simultaneously, resulted in a substantial but delayed increase in biomass with P only (Figure 6.33). Presumably time was required for the necessary N to be supplied by fixation. Subsequent dilution of the lake with large amounts of low nutrient water showed N to be the controlling nutrient; summer chlorophyll *a* was directly related to the inflow NO_3 concentration (Welch *et al.*, 1984). Although N-fixing *Aphanizomenon* was the dominant blue-green, its N-fixation rate was apparently not fast enough to utilize substantially the available P at the relatively high, dilution-imposed, flushing rate (average $0.1\,day^{-1}$). Horne and Goldman (1972) have suggested that the maximum fixation rate is $0.05\,day^{-1}$.

The decreasing tendency for P to limit as eutrophication proceeds (trophic state increases) was shown by Miller *et al.* (1974) from bioassays using *Selenastrum* (AGPs) in water from 49 lakes in the United States. As

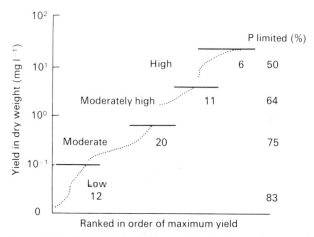

Figure 6.34. Yield of algal biomass in algal growth potential (AGP) experiments from 49 lakes in the USA showing a decrease in the percentage that were P-limited as the yield (and productivity) increased (modified from Miller *et al.*, 1974)

the yield of *Selenastrum* increased, as an indication of the increasing trophic state, the percentage of P-limited lakes decreased (Figure 6.34). A similar pattern has been shown by a multiple regression analysis of TP, TN and chlorophyll *a* data from 127 North American lakes (Smith, 1982). Although the regression and N:P ratios showed that P limited in most lakes, a better fit of the data was obtained when TN was included. That is obvious from the plot of TP on chlorophyll *a* where ≈20 of the high chlorophyll *a* observations are skewed substantially to the right of the prediction line and these represent lakes with TN:TP ratios <10, suggesting N limitation.

Explanations of why P limitation in lakes gives way to N limitation as eutrophication proceeds involves a consideration of the N and P cycling processes and the character of the inputs. According to Thomas (1969), NO_3 is always observable in oligotrophic lakes, but as PO_4 is added culturally, the 'NO$_3$ reserve' is depleted. One reason for this is probably that the ratios of N:P in wastewater sources are low, e.g. N:P in sewage effluent is about 2:1 to 4:1. Addition of effluent with this ratio would lower the lake ratio. Also, as eutrophication increases, light availability becomes less, due to attenuation from the increased organic matter. As light, rather than nutrients, becomes more limiting there are higher concentrations of unused nutrients remaining. Associated with that process is an increased potential for N loss from the lake through denitrification. As eutrophication increases, oxygen depletion, especially at the sediment–water interface becomes more pronounced (see P cycle section). This should result in higher rates of denitrification, because that is an anaerobic process. As more nutrients remain unused due to increasing light limitation, and N is preferentially lost due to denitrification, the N:P ratio, especially the soluble fractions, would decline. In addition, the more extensive anoxic conditions at the sediment–water interface result in higher internal loading of P, which would lead to further lowering of the N:P ratios since N release does not increase with anoxic conditions. N fixation tends to restore the N:P ratio to a level of equal limitation, i.e. the ratio in unstarved cells. However, fixation is a rather slow, energy-demanding process compared to those processes that contribute high P relative to N.

P is most apt to be limiting in oligotrophic through mesotrophic lakes because sediments are usually still oxic and represent an effective, relatively permanent sink for P because of the insoluble complexes formed with metals and its tendency to sorb to particulate matter. Soluble inorganic N, on the other hand, is not susceptible to such complexation and sorption and thus tends to be more conservative with respect to physical processes. Furthermore, precipitation represents a relatively greater and more consistent source of N than P. Although rainwater contains some P (largely associated with dust) and has been shown to be a relatively important contributor in phosphorus-poor watersheds (Schindler, 1974), the NO_3 in rain is a rather constant phenomenon, having been transformed from the N_2 reservoir in the atmosphere by lightning.

Overall, there are obviously fewer sources for P than for N and sedimentation is a more efficient remover of P than N from the water column. With few exceptions, N usually becomes limiting only in highly eutrophic or hypereutrophic lakes and even then fixation by blue-green algae usually becomes an important source for N. With these processes operating, limitation of algal biomass to relatively low levels should be more easily attained in most lakes by restricting the availability of P rather than N, even if N is limiting in highly eutrophic systems. However, in lakes with high natural inputs of P, results may be more promising from N control. This is apparently the case in many lakes in New Zealand (White, 1983).

This reasoning is not supported in marine systems where N is usually the limiting nutrient, whether the system is oligotrophic or eutrophic (Ryther and Dunstan, 1971; Smayda, 1974). Why N is more consistently limiting in marine waters is not entirely clear. A partial explanation may relate to water column recycling for P, which is apparently faster than for N. Particulate matter, therefore, becomes richer in N than P as depth increases in the sea (Ryther and Dunstan, 1971), while the reverse occurs in lakes (Schindler, 1974; Birch, 1976; Gachter and Mares, 1985). Thus, as particulate matter sinks through a much deeper mixed layer in the sea, consistently lower soluble N:P ratios would result than in shallower lakes, where the high-P particulate matter would tend to be trapped by bottom sediments before recycling could occur. The N:P ratio would be lowered even more if N fixation were relatively unimportant locally or in the short term in the sea as suggested by Ryther and Dunstan (1971). Smith (1984) contends that there really is not that much difference between marine and freshwater environments with respect to which nutrient is limiting. Rather than basing the designation of limitation on bottle experiments, which is usually the case, whole ecosystem nutrient budget information should be used. He suggested that where N limits in the sea, the explanation may be due to high rates of hydrographic exchange that dominate nutrient availability permitting insufficient time for N fixation to restore N losses.

Thus far the discussion of N and P limitation has considered conditions under which yield (biomass) is limited. That is, if the N:P ratio in the ambient environment indicates that N is ultimately limiting, then growth will stop when N is exhausted and the final yield will be proportional to the initial N concentration. This concept was demonstrated for dynamic systems in the section on growth kinetics, such that the steady-state biomass concentration is directly proportional to the inflow concentration of the limiting nutrient. Uptake rate, on the other hand, is dependent on the ambient concentration of nutrient, according to Michaelis–Menten kinetics. Growth rate is dependent on the cellular concentration, which may be higher than ambient if there is luxury storage and, therefore, the growth rate would be higher than the uptake rate. If cells have below quota concentrations, uptake could be higher than growth rate. If one assumes that cell concentrations are at quota amounts, then growth rate can be predicted, according to Michaelis–Menten kinetics, based on the ambient concentration.

There are several approaches to predicting growth rate of phytoplankton if there are changes in limitation between N and P, assuming a dependence on ambient concentrations. The first formulation is the 'multiplicative' model (Chen, 1970):

$$\mu = \mu_{max}[P/(K_P + P)][N/(K_N + N)] \qquad (6.32)$$

This model produces a low estimate of μ when the concentrations of both nutrients are near the limiting level. For example, if K_P and K_N each $= 0.05 \, \text{mg} \, l^{-1}$, $P = 0.05 \, \text{mg} \, l^{-1}$ and $N = 0.25 \, \text{mg} \, l^{-1}$, $\mu = 0.42 \, \mu_{max}$.

The second is an additive model (Patten et al., 1975):

$$\mu = \mu_{max}/2[P/(K_P + P) + N/(K_N + N)] \qquad (6.33)$$

Using the same values for N and P concentrations and the half saturation constants, $\mu = 0.67 \, \mu_{max}$. By averaging the reduction ratios to account for the possible changing limitation, instead of multiplying them together, the estimate of μ increases.

Another form of averaging treats the reciprocal of the reduction terms (Bloomfield *et al.*, 1973):

$$\mu = \mu_{max}2/[1/(P/(K_P + P)) + 1/(N/(K_N + N))] \qquad (6.34)$$

In this case, μ is estimated at $0.62\mu_{max}$. The final and possibly more supportable formulation is the 'threshold' model (Bierman and Dolan, personal communication, 1976):

$$\mu = \mu_{max}P/(K_p + P) \ldots \quad \text{if} \quad N:P > 13.5:1 \quad \text{(by weight)} \quad (6.35)$$
$$\mu = \mu_{max}N/(K_N + N) \ldots \quad \text{if} \quad N:P < 13.5:1 \quad \text{(by weight)} \quad (6.36)$$

With the same K_s and concentrations, this approach predicts a μ of $0.83\mu_{max}$, because $N:P$ is $5:1$ and N should limit. Bierman and Dolan (personal communication) found that the threshold model predicted a μ closer to observed than the multiplicative model, when compared with chemostat culture data using *Scenedesmus*. They observed that the predicted ratios of μ/μ_{max} were 0.29 ± 0.3 for the multiplicative model and 0.43 ± 0.07 for the threshold model. The actual ratio, determined in cells from chemostat experiments, was 0.44 ± 0.02, thus supporting the threshold model. The data and threshold $N:P$ ratio ($30:1$ by atoms; $13.5:1$ by weight) were taken from Rhee (1978).

The light model of Steele and the temperature model of either Eppley or Bělehrádek (see Ahlgren, 1987) are often linked together with a multiple nutrient model to predict growth rate with depth and time. However, light, temperature and other nutrients may interact with N and P to produce variable growth responses, thus altering the half-saturation coefficients for N and P. For example, Young and King (1980) showed that growth rate of *Anacystis nidulans* tended to saturate at lower CO_2 concentrations as the combination of light and P content increased. Also, Hughes and Lund (1962) showed that the critical level for SiO_2 of $0.5\,mg\,l^{-1}$, with respect to *Asterionella* growth, could be effectively lowered with increased P addition.

6.5.3 Trace elements and organic growth factors

Many trace elements are required by algae for growth as co-factors for enzymes and there is a sound rationale as to how growth rate is affected by trace-element deficiency (Dugdale, 1967). Trace-element limitation in natural systems apparently does not occur frequently, however, in view of the fact that full response is nearly always attained with the addition of N and P. Experiments in four lakes, two of which were oligotrophic, in the Lake Washington drainage showed that trace-element additions alone never increased growth, although in some instances an additional enhancement was achieved by a complete medium addition over that of N and P (Hendrey 1973). In view of these results, it would seem that trace elements are usually in adequate natural supply, particularly in eutrophic waters. A notable exception is the productivity response to Mo addition in oligotrophic Castle Lake, California (Goldman, 1960a) and some Alaskan lakes (Goldman, 1960b). Although many organisms have vitamin demands and may respond to dissolved organic compounds (e.g. glycolic acid, Fogg, 1965), these factors have seldom been demonstrated as important limiters to growth under natural conditions.

6.5.4 Chemical complexers

Organic complexers can either remove trace elements from solution and reduce inhibition or make them more soluble and available for growth. The effect of complexers, such as EDTA, has been demonstrated by several workers (Wetzel, 1975).

Another factor linked to productivity control is the ratio of monovalent to divalent ions. As that ratio increased, productivity was observed to increase (Wetzel, 1972).

6.5.5 Carbon

Carbon is the macronutrient required by plankton in the greatest quantity, although it was not considered seriously as a principal cause for increasing eutrophication until 1969–1970. Essentially, the argument centred on the possibility that the organic carbon in waste effluents could cause eutrophication in many carbon-limited situations when the organic C was released by bacterial decomposition as CO_2 (Kuentzel, 1969; Kerr et al., 1970). This idea was confronted by those who believed strongly that organic C from sewage or other waste effluents was not a principal cause for eutrophication, notwithstanding the potential limitation by C in some situations. Much was learned from this controversy and related experiments. The points and counterpoints made by the two schools of thought in 1970 are listed in Table 6.6.

To best clarify the problem, it would be most instructive to retrace the steps of this controversy, which originated in 1969 in the article by Kuentzel. He made the following points.

1. Inorganic C in the alkalinity system is insufficient to form a large algal bloom, which frequently occurs in enriched lakes. At pH of 7.5–9, the atmospheric input of CO_2 would amount, at most, to $1\,mg\,l^{-1}$ after several days of transport.
2. The $56\,mg\,l^{-1}$ (dry weight) of algae observed in a bloom in Lake Sebasticook, Maine would require $110\,mg\,l^{-1}$ CO_2, which would only come from a supply of organic matter decomposed by bacteria as a ready

TABLE 6.6.   Two schools of thought clashed on many points as to whether C or P is the principal key to eutrophication control

School: carbon is key	School: phosphorus is key
Carbon controls algal growth	Phosphorus controls algal growth
Phosphorus is recycled again and again during and after each bloom	Recycling is inefficient: some of the phosphorus is lost to bottom sediment
Phosphorus in sediment is a vast reservoir always available to stimulate growth	Sediments are sinks for phosphorus, not sources
Massive blooms can occur even when dissolved phosphorus concentration is low	Phosphorus concentrations are low during massive blooms because phosphorus is in algal cells, not water
When large supplies of CO_2 and bicarbonate are present, very small amounts of phosphorus cause growth	No matter how much CO_2 is present, a certain minimum amount of phosphorus is needed for growth
CO_2 supplied by the bacterial decomposition of organic matter is the key source of carbon for algal growth	CO_2 produced by bacteria may be used in algal growth, but the main supply is from dissociation of bicarbonates
By and large, severe reduction in phosphorus discharges does not result in reduced algal growth	Reduction in phosphorus discharges materially curtails algal growth

Source: Great Phosphorus Controversy (1970)

source of CO_2. About $30\,mg\,l^{-1}$ of organic carbon would be sufficient for that quantity.

3. The organic C in sewage effluents furnishes the needed nutrient (C) to form large algal blooms, because such low concentrations of dissolved PO_4-P ($<10\,\mu g\,l^{-1}$) are associated with large blooms. Therefore, it is obvious that sufficient PO_4 exists even at these low concentrations!

Shapiro (1970) responded to this article with the following points.

1. PO_4^{3-} is frequently, but not always, the cause of eutrophication.
2. Algae have an absolute requirement of P; $10\,\mu g\,l^{-1}$ dissolved at the time of a bloom does not indicate the supply, but only the difference between uptake and supply.
3. Sediments are not usually a source for P, but rather a sink.

A report by Kerr *et al.* (1970) was the next input to the controversy, demonstrating under experimental conditions that C could limit algal production in naturally soft water and that bacteria-originated CO_2 from organic matter serves as an effective source of C in such environments. More specifically, Kerr's findings were briefly as follows.

TABLE 6.7.   Relative response of Anacystis nidulans grown under various conditions

| | Natural populations | | Effect of E. coli | |
Conditions	Bacteria	Algae	$CO_2 + HCO_3$ increase $(mg\,l^{-1}\,72\,h^{-1})$	CO_2 produced (mg trapped)
Pond water	Decrease	No change	3.6	0
+ N and P	Large increase	3 × increase	10.0	0.003
+ Glucose	2 × increase	2.5 × increase	9.2	0.005
+ Glucose	Large increase	6 × increase	12.0	
+ N and P				0.030

Source: Kerr *et al.* (1970)

1. *Anacystis nidulans*, an obligate photoautotroph, was used as the test organism, which attained concentrations of 60×10^6 cells ml^{-1}.

2. Algae that received CO_2 from a dialysis bag with bacteria and organic matter grew to double the mass of that in the flask without such a CO_2 source; 6.5 versus 3.7×10^7 cells ml^{-1}. The same results were produced using CO_2-free air and 5% CO_2/air bubbled into flasks.

3. To determine if such bacteria-originated CO_2 was important in natural systems and, consequently, if the C from organic matter added to natural systems was important in bloom production, various enrichments with 'infertile' pond water ($<5\,\mu g\,l^{-1}$ P and N) were tested and population increases were recorded after 30 h of 'incubation' (Table 6.7).

Kerr showed that CO_2 probably limited growth rate in the water tested, which was soft (alkalinity only $14\,mg\,l^{-1}$ CaCO$_3$; dissolved inorganic carbon (DIC) = $5\,mg\,l^{-1}$), because the addition of organic C to this natural infertile water had about as significant an effect as N and P addition. Heterotrophs and autotrophs were stimulated by both N + P and glucose, but the greatest stimulation occurred with the addition of all three nutrients. Limitation of growth rate in these waters was clearly C > P + N. However, a question remained. Because glucose addition produced the smallest increase of any combination, would glucose addition produce the greatest long-term effect over one or more years in such infertile water? That is, which nutrient is most scarce (controlling) in the long term?

Schindler (1971b, 1974) answered the latter and most important question from the standpoint of lake management by a fertilization experiment in a low alkalinity ($15\,mg\,l^{-1}$ CaCO$_3$) lake (Lake 227) in Ontario, Canada. To that system, N and P were added weekly for three summers

Figure 6.35. The response of phytoplankton standing crop (as chlorophyll *a*) in the epilimnion of lake 227, fertilized with 0.48 g P and 6.29 g N m^{-2} annually beginning in 1969. Chlorophyll concentration from six other unfertilized lakes never exceeded 10 μg l^{-1}. All lakes are less than 13 m deep. Note that lake 227 had a low standing crop (+) similar to that of the other lakes prior to fertilization. Large inputs of P and N will cause eutrophication problems regardless of how low carbon concentrations are, because the necessary carbon is drawn from the atmosphere (modified from Schindler and Fee, 1974)

in an amount to reach 10 μg l^{-1} P. There was no change in total inorganic C in the water and the greatest change in lake algal content did not occur until the third year (Figure 6.35).

Dissolved P concentration showed very little change in response to fertilization, indicating the supply rate as the significant factor. Even in this low-carbon lake, in which short-term (hours) *in situ* photosynthesis experiments showed C to limit (Schindler, 1971b), a large mass of blue-green algae (*Lyngbya* and *Oscillatoria*) can be produced if the N and P input is maintained long enough.

By following the CO_2 diffusion pattern with radium, Schindler (1971b, 1974) showed that the C supply originated in the atmosphere and that CO_2 was made to be limiting in the short term by adding N and P (as Kuentzel hypothesized and Kerr demonstrated). The pattern of CO_2 concentration in the fertilized and unfertilized lake over the summer can be illustrated as in Figure 6.36.

The arrows, which indicate the determined direction of CO_2 transfer in the fertilized lake, show that additional amounts of N and P increased the supply of CO_2 from the atmosphere and eventually a large nuisance level of algae resulted. In a lake with higher alkalinity, the problem of a large algal mass may have occurred sooner because C would not have limited growth rate to such an extent in the short term.

In another lake (Lake 304), the same result occurred; that is, a gradual

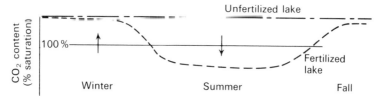

Figure 6.36. CO_2 level in fertilized and unfertilized lakes compared with the air saturation level of 100% (idealized from Schindler, 1971b)

Figure 6.37. Phytoplankton standing crop (as chlorophyll *a*) in the epilimnion of lake 304 before fertilization (1968, 1969 and 1970); during fertilization with P, N and C (1971 and 1972); and with the elimination of P and continued fertilization with N and C (1973). Results show the effectiveness of phosphorus control in eutrophication management (Schindler and Fee, 1974)

increase to a large algal biomass from enrichment with P, N and C. In this instance, however, biomass rapidly declined when P input was stopped in spite of continually added N and C (Figure 6.37), indicating that P was the key nutrient.

The conclusion with regard to the roles of C, P and N in eutrophication is that P is usually more significant than N or C in the long term because the P supply relative to the needs for growth is usually less than for N or C. This is because the sources of P are more limited (with no atmospheric source), and PO_4 is more readily bound chemically, resulting in sediments usually being more of a sink for P than for N or C. The long-term effect would be more evident from biomass (Figure 6.37).

However, any nutrient (usually C, N or P) may limit growth rate in the short term. In this consideration the most limiting (controlling) nutrient can usually be indicated from the ratio of one concentration to another as indicated previously.

6.5.6   Species succession and dominance (why blue-greens?)

One of the most interesting and challenging subjects in limnology is understanding the succession of species in the phytoplankton. Althouth this is a problem in its own right, it is also a problem in water-quality management. Succession occurs seasonally in lakes, as has been discussed in general and previously in relation to temperature and light. Usually diatoms dominate in the spring, greens in summer, blue-green algae (cyanobacteria) in later summer and possibly diatoms again in the autumn. However, there is much variability in this pattern. The variability that is of particular interest here is the tendency for blue-green algae to dominate an increasing part of the growing season as eutrophication increases algal biomass and phosphorus content (Smith 1990a,b). The objections to the blue-greens are many, but particularly they include: an objectionable appearance because they float and clump together causing scums, objectionable odours around beaches and taste/odours in drinking water supplies, a reduction in efficiency in predator–prey food chains because the filamentous and colonial nature of blue-green algae preclude effective grazing by filter-feeding zooplankton and a direct toxicity to mammals and fish when the blooms are very dense. (See Chapter 7 for more on toxicity.)

The challenge to water-quality management is to understand the cause(s) for the domination by blue-green algae when eutrophication increases. If something can be added biologically or chemically to more efficiently utilize the nutrients added to the system, the effects of eutrophication (obnoxious blooms of blue-greens) can be alleviated. This would be particularly valuable in situations where for practical or economic reasons the external nutrient supply cannot be reduced (Shapiro *et al.*, 1975).

The role of light and temperature in the succession of phytoplankton, discussed previously, concluded that those factors, while important, could not begin to explain the phenomenon of blue-green dominance either seasonally or with an increase in eutrophication. This was also concluded by Shapiro (1990).

Varying requirements for certain nutrients was thought to contribute to succession. The succession from *Asterionella formosa* to *Dinobryon divergens* in the English lakes has been related to the different P requirements of the two organisms being met successively as the P content declines in the spring (Lund, 1950). The following ranges of dissolved P in $\mu g l^{-1}$ for optimum cell growth were suggested by Rodhe (1948):

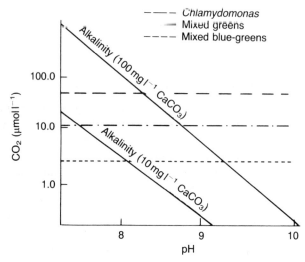

Figure 6.38. Boundaries for indicated algal communities in sewage lagoons in relation to alkalinity, pH and free CO_2 (modified from King, 1970)

Low P, *Dinobryon divergens* and *Urogena americana*, lower limit < 20, upper limit < 20; Medium P, *Asterionella formosa*, lower limit < 20, upper limit > 20; High P, *Scenedesmus quadricauda*, lower limit > 20, upper limit > 20.

The Chlorococcales (e.g. *Chlorella, Scenedesmus*) apparently require high, extracellular concentrations of phosphorus to grow at maximum rates (see above) and probably also nitrogen and CO_2. The blue-green algae, or cyanobacteria (at least the nuisance-forming species), on the other hand, appear to grow well at low soluble concentrations, but require high supply rates (i.e. total concentrations) in order to produce a high nuisance biomass. These points will be apparent in the following discussion of the hypotheses that have been suggested to explain partially or wholly the dominance of blue-greens.

Hypothesis 1 is based on blue-greens being able to dominate at low CO_2 and high pH, which are conditions that occur more frequently as lakes become eutrophic. If CO_2 is increased and pH lowered, by adding either CO_2 or HCl, then the phytoplankton community should shift from blue-greens to green algae. Furthermore, by again raising pH (and lowering CO_2) blue-greens should once more dominate. This hypothesis was first proposed by King (1970) and further substantiated by Shapiro (1973; 1984; 1990).

Through laboratory and field work with sewage lagoons (King, 1970; 1972) originated the hypothesis that blue-green/green algal dominance is controlled by CO_2. The hypothesis is easily explained by Figure 6.38.

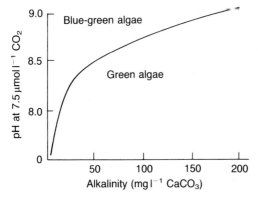

Figure 6.39. The pH at which the critical CO_2 concentration is reached, separating blue-green from green algae as a function of alkalinity (King, 1972)

Figure 6.40. Hypothetical amount of P that must be added to extract enough CO_2 through photosynthesis to, in turn, raise pH to the level providing the critical CO_2 concentration as a function of alkalinity (King, 1972)

Given a particular water with a constant alkalinity, the algal community would progress from one dominated by *Chlamydamonas* to a mixed assemblage of greens and finally to blue-green dominance as high photosynthetic rates and CO_2 consumption (stimulated by high N and P) raised the pH and lowered the CO_2 concentration. Furthermore, King contended that community shifts were associated with particular levels of CO_2, the critical level for blue-green dominance being $7.5 \mu mol\,l^{-1}$ ($0.33\,mg\,l^{-1}$). Figure 6.38 also indicates that for a water of lower alkalinity the critical CO_2 concentration would be reached at a lower pH.

The concept is clarified further in Figures 6.39 and 6.40. Figure 6.39 shows the relationship between the pH, which provides the critical CO_2 level, and the water's alkalinity. Blue-greens should dominate at pH

TABLE 6.8.   The ratio of blue-green cell counts to greens, maximum pH and minimum CO_2 in 1 m deep plastic-bag experiments lasting 1.5 weeks in Moses Lake, Washington

NO_3N $(mg\,l^{-1})$	Max pH	Min CO_2 $(mg\,l^{-1})$	Blue-green:green
0.74	10.1	0.02	1.9:1
0.5	9.8	0.07	1.4:1
0.23	9.6	0.1	0.9:1

Source: Buckley (1971)

values above the line and green algae at pH values below the line. The concept is expanded further in Figure 6.40, which relates the amount of P required stochiometrically to remove enough CO_2 to raise the pH to the level providing the critical concentration. The amount of P required would vary with the alkalinity of the water. Waters of low alkalinity theoretically need much less P enrichment to drive the pH up and, consequently, the CO_2 concentration down to the critical level than would be the case in high alkalinity waters. This is because higher alkalinity provides more buffering, i.e. less pH change for a quantity of CO_2 removed.

An experiment was performed *in situ* in 1 m deep bags in Moses Lake, Washington that tended to support King's hypothesis (Buckley, 1971). The experiment was performed to determine the effect of low-nutrient dilution water on the biomass and species composition of algae in the lake and, by adding back N and P, to determine which nutrient was responsible for the observed proportional reduction in biomass with increased dilution water. When N (as NO_3) was replaced, pH increased and CO_2 decreased as N increased (Table 6.8). Added N was expected to favour green algae, because green algae dominated in another portion of the lake where sewage effluent entered and the concentration of all macro-nutrients were high. However, the ratio of blue-greens (primarily *Aphanizomenon*) to greens increased in relation to added N (Table 6.8). Note that CO_2 was below King's critical level in all three treatments. The benefit to *Aphanizomenon* from increased N, when CO_2 is low, may be explained in later experiments by Shapiro.

Shapiro's experiments were initially conducted *in situ* in polyethelene cylinders 1 m in diameter and 1.5 m deep, utilizing the blue-green dominated phytoplankton community in eutrophic Lake Emily, Minnesota (Figure 6.41). Nutrients were added where indicated at $700\,\mu g\,N\,l^{-1}$ and $100\,\mu g\,P\,l^{-1}$. Initially, pH was lowered to 5.5, but later to various levels. Of 20 *in situ* experiments in which N, P and CO_2 were added to obtain pH 5.5, shifts to greens occurred in all except one. Shifts were observed

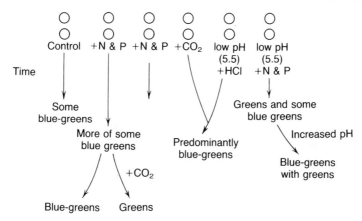

Figure 6.41. Shapiro's (1973) experiment showing the significance of CO_2 in causing a shift between green and blue-green algal dominance (schematic from personal communication)

consistently in ten other lakes and usually when pH was lowered by adding HCl. However, the shift usually did not not occur if N and P were not also added and shifts were inconsistent if pH lowering were stopped at a level higher than 5.5. The shift to greens and back to blue-greens was also accomplished to some extent by adding KOH after the shift had occurred by lowering pH (see far right, Figure 6.41).

The overall conclusion seemed to be clear. Increased eutrophication with N and P (or whichever one was limiting) could result in dominance by either green algae, which are more easily consumed in the food web, or blue-green algae, which are less easily consumed and create nuisance conditions, depending on the level of CO_2 reached. The mechanism for the shift may be that blue-green algae in general have lower half-saturation uptake constants than do green algae, and therefore, can outcompete greens as CO_2 decreases. However, Shapiro (1984) was doubtful of this, even though such uptake kinetics had been determined by Long (1976). The reason was that the shift did not occur as a gradual change in dominance, but rather the blue-greens crashed abruptly, followed by an increase in N and P in the water, possibly as a result of cell lysis due to cyanophage.

The idea that the concentration of free CO_2 is a key factor to the dominance of blue-green or green algae was challenged by Goldman (1973), who maintained that the free CO_2 concentration is not critical to growth rate. Rather, algal growth is related to C_T. That contention was also challenged by King and Novak (1974). Although the debate was very stimulating, no consensus was reached. However, from Shapiro's experiments (Shapiro, 1973; Shapiro *et al.*, 1975), it does appear that algal growth responds to free CO_2 because CO_2 was increased by lowering pII with acid, which in the short term should not have resulted in decreased C_T.

Hypothesis 2 is related to the buoyancy mechanism possessed by bluegreens, but is interrelated with the CO_2/pH/N and P phenomenon just described. Reynolds *et al.* (1987) have reviewed the state of knowledge on this phenomenon. Blue-green algae have gas vacuoles, which are composed of stacks of proteinaceous gas vesicles that provide rigidity to the vacuoles. Production of gas vesicles allows blue greens to migrate vertically in the water column, while collapse of the vesicles will result in sinking. *Oscillatoria* and *Lyngbya* are examples that tend to remain neutrally buoyant to occupy intermediate depths in deep mesotrophic lakes where nutrient content is more available than in the relatively nutrient-depleted, but lighted and mixed surface layer. Their filaments are small, providing a high 'coefficient of form resistance' to sinking. With increased nutrient content, the production of gas vesicles will increase and cells will rise into the lighted zone (Klemer *et al.*, 1982; Booker and Walsby, 1981). As photosynthesis increases the gas vesicles collapse due to increased cellular turgor pressure resulting in increased cell density, which is caused by the accumulation of carbohydrate (photosynthate) (Grant and Wallsby, 1977) and potassium (Reynolds, 1975; Allison and Wallsby, 1981; Reynolds, 1987). Increased cell density can also cause sinking through the accumulation of polysaccharide as ballast (Klemer *et al.*, 1988; Klemer and Konopka, 1989).

At depth, where CO_2 and/or light limit photosynthesis, but N and P are available, previously accumulated photosynthate can be assimilated into proteins and cause cell division. Turgor pressure is reduced by 'dilution' of photosynthate when the cells divide or, alternatively, by polysaccharide utilization and depletion of the ballast. Such buoyancy response to CO_2 limitation and N enrichment has been shown with *Oscillatoria* (Klemer *et al.*, 1982), with N and P enrichment in *Anabaena* (Booker and Wallsby, 1981), and with CO_2 (by pH adjustment) limitation in *Anabaena* and *Aphanizomenon* (Paerl and Ustach, 1982). The latter experiment clearly shows the effect of increasing pH (and hence reduced free CO_2 content) on buoyancy of blue greens (Figure 6.42). More recently, Klemer and Konopka (1989) have argued that low CO_2 can promote buoyancy only if other macronutrients are limiting. In their experiments, if N and P were not limiting, then buoyancy was found to be indirectly related to increased inorganic C. That is, addition of N, P and C promoted more buoyancy than did N and P without C.

Large *Microcystis* colonies can sink or float through a 15 m water column in just a few hours while many days would be required for such an excursion by small *Oscillatoria* filaments (Reynolds, 1987). Thus, species like *Microcystis* (and apparently *Anabaena* and *Aphanizomenon*)

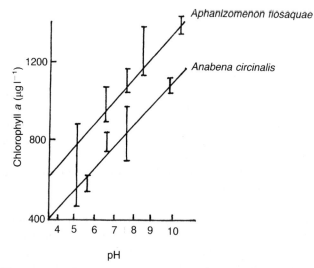

Figure 6.42. Buoyancy response of two blue-green algae from exposure to increasing pH and consequent low CO_2 concentration (Paerl and Ustach, 1982). Chlorophyll *a* content is from the upper 10 ml of 100 ml graduated cylinders after 2 h

are best adapted to rather turbid, stable (or alternately mixed/stable) water columns where they can make rather quick vertical excursions for nutrients and light, while *Oscillatoria* is adapted more to deeper, clearer waters, where even if well mixed they can maintain themselves below the mixed layer in low light and resist transport from depths where nutrients are available. As lakes become enriched and increased particulate matter restricts light penetration, the blue-green algae, with their buoyancy regulation mechanisms, are favoured under conditions of strong stratification or diel partial mixing. Those conditions are not favourable to diatoms and green algae, both of which grow faster and require higher light levels than blue-greens, and lack the buoyancy mechanism (Reynolds, 1987). Spencer and King (1987) showed that even in a 1.8 m pond the light gradient was sufficient to promote strong buoyancy/sinking behaviour in *Anabaena* that led to its dominance.

Knoechel and Kalff (1975) used buoyancy regulation to explain why a blue-green (*Anabaena*) became dominant in late summer, replacing a diatom (*Tabellaria*). Because specific growth rates of the two species were not different, the authors concluded that reduced mixing as thermal stratification developed and the epilimnion deepened was the principal reason for the species shift. That is, the blue-green was favoured because its sinking rate was less than that of the diatom, which requires mixing to keep it suspended. Reduced sinking or buoyancy of the blue-green may have been caused by changes in CO_2/pH, but the authors did not investigate that aspect.

The buoyancy mechanism is apparently unsuccessful in favouring blue-green dominance in lakes and reservoirs that are turbid from high concentrations of non-algal particulate matter. Clay particles are especially effective in attaching to the musciglaginous covering, which results in flocculating out the blue-greens and favouring greens and diatoms (Cuker *et al.*, 1990). There are several accounts of blue-green algal abundance or fraction being much less than expected in turbid systems (Smith, 1990b; Søballe and Kimmel, 1988; Cuker, 1987).

Hypothesis 3 deals with zooplankton and their ability to promote the dominance of inedible colonial and filamentous blue-green algae. They may do that by either grazing the small edible species and leaving the inedible ones to dominate or by recycling nutrients when the edible species are consumed, which favours the large inedible blue-greens (Porter, 1972; 1977). Toxicity of some blue-greens to animals is also a protective device against grazing loss. The literature involving the testing of these hypotheses has been reviewed recently by Burns (1987), with respect to *in situ* enclosures and by Lampert (1987) for laboratory experiments.

Even large efficient grazers (e.g. *Daphnia*) usually avoid eating most blue-greens, which are either too large to handle (*Microcystis*, *Aphanizomenon* and *Anabaena*), or if small filaments (*Oscillatoria* and *Lyngbya*) they interfere with the filter-feeding mechanism and, therefore, growth and survival of zooplankton. Also, some blue-greens may produce toxins that inhibit zooplankton (especially filter feeders) growth and reproduction (Arnold, 1971; Lampert, 1981). All of these mechanisms should lead to greater dominance by blue-greens. However, decreases in blue-greens (especially small colonies) in enclosures have been related to grazing by large zooplankton (Lynch and Shapiro, 1981; Ganf, 1983), while other results show that zooplankton have little or no direct effect on blue-green abundance (Porter, 1977; Reynolds *et al.*, 1987). The degree to which blue-greens are consumed by large filter-feeding zooplankton depends on colony size, and whether there is an adverse effect of algal toxicity depends on the alga's growth phase. Thus, while many interactions between zooplankton and blue-greens have been observed, there is no consistent pattern of response, so this hypothesis cannot be regarded as a general case. In fact, grazing of diatoms and green algae has been shown to discourage the blooming of *Anabaena elenkinii* in eutrophic ponds because it prevented the low light and low CO_2, which would normally favour blue-greens through buoyancy stimulation. The low light and CO_2 were brought on by high densities of diatoms and greens (Spencer and King, 1987).

Hypothesis 4 deals with the requirements for N and P. On the one hand, there is some evidence that blue-greens may be favoured by low concentrations of P, as seems to be the case with C. The uptake rate for $^{32}PO_4$-P has been shown to be lower for blue-greens than for greens as reported by Shapiro (1973). Also, Ahlgren (1977) found that the K_s for growth of *Oscillatoria agardii* to be only $\approx 1\,\mu g\,l^{-1}$ in continuous-flow culture experiments, with $\approx 10\,\mu g\,l^{-1}$ being required for maximum growth rate.

On the other hand, Smith (1982, 1983, 1986) has set forth the hypothesis that blue-greens are favoured by low N:P ratios, which could mean that high P concentrations are beneficial. Smith showed in 17 lakes studied that the percentage of blue-green algae in the phytoplankton was <10% in lakes with TN:TP ratios >29:1, but various percentages were observed at low ratios. The rationale is that cellular N:P required by blue-greens is lower than for other algae and as the ambient N:P ratio decreases (as eutrophication proceeds) blue-greens would be better able to exploit the available nutrient resources. The ability of some of the blue-greens to fix N could also be the reason for some of these observations. He also presented some evidence that low transparency, in addition to low TN:TP ratios, promoted the dominance by blue-green algae.

The absence, or low concentrations, of trace metals (required for growth) was also suggested to affect phytoplankton dominance. In artificially fertilized Fern Lake, Washington (Olson, personal communication), 'complete medium' fertilization (including all trace elements) resulted in diatom blooms and subsequent zooplankton increase, whereas ammonium superphosphate fertilization produced an *Anabaena* bloom and relatively little zooplankton.

Hypothesis 5 deals with allelopathic excretory products from blue-green algae that inhibit other algae and therefore favour the dominance of blue-greens. This has been shown by Vance (1965), who found that only *Microcystis* grew well in pond water from which that organism dominated. Keating (1977) found that summers with large blue-green blooms in Linsley Pond, Connecticut and winters with low flushing, were followed with poor diatom growth in the spring and vice versa.

Hypothesis 6 is the silica depletion concept. Increased enrichment with N and P depletes silica, which is slow to regenerate in the water column, thus limiting diatom growth, leaving more nutrients for blue-greens. Schelske and Stoermer (1971) suggested that southern Lake Michigan gradually shifted to blue-green algae because silica was depleting. Schelske *et al.* (1986) have shown that the reserve of Si remaining after the stratified period is inversely proportional to the TP content.

Hypothesis 7 suggests that blue-green algal blooms may originate from the sediments (Fitzgerald, 1964; Silvey *et al.*, 1972; Osgood, 1988b) and rise into the water column, possibly by their buoyancy mechanism discussed earlier. Viable populations of blue-greens can exist in the sediments throughout the year (Roelofs and Ogelsby, 1970; Kappers, 1976; Fallon and Brock, 1984; Reynolds et al., 1981; Ahlgren et al., 1988a). When conditions are favourable, although initiating factors are not well known at present, a portion of the viable population rises into the water column.

Trimbee and Harris (1984) have trapped *Microcystis aeruginosa*, *Gomphosphaeria lacustris* and *Lyngbya Birgei* rising from the sediments in an amount equivalent to 2–4% of the water column maximum biomass. Stimulation of *Microcystis* populations in the sediment to rise into the water column may be related to the onset of low DO and increasing light in the spring, which was indicated by the growth response of sediment populations (Reynolds et al., 1981). More directly, Trimbee and Prepas (1988) have shown that the fraction of blue-greens in the phytoplankton in 39 North-temperate lakes was inversely related to an OXYD index, which is the product of mean depth and the proportion of sediments that are oxic. This supports the hypothesis that recruitment of blue-greens from the sediments is more apt to occur in shallow lakes and is favoured by anoxic conditions. One the other hand, anoxic conditions may indicate high internal loading of P, which may be the factor favouring increased blue-greens.

Other species, such as *Aphanizomenon* and *Anabaena flos-aquae*, are not as durable as *Microcystis* in the vegetative form, but do have resting spores (akinetes) that may be important in overwintering and subsequent bloom formation. However, the role of akinetes in initiating blue-green blooms is not firmly established (Wildman et al., 1975; Rother and Fay, 1977; Jones, 1979). Barbiero (1991) has observed that migration of sediment populations of *Gloeotrichia* (caught in traps) contributed well over 50% of the water column bloom.

As the above hypotheses demonstrate, there are many factors that have been implicated in promoting dominance by blue-green algae, e.g. nitrogen fixation, low soluble P concentration, low CO_2/high pH, low N:P, water column stability, low transparency, buoyancy, reduced grazing, high temperature, depleted silica and allelopathy. A case can be made for each of these factors under the conditions of the particular experiments or test site. However, some of these factors are highly correlated with each other in eutrophic lakes. For example, a highly eutrophic lake during summer would usually have the following conditions: a high temperature in the epilimnion; a stable water column; low CO_2/high pH in the epilimnion; low transparency due to high algal biomass; a low N:P ratio resulting from P gain from anoxic sediments and N loss due to denitrification under anoxia; and possibly low silica due to high productivity and loss during the spring diatom bloom. Except for CO_2/pH (Shapiro, 1973; 1984), the case for each of the factors has been made largely through correlation, which does not prove cause and effect.

Cause and effect are most convincing through experimentation where the causitive factor is altered and the algal dominance changes in response to the altered conditions. Recently, however, Stockner and Shortreed (1988) promoted the bloom of *Anabaena* in one arm of a lake that was treated with a low N:P fertilizer in contrast to the other arm. Subsequently the N:P ratio was raised in the affected arm and *Ababaena* declined.

Doubts can be raised about the exclusive effect of some of these factors on blue-green dominance. Most blue greens have relatively high thermal optima which have been determined experimentally, yet blooms have occurred at low temperature and many green algae have optima as high or higher than blue-greens (see Chapter 6.4). If blue-greens are favoured by low soluble P and high temperature, then those conditions exist in oligotrophic lakes as well as eutrophic lakes. N:P ratios are very low in Lakes Tahoe and Crater, but blue-green algae are not dominant there. Zooplankton feeding, growth and reproduction usually decrease with an increase in large colonial and filamentous blue greens. This should contribute to blue-green dominance, but there is conflicting evidence on the importance of that factor (see above).

Some combination of these factors, which may vary from time to time and place to place, would seem to be the most probable explanation for the dominance of blue-green algae. The following general scenario is proposed as an explanation for blue-green dominance under some conditions. Increased productivity results from increased N and P and causes raised pH, lowered CO_2 and reduced light transmission in the surface water. These conditions alone would favour blue-greens because of their ability to grow at low CO_2 and light. In addition, water-column stability promotes anoxic conditions in the hypolimnion leading to increased nutrient content, especially in the hypo- and metalimnia. High surface temperature, which favours blue-greens directly, is also consistent with stability. These conditions would be favourable to blue-greens, whose buoyancy regulation could exploit such a physically–chemically stratified environment. Limitation by N and/or P at the surface where light is not limiting would allow an accumulation of polysaccharide ballast in excess of growth demands causing loss of buoyancy. The algae could then utilize the high nutrient content at depth, causing buoyancy to increase once the ballast is consumed through growth. That environment, with low nutrient content in the surface water and reduced mixing, would not be hospitable to non-buoyant species, which would tend to sink from the photic zone permanently.

7
EUTROPHICATION

7.1 DEFINITION

Eutrophication is the process by which water bodies become more productive through increased input of inorganic nutrients. In a strict sense, the term refers to nutrients only and not necessarily to a response in production (Beeton and Edmondson, 1972). Thus, a lake could become eutrophic from increased nutrients even though productivity did not increase, as for example if algal growth were limited by light due to high suspended solids. Normally, however, increased plant productivity and biomass are considered part of the eutrophication process. Increased input of sediment can also cause eutrophication by decreasing depth, which could expand the area suitable for macrophytes, as well as encourage a more effective exchange of nutrients between sediments and water. Sediment can be allochthonous organic or inorganic and autochthonous organic.

Depending on the source of the increased nutrient supply, eutrophication can be either natural or cultural. Nutrient supply can increase naturally as a result of forest fires, earthquakes or simply increased erosion following a dramatic increase in precipitation. In these cases, the watershed and lake can stabilize when the rate of nutrient input (eutrophication) returns to earlier levels. Lakes can also age naturally and become more eutrophic. They will eventually become shallower through normal rates of sedimentation, which are typically of the order of a few millimetres per year. Assuming a sedimentation rate of 4 mm year^{-1}, 250 years would be required to reduce lake depth by 1 m. As the lake became shallower the sediment area to lake volume would increase, providing more opportunity for nutrient recycling (see internal loading) from sediments, as well as greater light availability for photosynthesis. However, this ageing process could take thousands of years and the lake may never become eutrophic, depending on the nutrient input.

The concern about eutrophication from a water quality standpoint is more probably due to cultural eutrophication. Increased nutrient input can originate from such man-caused sources as: direct discharge of treated or untreated domestic sewage (including phosphate detergents); industrial wastes, such as food-processing plants and dairies, leaching from fertilizer applied to forests, lawns, pastures or cultivated land; leaching from failing on-site sewage treatment systems; and stormwater runoff from urbanized land. Increased eutrophication in a water body occurring over a period of only 10 to 50 years usually indicates anthropogenic rather than natural processes.

7.2 PHOSPHORUS MASS BALANCE MODELS

Since Vollenweider's initial work (Vollenweider, 1969a), many models have been developed that describe the mass balance of phosphorus in lakes. Notable among these are Dillon and Rigler (1974a), Vollenweider (1975, 1976), Chapra (1975), Nurnberg (1984) and Walker (1977). Walker (1987a) has adapted the P mass balance model for reservoirs. All P mass balance models are based on the kinetics of continuously stirred tank reactors (CSTR) commonly used in chemical engineering (Reckhow and Chapra, 1983). By continuously mixing the volume in such a reactor, holding that volume constant and maintaining the water inflow rate equal to the water outflow rate, the following mass balance equation applies with units of mass/time:

$$dCV/dt = C_i Q - CQ + rCV \tag{7.1}$$

where C is substance concentration in the reactor and C_i is concentration in the inflow, Q is flow rate, V is reactor volume and r is the reaction rate coefficient. If K is taken as a first-order depletion reaction and both sides are divided by V, so that $Q/V = \rho$, the flushing rate in $1\,t^{-1}$, the equation becomes:

$$dC/dt = \rho C_i - \rho C - KC \tag{7.2}$$

and at steady state it becomes:

$$C = C_i/(1 + K/\rho) = \rho C_i/(\rho + K) \tag{7.3}$$

This is essentially the same as the mass balance for TP in lakes proposed by Vollenweider (1969a):

$$dTP/dt = L/\bar{z} - \rho TP - \sigma TP \tag{7.4}$$

where L is areal loading of TP in $mg\,m^{-2}\,year^{-1}$ ($L/\bar{z} = \rho C_i$ in Equations 7.2 and 7.3), $\bar{z}$ is mean depth in m, ρ is flushing rate in $year^{-1}$ and σ is the sedimentation rate coefficient in $year^{-1}$. At steady state the equation becomes:

$$TP = L/[\bar{z}(\rho + \sigma)] \tag{7.5}$$

which is the same as Equation 7.3 because $L/\bar{z} = \rho TP_i$.

Equation 7.4 shows that as each new concentration of TP enters the lake and is immediately mixed throughout the lake, a new concentration is produced after a fraction leaves through the outlet and a fraction sediments to the lake bottom, both as a function of the new, slightly changed concentration. Equation 7.5 shows that over the long term the lake will equilibrate to the given loading. If the loading is changed then a time will be required for equilibration to the new loading and, if a first-order rate reaction is assumed, the times to 50% and 90% of equilibrium will be, respectively:

$$t_{50} = \ln 2/(\rho + \sigma) \quad \text{and} \quad t_{90} = \ln 10/(\rho + \sigma) \qquad (7.6)$$

The difficult problem with this model is the determination of the sedimentation rate coefficient; all other parameters could be determined directly. This means that for a lake with loading, mean annual TP concentration and flushing rate known, σ could be estimated from Equation 7.5:

$$\sigma = L/(TP\bar{z}) - \rho \qquad (7.7)$$

However, to develop a model to describe a large number of lakes, it is useful to have some general way of estimating sedimentation. One approach is to use a unitless retention coefficient R_{tp} (Vollenweider and Dillon, 1974; Dillon and Rigler, 1974a), which can be determined directly as:

$$R_{tp} = 1 - TP/TP_i \qquad (7.8)$$

where TP_i is inflow concentration and TP is the lake concentration if assumed to equal the outflow concentration. With some manipulation (see Vollenweider and Dillon, 1974), Equation 7.5 can be written as:

$$TP = L(1 - R_{tp})/\bar{z}\rho \qquad (7.9)$$

R_{tp} has been related to hydraulic variables in several empirical formulations, one of which is $1/(1 + \rho^{0.5})$ (Larsen and Mercier, 1976; Vollenweider, 1976). Thus, as flushing rate increases, R_{tp} decreases. As a result, there are three algebraic forms of the steady-state equation (Equation 7.5) that are based on this dependence of retained TP on flushing rate. Using $TP_i = L/\bar{z}\rho$ for simplicity, they are:

$$TP = TP_i(1 - R_{tp}) = TP_i/(1 + 1/\rho^{0.5}) = L/[\bar{z}(\rho + \rho^{0.5})] \quad (7.10)$$

The negative relation between flushing rate and R_{tp} is logical, i.e. as flushing rate increases there is less time for TP to settle, so R_{tp} decreases. On the other hand, the sedimentation rate coefficient is positively related to the flushing rate ($\rho = \sigma^{0.5}$), which seems illogical. However, to calculate actual sedimentation, which is flux rate to the sediment, R_{tp}

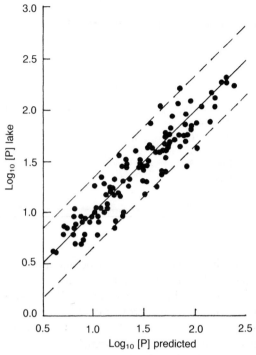

Figure 7.1. Comparison of lake TP [P] with that predicted from Vollenweider's (1976) improved loading criteria model according to a slightly modified version: $\log[P]$ lake $= \log[P]_i/(1 + 1.17^{0.45})$ (modified from Chapra and Reckhow, 1979). Dashed line is 95% confidence interval

is multiplied by L, but σ is multiplied by lake TP. Therefore, one can readily see that if L is held constant, increasing the flushing rate will give increasingly smaller TP_i, so σ must increase in order that the flux rate to the sediments does not decrease too rapidly. Ahlgren et al. (1988b) have modified a relationship found by Canfield and Bachmann (1981) that shows such a relationship between σ and both flushing rate and TP_i:

$$\sigma = 0.129(TP_i \, \rho)^{0.549} \tag{7.11}$$

At steady state, the mass balance model illustrated by Equation 7.10 has been verified for a large population of lakes (Chapra and Reckhow, 1979), which suggests that the general form of the sedimentation term is reasonable. However, the error for predicting the TP content in any given lake may be quite large (about $\pm 50 \, \mu g \, l^{-1}$, Figure 7.1). For example, Equation 7.10 predicts pre- and post-diversion TP in Lake Washington quite well, but provides a much poorer fit for Lake Sammamish and post-diversion levels in Shagawa Lake (Table 7.1). The poor agreement for all

TABLE 7.1. Annual mean TP predicted before and after P control using three models compared with observed. Data from Lake Washington and Shagawa Lake are from Edmondson and Lehman (1981) and Larsen et al. (1979), respectively

| Lake | Treatment | | TP ($\mu g\ l^{-1}$) | | | |
			Obs	Vollenweider	Nurnberg	OECD
Washington	Diversion	Pre	64	67	84	49
		Post	17	17	21	16
Sammamish	Diversion	Pre	33	43	55	34
		Post	19	29	33	25
Shagawa	P removal	Pre	51	50	54	39
		Post	30	7	8	11

models with post-treatment values in Shagawa Lake may be due mostly to insufficient time for internal loading to have declined. Several years were required in Lake Sammamish. The Nurnberg (1984) and OECD (1982) models show variable results, with the latter providing a better fit for Lake Sammamish (see Chapter 10 for more on the response of these lakes). This comparison suggests an alternative to calibrating a model to a particular lake data set, i.e. to compare several models and select the one that provides the best fit.

If internal loading is important, as may be the case in either un-stratified oxic or stratified anoxic lakes (see Chapter 4.1), then the model may need to be modified to account for the two sources. Nurnberg (1984) suggested using the following model to account for internal load (L_{int}):

$$TP = TP_i(1 - R_{pred}) + L_{int}/\bar{z}\rho \qquad (7.12)$$

where R_{pred}, the predicted TP retention coefficient, was found in 54 oxic lakes to be best represented by:

$$R_{pred} = 15/(18 + \bar{z}\rho) \qquad (7.13)$$

There are several other formulations where values refer to settling rates that range from 10 to $16\,\mathrm{m\,year^{-1}}$ (Chapra, 1975; Kirchner and Dillon, 1975; Vollenweider, 1975). Internal loading can also be added to the previous model versions (Equations 7.9 and 7.10). However, there was no attempt to treat oxic and anoxic lakes separately in the development of these other approaches of estimating sedimentation.

Using observed TP, and solving Equation 7.12 for L_{int}, makes it possible to calibrate Nurnberg's model for a particular stratified, anoxic lake. L_{int} so calculated can be compared with other estimates of internal loading for the lake in question, such as P release rates determined from sediment cores in the laboratory or observed rate of increase in hypolimnetic P concentrations. Nurnberg (1987b) has shown rather good agreement between these two methods of estimating internal loading in anoxic lakes. If there is good agreement among these different estimates of internal loading for a particular lake, the model is verified for that lake. If not, there may be an error in the estimate for sedimentation and a different modelling approach should be taken. Also, the lake may not be in equilibrium with its external loading.

Example problem 1
(a) Given a lake with a mean depth of 15 m, a flushing rate of 1.5 year^{-1} (outflow rate/lake volume) and a mean inflow TP concentration of 80 μg l^{-1}, calculate the lake's expected external TP loading in mg m^{-2} year^{-1}.

From the conversion of Equation 7.9 to 7.10 we know that $L/\bar{z}\rho = TP_i$, so

$$L = TP_i \, \bar{z}\rho = 80 \, \text{mg m}^{-3} \times 15 \, \text{m} \times 1.5 \, \text{year}^{-1} = 1800 \, \text{mg m}^{-2} \text{year}^{-1}$$

(b) If the lake concentration is actually 70 μg l^{-1}, calculate its expected internal loading of TP.

Using a modified Equations 7.10 and 7.12, gives

$$L_{int} = \bar{z}\rho[TP - TP_i/(1 + 1/\rho^{0.5})]$$
$$= 15 \times 1.5[70 - 80(0.55)] = 585 \, \text{mg m}^{-2} \text{year}^{-1}$$

Or using Equation 7.12, gives

$$L_{int} = 15 \times 1.5[70 - 80(1 - 15/(18 + \bar{z}\rho))] = 441 \, \text{mg m}^{-2} \text{year}^{-1}$$

Thus, the difference between internal loading with the Vollenweider and Nurnberg versions of the sedimentation term is 144 mg m^{-2} year^{-1}. Formulated in this way, L_{int} is net internal loading, because there is no loss of L_{int} due to sedimentation. Being net internal loading, it should be comparable with the observed increase in hypolimnetic TP (Nurnberg, 1984). This is reasonable, because L_{int} results in TP build-up in the hypolimnion, which is essentially net internal loading. P released into the hypolimnion should not sediment at the same rate as P in the epilimnion because being in the dark it is not as subject to uptake and sinking by algae. That is supported by the similarity between sediment P release determined in latoratory cores and field observations of hypolimnetic P build-up (Nurnberg, 1987b).

Even if one can verify a particular steady-state model for a lake, there are problems in using the steady-state version. Firstly, it is often difficult to determine an appropriate time interval (most often annual) in which the lake mean TP represents a steady state, especially if flushing rate exceeds once per year. Secondly, internal loading usually occurs during the productive period (whether stratified or unstratified) and may contribute proportionately more to growing-season TP and biomass than would external loading, especially if the latter occurs primarily during the non-productive period (e.g. winter in the Pacific northwest). These problems can be averted by calibrating and verifying a non-steady-state version of the mass balance model (Equation 7.4), but including L_{int} as follows:

$$dTP/dt = L_{ext}/\bar{z} - \rho TP - \sigma TP + L_{int}/\bar{z} \qquad (7.14)$$

Note that TP resulting from L_{ext} and L_{int} in Equation 7.14 is subject to sedimentation and, therefore, L_{int} is a gross rate. In this case, the numerator in Equation 7.5 would be $L_{ext} + L_{int}$.

No more data are usually required for the non-steady-state version, because TP loading and lake concentration data are usually collected on at least a monthly basis. While the data are, however, usually reduced to annual means for a steady-state approach (or some interval consistent with ρ), TP is computed for each time interval with the non-steady-state version. In fact, weekly time steps are recommended to obtain more realistic smooth curves even if data were collected less frequently. The model can be calibrated by determining the sedimentation rate coefficient (σ) that gives the best fit between predicted and observed data for the oxic period. Although Larsen et al. (1979) used a constant σ among years in Shagawa Lake, with good success, the model could be verified in Lake Sammamish from year to year only if σ were allowed to vary as a function of flushing rate, i.e. $\sigma = \rho^x$ (where $0 < x < 1$; $x = 0.78$, Figure 7.2, Welch et al., 1986). This is analogous to Equation 7.10 where $x = 0.5$. A formulation such as $\sigma = y\rho^x$ may be necessary where sedimentation rates are low, because as x approaches zero the sedimentation rate remains around 1.0 regardless of the flushing rate.

There still may be a problem with the non-steady-state model in stratified lakes even if there is a good fit between observed and predicted whole-lake TP. Chlorophyll a and transparency are a function of TP in the productive zone, probably the epilimnion, and not a function of the whole-lake TP. Epilimnetic TP usually declines during the stratified period while hypolimnetic TP increases. Thus, either the epilimnion and hypolimnion must be modelled separately with a function included to account for exchange of TP between the two, or mean epilimnetic TP must be estimated from a relationship between epilimnetic TP and whole-lake TP. The latter may be adequate, because relationships among chlorophyll a, TP and transparency are usually based on summer means, which are in turn most often used for management purposes.

There have been several approaches to dealing with the uncertainty in TP predictions for individual lakes. For example, in using Equation 7.14 to predict future TP concentrations in Lake Sammamish resulting from increased development in the watershed, uncertainty was included by

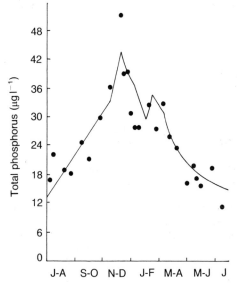

Figure 7.2. Predicted (—) versus observed TP (whole-lake) (●) in Lake Sammamish, WA

choosing a range in TP yield coefficients (areal loss rates for different land uses) and the 5 and 95% flow probabilities for the principal inflow stream (Shuster *et al.*, 1986). TP sedimentation was a function of p and increased/decreased flow resulted in dilution/concentration of the estimated TP loading. By this procedure, the prediction of $31\,\mu g\,l^{-1}$ TP by the year 2000 had a ±10% error due to TP yield and a ±20% error due to flow. Flow was found to cause most of the year-to-year variation in loading.

Another approach is to use first-order error analysis to calculate the uncertainty in the model and loading by utilizing low, high and most probable loading estimates from yield coefficients (Reckhow and Chapra, 1983). For a model of the type of Equation 7.13, Reckhow and Chapra (1983) determined an error of ±30%, which is added to the loading uncertainty. Once those uncertainties are summed, confidence intervals for a single model estimate for TP can be calculated. In order to evaluate small changes in TP, predicted from relatively small changes in loading, uncertainty can be applied to the TP concentration change, rather than the before and after concentration.

7.3 TROPHIC STATE CRITERIA

Criteria exist to describe a lake's quality and its trophic state. They include the concentration and loading rate of nutrients which are the

TABLE 7.2.   Qualitative characteristics of oligotrophic and mesotrophic lakes

	Oligotrophic	Eutrophic
Depth	Deep	Shallow
Hypolimnion: epilimnion	>1	<1
Primary productivity	Low	High
Rooted macrophytes	Few	Abundant
Density of plankton algae	Low	High
Number of plankton algal species	Many	Few
Frequency of plankton blooms	Rare	Common
Depletion of hypolimnetic oxygen	No	Yes
Fish species	Cold water, slow growth, restricted to hypolimnion	Warm water, fast growth, tolerate low O_2 in hypolimnion and high temperature of epilimnion
Nutrient supply	Low	High

cause, as well as biological and physical indices, which are the effect. The value of numerical criteria is to allow the quality of the lake to be defined or the lake classified. Criteria can be used to chart accurately the course of a lake as it becomes more or less eutrophic or to judge if lake quality is suitable or unsuitable for recreational or water supply use.

Some qualitative characteristics of eutrophic, compared to oligotrophic lakes, are shown in Table 7.2. Shallow lakes are more likely to be eutrophic than deep lakes, because of the greater potential of nutrient recycling from sediments, but shallow lakes can also be oligotrophic if they are naturally flushed at a fairly high rate with low nutrient water. As a corollary of depth, the relative size of the hypolimnion is apt to be less in a shallow lake and, thus, is more apt to become anoxic, facilitating internal loading of phosphorus. The amount of plankton, its productivity, frequency of blooms and resulting low water transparency are all typical of eutrophic lakes and atypical of oligotrophic lakes.

Blue-green algae frequently dominate the phytoplankton in eutrophic lakes. They can form scums on the water surface that can be very unsightly and odiferous. They can taint the flesh of fish with a musty odour or taste, a problem that is probably more widespread than appreciated, and also cause taste and odours in water supplies. Finally, some blue-greens are at times toxic and frequently result in deaths of domestic animals. The toxins from blue-greens have been grouped under neurotoxins, hepatotoxins and contact irritants (Carmichael, 1986). *Anabaena* and *Aphanizomenon* are usually associated with neurotoxins, while *Microcystis* produces hepatotoxins and species of several genera have been linked with irritants; *Gleotrichia*, *Anabaena*, *Aphanizomenon* and *Oscillatoria* (Carmichael, 1986). Surveys of lakes with blue-green blooms have shown that blue-green toxicity is probably a rather common occurrence. Repavich *et al.* (1990) sampled 102 sites in Wisconsin, USA and found that 25% contained toxic algae as determined by the mouse bioassay. Sivonen *et al.* (1990) showed toxicity by the same method in 44% of the 188 samples from 125 sites in Finland and site results from Sweden showing >50% incidence of toxicity.

Rooted macrophytes are also more typical of shallow lakes, which are more apt to have larger littoral areas (<3 m) with richer sediment. Factors controlling macrophyte distribution and abundance will be discussed in Chapter 9.

Combining the higher productivity and relatively smaller hypolimnion results in greater rates of oxygen depletion and lower oxygen concentration as hypolimnetic volume declines. The kinds and abundance of fish are affected mostly by the combination of oxygen and temperature. As lakes become more eutrophic, oxygen becomes depleted in the hypolimnion (see Chapter 7.3.3) limiting the suitable fish habitat to the epilimnion and metalimnion. The shallower the lake, the smaller the hypolimnion and the less is its cooling effect on the epilimnion. If the epilimnion is too warm (>20°C) for cold-water fish, then they will disappear as the hypolimnion with more suitable temperature becomes oxygen deficient.

There are many indices that have been used to classify trophic state and lake quality. Thirty different sources for trophic state criteria were listed by Porcella *et al.* (1980) and there are still others. There are also many goals for the quality of lakes, some of which may be in conflict. The aesthetically pleasing, clear blue water of ultraoligotrophic lakes does not yield large amounts of fish. Compromises may be called for between lake quality more favourable to fish production (meso or meso-eutrophic) and that preferred for swimming, boating and viewing. However, for cold-water fish species, there may be little difference between the appropriate trophic state for fisheries and recreational uses (see Chapter 7.3.3).

7.3.1   Nutrients, productivity, biomass and transparency

Beginning with Sakamoto's chlorophyll *a* (Chl*a*) versus TP relationship (Sakamoto, 1971) and Edmondson's observation (Edmondson, 1970) in Lake Washington's recovery from sewage diversion that chlorophyll *a* was closely tied to phosphorus concentration, TP and chlorophyll *a* have become two of the three most widely used indicators of trophic state. Transparency, determined by a black and white Secchi disc (SD), can

Figure 7.3. Relationship between summer levels of chlorophyll *a* and measured total phosphorus concentration for 143 lakes (Jones and Bachmann, 1976)

often be empircally related to chlorophyll *a* and TP, if water colour and inorganic suspended sediment are low (see Secchi disc in Chapter 6.3).

There are several empirical relationships between chlorophyll *a* and TP (see Ahlgren *et al.*, 1988b), but probably the two most often used are by Dillon and Rigler (1974b) (Equation 7.15) and Jones and Bachmann (1976) (Figure 7.3, Equation 7.16), which are respectively:

$$\log \text{Chl}a = 1.449 \log \text{TP} - 1.136 \qquad (7.15)$$

$$\log \text{Chl}a = 1.46 \log \text{TP} - 1.09 \qquad (7.16)$$

Dillon and Rigler used data from 46 lakes, mostly in eastern Canada, and Jones and Bachmann used data from 143 lakes covering a broad range in trophic state. The former data set contained TP values from spring turnover and mean summer chlorophyll *a* while the latter was composed of summer mean values. The equations agree rather closely, notwithstanding the difference in data-averaging times. Summer means for chlorophyll *a* and TP are most often used to define the trophic state of a lake. Therefore, sampling intensively throughout the non-growing season, simply to determine the lake's trophic state, may be unjustified. While TP may be higher when inflows are greater during the winter and spring, the summer mean represents the residual after sedimentation and, therefore, should be closely tied to the P in algal cells.

Because most chlorophyll *a*–TP relationships using large data sets are

TABLE 7.3. Some suggested trophic state threshold values

Variable	Oligotrophic	Eutrophic	Reference
TP^a (μg l^{-1})	<14	>25	Porcella *et al.* (1980)
Chla^a (μg l^{-1})	<2.8	>8.7	Chapra and Tarapchak (1976)
Secchi depth^a (m)	>3.6	<1.9	Equations; Carlson (1977)
ODR (mg m^{-2} day^{-1})	<250	>550	Mortimer (1941)
Net DO (mg l^{-1})	<4.5	>5	Porcella *et al.* (1980)
Min DO (mg l^{-1})	>7.2	<6.2	Porcella *et al.* (1980)
TN^a (μg/l^{-1})	<140	>180	Porcella *et al.* (1980)

^{a} Summer mean or median

log-log, the accuracy for prediction is not great for any single lake. In Equation 7.15, for example, errors of prediction for a chlorophyll *a* concentration of 5.6 μg l^{-1} (10 μg l^{-1} TP) can be $\pm$60–170% and 30–40% for 95% and 50% confidence limits, respectively. The high correlation coefficients (0.95) tend to hide the accuracy problem which may be due to lake-to-lake variations in cellular chlorophyll *a*, zooplankton grazing and/or other limiting factors such as light and N (Ahlgren *et al.*, 1988b). Sometimes it is possible to develop a relationship for the individual lake of interest that provides much greater accuracy of prediction (Smith and Shapiro, 1981). Usually, however, this requires too much data and one must rely on the published relationship that provides the best agreement with the individual lake data.

The chlorophyll *a*–TP relationship was used by Carlson (1977) to develop a numerical trophic state index (TSI), which is probably the most commonly used index. However, the three indicators represent a rather narrow range within the spectrum of trophic state indicators. While Carlson's TSI and Porcella's LEI (lake evaluation index, Porcella *et al.*, 1980) reduce lake trophic state to one or more numbers, in an attempt to remove the subjectivity inherent in the terms oligotrophic, mesotrophic and eutrophic, etc., it is still necessary to use those terms in communications about lake quality. In contrast to indices in which the values are dependent on regression analyses with a particular data set, Carlson's TSIs (and LEIs) represent absolute values for chlorophyll *a*, TP and SD. Log$_2$ transformations were used to interrelate these three indices, so that a doubling in TP is related to a reduction by half in SD. Values of TP, chlorophyll *a* and SD from the following equations for TSI of 40, 50 are respectively: 12; 24; 2.6; 6.4; 4.2.

$$TSI = 10(6 - \log_2 SD) \tag{7.17}$$

$$= 10(6 - \log_2 7.7/\text{Chl}a^{0.68}) \tag{7.18}$$

$$= 10(6 - \log_2 48/\text{TP}) \tag{7.19}$$

Figure 7.4. Relationship of Secchi-disc depth to chlorophyll according to equation from Carlson (1977)

If annual mean values are used for TP in Equation 7.19 then 64.9 is used as the numerator instead of 48. The non-linear relationship between SD and chlorophyll a is shown in Figure 7.4 (also see Figure 6.12). Note that the greatest change in SD occurs below a concentration of chlorophyll a of $\approx 30\,\mu g\,l^{-1}$. Above $30\,\mu g\,l^{-1}$, there is relatively little change in SD with increasing chlorophyll a. The significance of this will be discussed later with respect to expectations in water quality improvement relative to the costs of various reductions in TP.

Threshold values for trophic states are shown in Table 7.3. These average values are generally agreed upon as satisfactory to indicate recreational water quality and have been derived from several sources (Porcella *et al.*, 1980). SD thresholds are averages of values predicted from the TP and chlorophyll a thresholds using Equations 7.17–7.19. Absolute thresholds work well for communicating lake conditions; moreover, these values are interrelated and have recreational and water supply significance. For example, a mean chlorophyll a threshold for eutrophy of $8–10\,\mu g\,l^{-1}$ has been supported in work by Walker (1985), who used three TP–chlorophyll a models to show that the frequency of blooms, defined as chlorophyll $a > 30\,\mu g\,l^{-1}$, begins to increase only after the summer average reaches about $10\,\mu g\,l^{-1}$ (Figure 7.5). This chlorophyll a level is related to a TP concentration of $30\,\mu g\,l^{-1}$ by Carlson's equations (Equations 7.18–7.19) and by Equation 7.15 (Equation 7.16 gives $27\,\mu g\,l^{-1}$). The absolute oligotrophic and eutrophic threshold values plotted on the LEI graph (chlorophyll a and SD slightly different from Table 7.3) show considerable consistency (Figure 7.6). The narrow range

Figure 7.5. Percentage of chlorophyll *a* concentrations >30 μg l⁻¹ (frequency of blooms) versus mean summer chlorophyll *a* for three data sets (Walker, 1985)

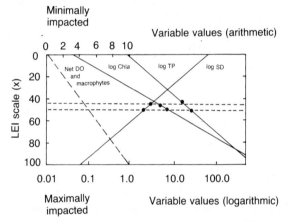

Figure 7.6. Lake evaluation index (LEI) values, where zero represents minimal impact and 100 is maximum, versus trophic state indicators (from Porcella *et al.*, 1980). Chl*a*, TP and SD are plotted on the logarithmic scale, while net DO and macrophytes are on the arithmetic scale

of mesotrophy reflects the logarithmic nature of the index, which follows from the non-linear relation of SD with TP and chlorophyll *a* as originally derived by Carlson (Equations 7.17–7.19). Although there is logic and consistency among these threshold values, they are nonetheless averages (or medians) and may not always be consistent when applied to an individual lake.

There is some advantage to expressing lake trophic state on a probability basis (OECD, 1982; Chapra and Reckhow, 1979). This approach acknowledges that there is a high degree of uncertainty in trophic state criteria. For example, from the OECD model an annual mean TP of 40 μg l⁻¹ has a 38% chance of representing eutrophy, a 56% chance of mesotrophy and a 6% chance for oligotrophy. The $25\,\mu g\,l^{-1}$ threshold (Table 7.3) represents a lake with a high probability of being mesotrophic, but has a low and equal chance of being either eutrophic or oligotrophic. Also, a lake with equal chance of being either eutrophic or mesotrophic, which could represent a meso-eutrophic threshold, would have a TP concentration of almost $50\,\mu g\,l^{-1}$. While overlap and uncertainty in trophic state are realities, a threshold value of $50\,\mu g\,l^{-1}$ represents a state that is far too degraded from the standpoint of recreational use to be interpreted as mesotrophy. This would more than double chlorophyll a from the eutrophic threshold and, according to Figure 7.5, produce blooms ($>30\,\mu g\,l^{-1}$) more than 20% of the time.

The OECD criterion, i.e. $50\,\mu g\,l^{-1}$ TP as a mesotrophic threshold, may be more appropriate for tropical than temperate lakes. Thornton (1987a) has suggested $50-60\,\mu g\,l^{-1}$ as the upper limit of mesotrophy in tropical lakes. The greater tolerance for P enrichment in the reservoirs of southern Africa may result from higher temperature and consequently higher metabolic rates, as well as higher flushing rates, than temperate lakes (Thornton, 1987b). The higher metabolic rates would tend to result in lower algal biomass even though productivity was higher, associated with higher TP. The reason for greater tolerance of P is probably not that N is limiting. Thornton (1987a) cites several studies showing that although there is some variability most workers found African lakes to be P limited usually.

Example problem 2
(a) Calculate the expected average summer chlorophyll a concentration and transparency in the lake from example problem 1.

Using Equation 7.15, which yields nearly identical results as Equations 7.18 and 7.19 combined, gives

$$\log \text{Chl}a = 1.449 \log 70 - 1.13$$
$$= 1.54$$
$$\text{Chl}a = 35.5\,\mu g\,l^{-1}$$

and using Equations 7.17 and 7.18 combined gives

$$SD = 7.7/35.5^{0.68}$$
$$= 0.68\,\text{m}$$

(b) By what percentage must internal loading be reduced in order to improve summer transparency to 2.0 m?

Using the same equations in reverse gives

$$\log \text{Chl}a = (\log 7.7/SD)/0.68$$
$$= (\log 7.7/2.0)/0.68$$
$$= 0.86$$

$$\text{Chl}a = 7.3\,\mu\text{g}\,\text{l}^{-1}$$

$$\log \text{TP} = (\log \text{Chl}a + 1.136)/1.449$$
$$= (\log 7.3 + 1.136)/1.449$$
$$= 1.37$$
$$\text{TP} = 24\,\mu\text{g}\,\text{l}^{-1}$$
$$L_{\text{int}} = 15 \times 1.5\,(24 - 40 \times 0.55)$$
$$= 45\,\text{mg}\,\text{m}^{-2}\,\text{year}^{-1}$$
$$\text{Reduction} = (585 - 45)/585 \times 100$$
$$= 92\%$$

TN (total nitrogen) is listed in Table 7.3, but has been used infrequently as an indicator. Except for unusual cases (e.g. Lake Tahoe, Goldman, 1981), TN would normally be a pertinent indicator only in highly eutrophic lakes where it could be expected to control productivity (see Chapter 6.5.2). Smith (1982) has presented a chlorophyll a predictive equation that includes TN and may be more useful in highly eutrophic systems than a TP–chlorophyll a relationship alone:

$$\log \text{Chl}a = 0.6531 \log \text{TP} + 0.548 \log \text{TN} - 1.517 \qquad (7.20)$$

Equation 7.20 predicted a chlorophyll a concentration of $21 \pm 9\,\mu\text{g}\,\text{l}^{-1}$ in Moses Lake, WA, while Equation 7.14, based only on TP, predicted $50 \pm 23\,\mu\text{g}\,\text{l}^{-1}$. The observed value in that N-limited system was $23 \pm 11\,\mu\text{g}\,\text{l}^{-1}$. Prarie et al. (1989) have presented equations for predicting chlorophyll a from TN and TP for different TN/TP ratios. This approach resulted in improved accuracy of prediction.

7.3.2 Primary productivity

Ranges in photosynthetic rate, as measured by radioactive-carbon assimilation, that are indicative of trophic states have been suggested by Rodhe (1969). These limits, with appropriate modifications, are given in Table 7.4. Similar values from two other authors are cited by Grandberg (1973). Although these ranges were determined for temperate lakes, recent comparisons of productivity at various latitudes suggest that a more general application may exist.

Although Table 7.4 shows quite a large range for eutrophy, Rodhe (1969) and others (Grandberg, 1973) have separated eutrophic from hypereutrophic at $\approx 40\%$ of the range $(250\,\text{g}\,\text{C}\,\text{m}^{-2}\,\text{year}^{-1}$ and $1000\,\text{mg}\,\text{C}\,\text{m}^{-2}\,\text{day}^{-1})$. The interval between oligotrophic and eutrophic in Table 7.4 is considered the mesotrophic state.

The main difficulty in using productivity as a trophic state indicator is that areal productivity depends as much on light availability as on

TABLE 7.4.   Ranges in rates of primary productivity as measured by total carbon uptake attributable to the trophic states of lakes

Period	Oligotrophic	Eutrophic
Mean daily rates in a growing season (mg C m^{-2} day^{-1})	30–100	300–3000
Total annual rates (g C m^{-1} year^{-1})	7–25	75–700

Source: modified from Rodhe (1969) with permission of the National Academy of Sciences

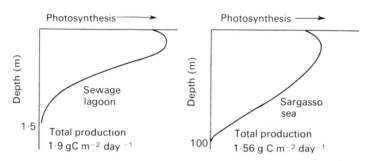

Figure 7.7. Comparison of productivity profiles in a sewage lagoon and the Sargasso Sea (modified from Ryther, 1960)

nutrients. Integration over the photic zone can give rates in an oligotrophic lake that are nearly as high as those in a eutrophic lake (see Figure 7.7). Moreover, of principal importance to the quality of a lake is its transparency, which is a function of the concentration of particles (i.e. algal biomass) in the water and not productivity per unit area.

7.3.3   Oxygen

Algal biomass as an organic matter source is really no different from sewage in its ultimate O_2 demand or BOD. The stoichiometry of nutrients versus BOD can be hypothesized from the photosynthetic equations (Stumm, 1963). CO_2 is fixed into organic carbon with light and nutrients according to a ratio of 106:16:1, C:N:P, as previously indicated:

$$106 \, CO_2 + 90 \, H_2O + 16 \, NO_3 + 1 \, PO_4 + \text{light energy} \rightarrow$$
$$C_{106}H_{180}O_{45}N_{16}P_1 + 154 \, 0.5 \, O_2 \tag{7.21}$$

The photosynthetic quotient, or PQ, = 154/106 = 1.45; experimentally this ratio averages 1.2. In the reverse reaction (respiration) O_2 is subsequently used to convert C_{106} to 106 CO_2, which results in O_2 deficits in lakes and sags (min. O_2 concentration downstream from BOD source) in streams. The potential for secondary BOD effect from treated sewage in the form of N and P can theoretically be estimated accordingly: $\approx 75\%$ of the assimilable C that is removed from sewage in secondary treatment can go back into decomposable organic matter and BOD in the form of plankton (fixed in photosynthesis) simply from the utilization of N and P that is not effectively removed by secondary treatment. In turn, the BOD potential of P, if completely utilized, can be estimated as follows:

$$\frac{154 \times 0.5 \times 32}{1 \times 31} = \frac{4950}{31} = 160 \, mg \; O_2 \, mg^{-1} \; P$$

thus $1 \, mg \; P = 160 \, mg \; O_2$ if all the P remaining from treatment is converted to organic C by photosynthesis. In order for that to occur, P must be limiting the plankton growth.

To illustrate the secondary BOD effect, an experiment by Antia et al. (1963) is appropriate (Figure 7.8). Measurements were made during and following a phytoplankton bloom in a 6 m diameter plastic sphere submerged in the sea. The top graph shows production of carbon by the oxygen method and particulate carbon measurements. The lower graph shows decomposition of produced organic carbon. The oxgen used is expressed as carbon. Note the release of phosphorus after blackout, which caused cessation of photosynthesis and death of the plankton. The greater oxidation of C than production of particulate C was explained as the bacterial breakdown of dissolved organic matter (difference between production as measured by particulate C and O_2). The authors suggested that oxygen consumption was large as a result of the bacterial breakdown of excreted dissolved C and that particulate matter was used more as bacterial substratum.

This process can be observed to cause O_2 problems in overfertilized waters. For example, Baalsrud (1967) showed that the oxygen depletion problem in the Oslo Fjord, Norway, was caused largely by the fertilizing effect of sewage added to inflowing fresh water (which also furnishes needed iron). The dominant organism was Skeletonema costatum, a marine diatom that frequently dominates in nearshore waters. This problem can occur because for each gram of phosphorus fixed into algal tissue from treated or untreated sewage at least 50 g of oxidizable carbon is fixed from naturally occurring CO_2 (Figure 7.9).

Indices of DO have included: oxygen deficit rate, ODR in $mg \, m^{-2}$ day^{-1} (or AHOD, areal hypolimnetic oxygen deficit); net DO; and minimum DO (Table 7.3). Neither minimum DO nor net DO has been correlated with TP loading or concentration, but AHOD was related to TP retention (Cornett and Rigler, 1979) and ODR to TP loading (Welch and Perkins, 1979):

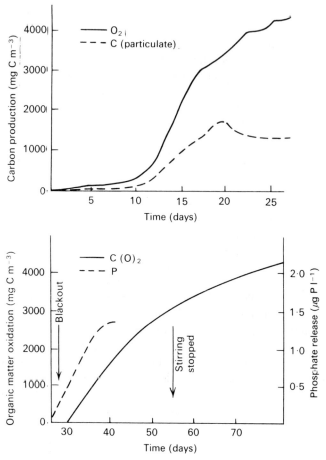

Figure 7.8. Results of a bloom and subsequent decomposition and nutrient release in a large plastic sphere (Antia *et al.*, 1963)

$$\log \text{ODR} = 1.51 + 0.39 \log L/\rho \quad (r = 0.73) \tag{7.22}$$

Criteria for ODR and trophic state were suggested by Mortimer (1941; Table 7.3).

ODR is usually calculated as the slope of the linear plot of hypolimnetic DO against time, multiplied by the hypolimnetic mean depth. DO may disappear too rapidly in some highly enriched lakes to give an accurate estimate of ODR, even if sampling is twice per month. In that case, twice weekly sampling may be necessary.

ODR indicated the increased eutrophication in Lake Washington (Edmondson, 1966) (Figure 7.10). ODR reached levels of 500–800 mg m^{-2} day^{-1} early in the eutrophication of the lake, but also

Figure 7.9. Algal growth potential experiment showing conversion of inorganic nutrients in sewage to oxygen-demanding organic matter. Top graphs show the potential of sewage in Oslo Fjord water at different distances from the sea (Baalsrud, 1967)

decreased between 1955 and 1962. The decrease in the rate was believed to be due to an increase in the quantity of blue-green algae, which would tend to be buoyant and be decomposed in the epilimnion.

Eutrophication has its greatest detrimental effect on fish populations through depletion of DO.

7.3.4   DO, eutrophication and fish

The principal effect of eutrophication on fish is one of DO depletion. An increase in plankton production to within the mesotrophic state may have a beneficial effect on the production of desirable species of fish, i.e. both

Figure 7.10. Hypolimnetic oxygen deficit below 20 m in Lake Washington in 1933, 1950, 1955, 1957 and 1962. For comparison of rates, constant slopes for deficit rates of 1, 2, and 3 mg $O_2 cm^{-2} month^{-1}$ are indicated (Edmondson, 1966)

cold- and warm-water sport fishes. However, once a state of eutrophy is reached and particularly if the lake stratifies in summer, there is a strong probability that DO in the hypolimnion may reach a critical level for survival toward summer's end.

As eutrophication increases, the minimum DO reached will continue to decrease and the minimum DO will occur earlier. This situation will be increasingly detrimental to particularly cold-water fishes, but may also adversely affect warm-water fish, because activity and growth will decrease progressively with decreasing DO (see Chapter 13). Both types will tend to evacuate the epilimnion when temperatures exceed their preferred level, although the problem would be faced more frequently by cold-water species. If adequate DO exists in the hypolimnion, it can be a healthy refuge during the warm summer period. If there is inadequate DO, the fish will be subjected to either stressful DO in the hypolimnion or, if excluded from the cooler but oxygenless hypolimnion, to stressful temperature in the epilimnion.

The higher the nutrient loading to the system, the greater the DO deficit and, depending upon the hypolimnetic depth, the lower the DO concentration and, consequently, the more inhospitable the entire lake will be to fish. Even shallow lakes that do not stratify will suffer from low DO in winter if they become ice covered. The problem of 'winter kill' in shallow lakes is a long-standing one in temperate areas (Halsey, 1968).

TABLE 7.5.   Comparative annual catch of commercial fish species in metric tons over a 40-year period in Lake Erie

Species	Early years	Recent years
Cisco	9000 (pre-1925)	3.2 (1962)
Sauger	500 (pre-1946)	0.45–1.8
Blue pike	6800	0.45 (1962)
Whitefish	1000	6.0 (1962)
Wall-eye	7000 (1956)	450
Total	24 300	
Drum, carp, perch, and smelt		≈22 240 (increased catch)
Total		22 700

Source: data from Beeton (1965)

Also, low DO can cause fish mortality following algal bloom die-off in summer (Ayles *et al.*, 1976).

The long-term effect of eutrophication will be one of changed species composition, a result largely of the changed DO status. However, increased food supply in the form of more detritus, which tends to lead to smaller zooplankton and a less diverse worm/midge-dominated bottom fauna, would tend to favour detritus/bottom-feeding fish such as suckers and carp. Haines (1973) has shown that the growth of carp was much greater (nealy four times) and small-mouth bass much less (factor of ≈2 to 6) in fertilized than in unfertilized experimental ponds. The fertilized ponds received phosphorus at a rather high rate of $\approx 2\,g\,m^{-2}\,year^{-1}$. DO was probably the principal factor detrimental to the bass because the diurnal range was from 18 to $<2\,mg\,l^{-1}$ in the fertilized ponds, but seldom exceeded $3.5\,mg\,l^{-1}$ range in the unfertilized ponds. (See Chapter 13 for the effect of diurnal DO range.)

Regardless of which factor is more important, DO or changes in the food supply, the results are similar. In Lake Erie, for example, the populations of cisco, whitefish, wall-eye, sauger and blue pike drastically declined over the 40-year period that loading of nutrients to the lake was increasing, as indicated by increases in major ions (Beeton, 1965). (The principal cause for their decline was a failure to reproduce.) The total fish catch, however, did not decline; rather, the catch of the desirable species was replaced with such species as carp, perch, buffalo, drum and smelt (Table 7.5). Although other factors, such as fishing pressure and sea lamprey predation, were involved in changes in fish production in the Great Lakes, Beeton and Edmondson (1972) suggested that the changes in Lake Erie, and in particular the cisco, were closely tied to the progressive eastward movement of polluted conditions.

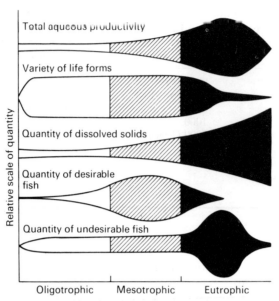

Figure 7.11. Suggested changes in various characteristics of lakes with eutrophication (modified from Lake Erie Report, 1968)

This pattern can be expected to occur in most lakes undergoing eutrophication, depending on the oxygen resources, which are a function of depth. Eutrophication is considered to be the principal problem that has limited salmonid fisheries in European subalpine lakes (Nüman, 1972).

The general pattern of change is shown in Figure 7.11. Increased abundance and productivity of 'desirable' fish species should occur in the mesotrophic stage. As eutrophy is approached, those species tend to diminish and are replaced by a dramatic increase in production of the more DO-tolerant and/or temperature-tolerant 'undesirable' species.

There is little doubt that a eutrophic state is definitely detrimental to the production of the more DO/temperature-sensitive species. The question is how much enrichment is detrimental? Are the criteria previously described for trophic state in relation to recreation also pertinent to fish propagation? Apparently the rate of phosphorus loading that is apt to cause a eutrophic state from the standpoint of algal biomass and transparency is also similar to the loading that will cause an ODR (oxygen deficit rate) that is representative of eutrophy (Table 7.4), and which may in turn begin to strain the oxygen resources from the standpoint of the fishery. Dillon and Rigler (1975) suggested very conservative guidelines for salmonid fisheries in Ontario lakes with respect to increased P loading and plankton algae.

To evaluate the connection between eutrophic threshold criteria for recreation and effects on cold-water fish, one can compare the TP loading for a threshold TP of $25 \mu g l^{-1}$ with the ultimate hypolimnetic DO concentration that loading should produce after some period of stratification. Using Equation 7.10 and an equilibrium TP of $25 \mu g l^{-1}$ gives a critical (eutrophic threshold) loading of $1145 \, mg \, m^{-2} year^{-1}$ for a lake with total and hypolimnetic mean depths of 12 m and 5 m, respectively, and a flushing rate of 2.3 $year^{-1}$ (means for 26 American OECD lakes, Welch and Perkins, 1979). After 60 days of stratification, average hypolimnetic DO content would have declined from saturation to $5.5 \, mg l^{-1}$, using Equation 7.22, and to $4.4 \, mg l^{-1}$, using the R_{tp} – AHOD model of Cornett and Rigler (1979), assuming 10°C and $R_{tp} = 1/(1 + \rho^{0.5})$. Those DO levels would produce stressful conditions for fish that need to reside in the cool waters of the hypolimnion during the summer, e.g. trout, striped bass and whitefish. This would be especially true if stratification lasted longer, which is usually the case. Theoretically then, an average stratified lake that had reached the threshold for eutrophy (Table 7.3), which would represent adverse conditions for purposes of recreation, would also produce adverse DO conditions for fish.

Even non-stratified lakes with high macrophyte density may show severely depleted DO within weedbeds (Frodge et al., 1987). Such DO depletion within macrophyte beds may be linked to observed die-offs of warm-water fish in shallow eutrophic lakes. Net DO may answer the dilemma of a suitable DO index for unstratified lakes, because ODR is appropriate for stratified lakes only. Porcella et al. (1980) developed net DO, with values ranging from 0–10, for stratified and unstratified lakes alike. Net DO is the absolute difference from an equilibrium condition (saturation) and is calculated by summing those differences (equilibrium DO – measured DO) over intervals of depth, thus incorporating the increasing tendency of supersaturation as well as deficiency in response to eutrophication.

7.3.5 Indicator organisms

While impairment of transparency (and the algal abundance that imparts it) and oxygen resources are important indicators of the suitability of lake water for recreation and water supply, there are other characteristics that are equally pertinent. The abundance of blue-green algae is readily observable because many of them are buoyant and cause surface scums (see Chapter 6.5.6 for discussion of causes). Thus, the fraction of the total biovolume that are blue-greens tends to increase with eutrophication and has been demonstrated to be a very useful indicator that can have different response times to enrichment than transparency, TP and chlorophyll a (e.g. see Edmondson and Litt, 1982). In an attempt to

TABLE 7.6.   Phytoplankton concentration before and after treatment indicated

Lake	Treatment	Blue-greens (%) Before	After
Moses Lake Parker Horn	Dilution	96	55
Moses Lake Pelican Horn	Diversion	27	47
Lake Washington[a]	Diversion	95	20
Lake Sammamish	Diversion	68	30
Long Lake	Alum	85	20
Lake Norrviken[b]	Diversion	95	92

[a] Edmondson and Litt (1982); [b] Ahlgren (1978)

develop a trophic state index using blue-green algae, Smith (1986) has related the fraction of phytoplankton as blue-greens to the ratios of TN:TP and SD/Z_{mix}. For example, given a mixed depth of 10 m, a TN:TP ratio of 10 and a eutrophic threshold SD of 2 m, Smith's model predicts 27% blue-greens.

The blue-green fraction has been known to respond sensitively to nutrient reduction as shown in Table 7.6. Prior to sewage diversion, Pelican Horn of Moses Lake, USA was dominated by small green algae, representative of sewage lagoons, and the increase in % blue-greens after diversion was actually associated with an improvement in quality. Parker Horn, on the other hand, was dominated by blue-greens initially and they declined following dilution (Welch and Patmont, 1980). The blue-green fraction in Lake Norrviken remained high after treatment, because TP was still very high, but at least the lake no longer contained a mono-culture of *Oscillatoria* (Ahlgren, 1978). The blue-green fraction responded slowly following diversion of sewage effluent from Lake Washington and did not reach the low level noted in Table 7.6 for about seven years (Edmondson and Litt, 1982).

Blue-green blooms are important only in freshwater and in some instances in brackish water, such as the Baltic Sea. Diatoms and dino-flagellates are most commonly the dominants in blooms in estuaries and nearshore marine areas.

Sediment diatoms, or the ratio (A/C) of planktonic pennate (long narrow cells–Araphidinae) to centrate (round cells–Centrales) species, is also a useful index of trophic state. The advantage of sediment core data is that the history of trophic state can be determined if sediment dating is available. The disadvantage of the diatom index is that it is restricted largely to stratified lakes, because of interference from littoral pennate species (Stockner, 1972). Results from Lake Washington, for example, show a rather clear picture, where the A/C ratio rose above the eutrophic threshold of 2 during the period of maximum enrichment and then recovered promptly after diversion (Stockner, 1972). There are equally good results from other lakes. Benthic invertebrate indicators of eutrophication are discussed in Chapter 12.

7.3.6 Nutrient loading

Although the trophic state of a lake is related to the concentration of TP, chlorophyll a, etc., it also useful to know the loading rate that produces that trophic state. This is true because it may be desirable to improve the quality of the lake and, to do that, the loading (either external or internal) must be manipulated. This was illustrated in example problem 2 above. Thus, the critical loading (L_c) to produce a mesotrophic or eutrophic state is often used as a recovery goal, compared to the current or worsening state of a lake.

There is an interesting chronology in the development of nutrient loading criteria that is worth reviewing, because it illustrates the evolution in thinking about lake behaviour. The initial attempt to define L_c involved only two variables; measured areal loading and mean depth (Vollenweider, 1968). This relationship is shown in Figure 7.12. Based on judgment about trophic state, guidelines were approximated for loading that represented thresholds or limits. Shallow lakes were seen as more sensitive to loading, because the load would presumably be more concentrated and recycling from the sediment would be greater. The same apparent effect of depth was seen in lake response (algal biomass and productivity) to the area/volume index in the Ontario experimental lakes (Figure 7.13).

However, such loading–mean depth relationships were unreliable where flushing rate varied greatly. The flushing effect was clearly demonstrated in two Ontario lakes (Cameron and Four Mile) by Dillon (1975). Although these two lakes were approximately the same trophic state, the flushing rate of Cameron was 70 times that of Four Mile, while the TP loading to Cameron was 20 times that of Four Mile. Correction of the loading for flushing rate explained the apparent anomaly. A consideration of residence time, τ (1/flushing rate) led to refinement of the loading graph (Vollenweider and Dillon, 1974; Vollenweider, 1975), which is shown in Figure 7.14. The equation for the straight lines in the figure is:

$$L_c = 100 \quad \text{or} \quad 200\,\bar{z}\tau^{0.5} \tag{7.23}$$

With that correction, some most disturbing points on the loading–mean depth graph could be clarified. For example, it appeared that

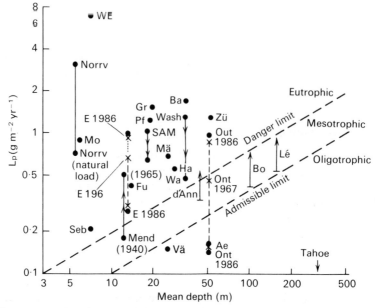

Figure 7.12. Loading graph for P in various lakes. Ae, Aegerisee (Switzerland); Ba, Baldeggersee (Switzerland); Bo, Bodensee, Lake Constance (Austria, Germany, Switzerland); d'Ann, Lake d'Annecy (France); E, Lake Erie (USA); Fu, Lake Fureso (Denmark); Gr, Greifensee (Switzerland); Ha, Hallwilersee (Switzerland); Lé, Lake Léman, Lake Geneva (France, Switzerland); Mä, Lake Mälaren (Sweden); Mend, Lake Mendota (USA); Mo, Moses Lake (USA); Norrv, Lake Norrviken (Sweden); Ont, Lake Ontario (USA); Pf, Pfäffikersee (Switzerland); Sam, Lake Sammamish (USA); Seb, Lake Sebasticook (USA); Tahoe, Lake Tahoe (USA); Wa, Lake Washington (USA); WE, Western Lake Erie (USA); Vä, Lake Vänern (Sweden); Zü, Zürichsee (Switzerland). For Lakes Erie and Ontario, dotted lines show P loading by 1986 (from 1967 estimates) without P control; dashed lines show P loading by 1986 (from 1967 estimate) with no P in detergents and with 95% P removed from all municipal and industrial wastes (Vollenweider, 1968)

ultraoligotrophic Lake Tahoe was completely insensitive to increased loading due to its very great depth. However, correction for flushing rate moved Lake Tahoe to the other side of the graph where it became much more sensitive to increased loading. The reverse was true for other lakes originally located high on the left-hand side of the graph (high flushing rates and shallow). Those were shifted to the right to a position more in line with their observed trophic state. The refined graph converted loading to inflow concentration, which, referring back to continuous flow culture kinetics, would be expected to be more directly related to the maximum algal biomass per volume that might develop.

The next step was to include sedimentation loss in the loading relation-

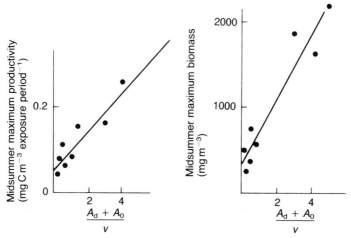

Figure 7.13. Relationships of lake response to watershed (A_d) plus lake (A_o) area divided by volume (v) (Schindler, 1971a)

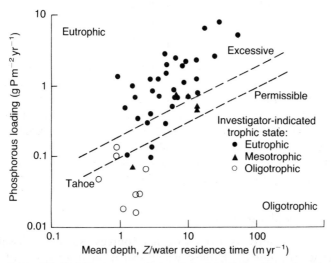

Figure 7.14. US OECD data applied to initial Vollenweider phosphorus loading and mean depth/hydraulic residence time relationship (Rast and Lee, 1978)

ship, as in Equation 7.10, to produce the following critical loading equation:

$$L_c = \mathrm{TP}_{e/m,m/o}\, \bar{z}(\rho + \rho^{0.5}) \tag{7.24}$$

where $\mathrm{TP}_{e/m}$ is 20 or $\mathrm{TP}_{m/o}$ is $10\,\mathrm{mg\,m^{-3}}$ for the eutrophic–mesotrophic or mesotrophic–oligotrophic threshold, respectively (Vollenweider, 1976).

Figure 7.15. Relationship of inflow concentration (TP$_i$) with water retention time (τ) for two lake concentrations (TP = 10 and 20 mg m^{-3})

The effect of sedimentation is indicated in Figure 7.15 in which the critical inflow concentration, TP$_{ic}$, is plotted against τ, or $1/\rho$, according to Equation 7.10. If $1/\tau$ is substituted for ρ in Equation 7.10:

$$TP_i = TP(1 + \tau^{0.5}) \qquad (7.25)$$

At low τ, the lines for lake concentrations of 10 and 20 mg m^{-3} become parallel with the abscissa indicating that sedimentation becomes minimal at short residence times (high flushing rates) and the lake concentration equals the inflow concentration. As τ increases, sedimentation becomes increasingly important permitting higher TP$_{ic}$ without exceeding the critical lake concentration. That is lakes become more tolerant of increased TP$_i$ as τ increases. See Reckhow and Chapra (1983) for a more detailed discussion of these loading relationships.

7.3.7   Sources of phosphorus

P and N enter waterways from external sources, such as: dissolution from rock; wastewater discharges, which can include a variety of waste types, e.g. treated or untreated sewage, dairies, slaughter houses, seafood canneries and other food processing plants; urban stormwater runoff; rural stormwater runoff from either fertilized or unfertilized pastures and croplands; forest; precipitation; and waterfowl. Only rock dissolution and forest runoff are considered natural background; all others are affected by man (actually forest runoff is if fertilized). The input from precipitation (even in remote areas) and waterfowl (attracted by feeding) are often affected by man. Only wastewater discharges are considered point sources, i.e. enter directly through some conveyance system. All others are considered non-point, or diffuse sources. While point sources are usually rather easy to estimate or determine directly, non-point sources are difficult to estimate. Non-point sources are often estimated from knowing the distribution of land-use types in a watershed and applying yield coefficients that are based on a variety of data sets. For example, one set of estimates for TP (Reckhow and Chapra, 1983) is, in $mg\,m^{-2}$ $year^{-2}$: forest, 2–45; precipitation, 15–60; agriculture, 10–300; urban, 50–500; and input to septic tanks from $0.3–1.8\,kg\,capita^{-1}\,year^{-1}$.

The ranges are large and it is usually difficult to select the appropriate value for a particular watershed. One approach is to select an appropriate portion of the range based on measured loading and land-use information in the watershed of interest. Assuming that ratios among the respective land-use types given above are appropriate for the watershed in question, yield coefficients can be selected by holding the ratios constant and scaling the coefficients to conform with actual loading and land-use areas (Shuster et al., 1986). Then a range in probable yields, or some method of expressing uncertainty should be employed (see Reckhow and Chapra, 1983).

Obviously the loading from septic tank drainfields can represent the most concentrated non-point source in a watershed. Assuming 2.5 members per family and a $4000\,m^2$ lot size, the yield would be $188–1125\,mg\,m^{-2}\,year^{-1}$ if the drainfield was not percolating adequately and removing P. Properly working drainfields ('on-site waste treatment') are actually very efficient at removing P. However, even if a few fail in the watershed the source can be significant. Moreover, P is probably in a soluble, readily available state.

Urban runoff can be a surprisingly significant source of TP as the values indicate. However, much of that TP can arrive as particulate P, sorbed to the high content of suspended matter in stormwater, and settle out shortly after entering the lake. If that happens, the contribution to productivity may not be great, even though the loading is high. This may be especially true if the loading arrives during the non-growing season, as it often does in areas with highly wet winters and dry summers, like the Pacific northwest (Stockner and Shortreed, 1985). However, the highest fraction of TP that was bioavailable arrived during the high runoff period in winter in Lake Sammamish, a Pacific Northwest lake (Butkus et al., 1988).

8
ZOOPLANKTON

Zooplankton are the primary consumers of phytoplankton in lakes, the ocean and in deep slow-moving rivers where the principal energy pathway is through a predator–prey oriented food web. In highly eutrophic or organically polluted waters, where energy is diverted through detritus pathways, zooplankton may also be the transporter of that energy by consuming bacteria and detritus. Assuming that temperature is optimum and there is minimal inhibition by low DO or toxicity, the effectiveness of zooplankton consumption and production will usually be controlled by the size, quality and abundance of food (the phytoplankton), as well as by planktivory – the predation by planktivorous fishes. Control of their consumption and production effectiveness by those factors has often been related to the resulting change in their size structure. The role of planktivory can be manipulated to increase consumption of phytoplankton, improve water clarity and reduce nutrients. Results of such attempts at 'biomanipulation' are covered in Chapter 10, but the ecological principals supporting that technique will be discussed in this chapter.

Zooplankton, especially large species of *Daphnia*, e.g. *D. magna*, are sensitive to inhibitors and are frequently used as test organisms in toxicity bioassays. Results of such toxicity tests are covered in Chapter 13.5.

8.1 POPULATION CHARACTERISTICS

The zooplankton are composed primarily of populations of microscopic or near-microscopic organisms from the groups Rotifera (rotifers or 'wheeled organisms'), and the Cladocera and Copepoda (Crustacea). Rotifers and cladocerans (e.g. *Daphnia*) can reproduce by parthenogenesis, i.e. adults produce eggs that have a full complement of chromosomes (diploid). Under crowded or stressed conditions, male cladocerans develop from a diploid egg to produce a haploid (half chromosome number) sperm to unite with a haploid egg. That union produces a

TABLE 8.1.   Maximum feeding rates of *Daphnia* using ^{32}P-labelled organisms

Organisms	Cells consumed (10^6 animal^{-1} h^{-1})	Particle size (μm^3)	Volume consumed (mm^3 animal^{-1} h^{-1})
E. coli (bacteria)	5.6	0.9	0.005
Chorella (green algae)	0.5	34.0	0.017
Yeast	0.25	66.0	0.016
Protozoan	0.0028	1.8×10^4	0.051

Source: modified slightly from McMahon and Rigler (1965)

fertilized diploid egg as a resting stage awaiting more desirable environmental conditions for population growth. The rapid population growth that can result from these adaptations partly account for the ability of the zooplankton to control the size of the phytoplankton. Generation times in cladocerans can be as little as 2–5 days or as long as a month or more. The copepods (e.g. *Diaptomus*) have no parthenogenic stage, but can also reproduce rapidly because the female can store sperm for several fertilizations.

Adult cladocerans such as *Daphnia* (e.g. *D. magna*, *D. pulex*) can be as large as 2 mm in carapace length, while others (*Moina*, *Polyphemus* and *Ceriodaphnia*) attain a length of 1.5 mm. Small cladocerans, such as *Bosmina*, are of the order of 0.4 mm. *Daphnia* are generally large and can consume a rather wide size range of food (Table 8.1), while small cladocerans (*Bosmina*) are more restricted in food size (Sarnelle, 1986). Size is a very important factor in relationship of zooplankton to phytoplankton, because not only does it determine the size of food particles that can be consumed, but the rate at which they consume food increases with size (Burns, 1968; 1969) and planktivores select the larger size zooplankton, such as *Daphnia*.

8.2   FILTERING AND GRAZING

Zooplankton feed by filtering the water as they move. The setae on their maxillary appendages serve to collect phytoplankton as water is circulated by swimming movements created by their antennae (see Russell-Hunter, 1970). The filtering, or clearing, rate can be determined by recording the change in food particle concentration over time according to the following equation:

$$\text{ml animal}^{-1}\,\text{day}^{-1} = \text{ml animal}^{-1}\,(\ln C_0 - \ln C_t)\,\text{days}^{-1} \quad (8.1)$$

where C_0 and C_t are, respectively, the initial and ending concentration of food particles. At low food concentrations, the rate is more accurately determined over much shorter time periods by tagging the algae with a radioisotope and measuring the activity of the zooplankton themselves. Results using the copepod *Diaptomus* and the cladoceran *Daphnia* have given filtering rates ranging from $0.1-5.5$ ml animal^{-1} day^{-1}, with higher rates for the larger *Daphnia* (see Jorgensen *et al.*, 1979). The grazing, or feeding rate, is in turn determined as the product of the filtering rate and the concentration of phytoplankton available:

$$L \text{ animal}^{-1} \text{ day}^{-1} \times \text{ mg } C_{\text{phyto}}/L = \text{ mg animal}^{-1} \text{ day}^{-1} \qquad (8.2)$$

The loss rate of phytoplankton through zooplankton grazing can be compared with the rate of primary productivity by multiplying the grazing rate times the concentration of zooplankton:

$$\text{mg } C_{\text{phyto}} \text{ animal}^{-1} \text{ day}^{-1} \times \text{ animal m}^{-3} = \text{ mg C m}^{-3} \text{ day}^{-1} \qquad (8.3)$$

which is the rate at which phytoplankton are consumed (grazed) by zooplankton and is the same unit as primary productivity. If the ratio of grazing rate to primary productivity averages around 1.0 for a growing season, it indicates that all of the phytoplankton produced is removed by grazers. Such equality of the two rates has been observed (Wright, 1958; Green and Hargrave, 1966) and is usually typical of more oligotrophic environments. Within more eutrophic environments, the ratio tends to fall below 1.0 because of the increasing presence of larger species of phytoplankton that are less easily consumed, as will be discussed later.

As discussed earlier (Figure 2.2), only a portion of food that is consumed is assimilated for growth and activity. Uhlmann (1971) cited results from several authors concluding that to produce one unit of *Daphnia* requires ≈ 5 units of phytoplankton. This represents a yield of 0.2; the values ranged from $0.15-0.4$ in continuous culture of *D. magna* (Uhlmann, 1971).

Equation 8.2 suggests that given a constant filtering rate, grazing rate will continue to increase as phytoplankton concentration increases. However, at some point filtering rate will decrease and digestion will decrease as well (Uhlmann, 1971). Thus, the relationship of grazing rate to phytoplankton biomass will take a Michaelis–Menten hyperbolic shape (Figure 7.14), with grazing rate becoming saturated at some phytoplankton concentration (Uhlmann, 1971).

To comprehend the potential for zooplankton grazing to control the biomass of phytoplankton, the results from continuous culture experiments with *D. magna* and *Chlorella* (as food) are instructive. The grazing rate, G, can also be expressed in units of day^{-1}, which if multiplied by the biomass of algae would give the same units as in Equation 8.3, and:

$$GX_{\text{phyto}} = \mu_D X_D / Y \qquad (8.4)$$

where X is biomass of phytoplankton and *Daphnia*, Y is the yield coefficient, taken as 0.25 (see above), and μ_D is the maximum growth rate of *Daphnia*, observed to be 0.3 in continuous culture (Uhlmann, 1971). Assuming that X_D and X_{phyto} have the same order of magnitude, the equation becomes:

$$G = \mu_D/Y \qquad (8.5)$$

which is 1.2 if the values given above are used. This means that *Daphnia* is capable of consuming 1.2-times the algal biomass per day if they are of equal magnitude. Capacity of the zooplankton to remove the phytoplankton would be, of course, less with more phytoplankton and vice versa.

A similar, but less specific example, is to assume that the grazing rate of zooplankton equals the growth rate of phytoplankton. If the grazing rate is doubled from that steady-state condition, the phytoplankton biomass would decline from an arbitrary initial level of 1×10^6 cells to 27×10^3 cells in 5 days. If the grazing rate is tripled the phytoplankton biomass would be eliminated in 5 days (Sverdrup *et al.*, 1942).

8.3 ZOOPLANKTON GRAZING AND EUTROPHICATION

As indicated above, zooplankton grazing usually has greater control on phytoplankton biomass in oligotrophic than eutrophic waters. As a result, predator–prey conversion efficiencies would tend to decrease with eutrophication. Observations of zooplankton–phytoplankton interrelationships across trophic gradients have shown that, along with a decrease in the ratio of zooplankton consumption and production to phytoplankton production, there is a decrease in the size of zooplankton (Gliwicz, 1975; Hillbricht-Ilkowska, 1972; Gliwicz and Hillbricht-Ilkowska, 1973; Pederson *et al.*, 1976) (Tables 8.2 and 8.3). Associated with this change in activity and energy transfer, there was also a decrease in the size of zooplankton and an increase in the size of phytoplankton to more colonial and filamentous forms. Also, Gliwicz showed that the filtering rate increased from ultraoligotrophy through mesotrophy, but decreased in eutrophic lakes. Zooplankton biomass increased through mesotrophy, but failed to increase further in eutrophic lakes. Thus, the efficiency of phytoplankton removal was poor, with eutrophy, resulting from an increasing amount of net phytoplankton and bacteria/detritus.

There are several processes, which have been further clarified in recent years, that can account for this observation. As discussed in the chapter on phytoplankton (Chapter 6), zooplankton are relatively ineffective at grazing large colonial and filamentous species, such as the nuisance-bloom-forming blue-green algae. The size and shape of colonial and filamentous blue-greens create problems for filter-feeding grazers. The filaments are either too large to consume or they interfere with their filtering mechanism resulting in added energy costs that reduce growth and survival (Porter, 1972, 1977; Burns, 1987; Infante and Abella, 1985;

TABLE 8.2. Expected character of phytoplankton–zooplankton relationships with respect to trophic state

Variable	Oligotrophic-mesotrophic, moderately fertilized, pelagic-marine	Eutrophic, heavily fertilized ponds, neritic-marine
Ratio of zooplankton consumption/primary production	≈100%	≈30%
Efficiency of energy conversion	≈20%	≈10%
Size of zooplankton	Large	Small and do not control phytoplankton
Size of phytoplankton	Nanoplankton <50μm	Nanoplankton not used, bacteria and detritus consumers dominate

Source: Hillbricht-Ilkowska (1972)

TABLE 8.3. The ratio of zooplankton to phytoplankton productivity as a measure of food-web efficiency in experimental ponds and three lakes

	Nutrient status	Zooplankton/ phytoplankton production
Ponds		
	High	0.07–0.05
	Medium	0.41–0.08
	Low	0.56–0.20
Lakes (2 years)		
Sammamish	Mesotrophic	0.04–0.04
Chester Morse	Oligotrophic	0.09–0.08
Findley	Oligotrophic	0.18–0.08

Source: Hall *et al.* (1970); Pederson *et al.* (1976)

Porter and McDonough, 1984). Even large colonial diatoms are not favoured (Infante and Litt, 1985). Also, toxicity produced by some nuisance blue-greens has been shown to inhibit growth and reproduction of zooplankton (Arnold, 1971; Lampert, 1981). Rather than zooplankton grazing causing the increase in larger colonial and filamentous algae (especially blue-greens) as eutrophication increases, the evidence suggests that changing physical–chemical conditions (CO_2/pH, N:P, stability, etc.) as they affect buoyancy, sedimentation and light availability (see Chapter 6) are the more probable causes for the changes in phytoplankton dominance. The decreased size of zooplankton, with associated reduced capacity to graze the phytoplankton, would then be a response to the greater availability of bacteria/detritus resulting from the relatively ungrazed phytoplankton crop.

On the other hand, *Daphnia* is considered to be more representative in eutrophic than oligotrophic lakes (McNaught, 1975; Patalas, 1972). An explanation is that *Daphnia* species are large and therefore have the capacity to select a wider size range of food particles (see above), thus being more versatile than the calanoid copepods. Except in cases of blue-green toxicity and filamentous forms that interfere with filtration (Edmondson and Litt, 1982; Porter and McDonough, 1984), the large grazing-efficient *Daphnia* should be abundant and able to control most phytoplankton in eutrophic lakes. However, there is another important process that would tend to reduce the abundance of large grazing-efficient zooplankton and that is planktivory. The predatory activity of small fish, such as yellow perch (*Perca flavescens*) and bluegill sunfish (*Lepomis macrochirus*) are size-selective and can effectively reduce populations of large cladocerans (Brooks and Dodson, 1965; Hall *et al.*, 1970; O'Brien, 1979; Shapiro *et al.*, 1975; Mills and Forney, 1983; McQueen *et al.*, 1986).

There are many examples of size-selective zooplanktivory by fish. Hall *et al.* (1970) showed in a series of experimental ponds that regardless of the level of N and P enrichment, the presence of bluegill sunfish determined the zooplankton composition; an average of 53% of the biomass was contributed by the relatively large *Ceriodaphnia* in ponds without fish, but it made up only 3% in ponds with fish. In Oneida Lake, New York, year-to-year and seasonal fluctuations in *Daphnia pulex* were controlled largely by the abundance of young yellow perch (Mills and Forney, 1983). Fish removal or exclusion experiments have also clearly demonstrated the adverse effect of planktivores on large cladoceran zooplankton (Shapiro and Wright, 1984; Lynch and Shapiro, 1981; Spencer and King, 1984; Post and McQueen, 1987; Raess and Maly, 1986). Preserving the large, grazing efficient zooplankton has also been accomplished by stocking piscivorous fish, such as large trout or bass, which is another way of removing the planktivores (Benndorf *et al.*, 1984; Carpenter *et al.*, 1987; Wagner, 1986). Without exception, such manipulations have resulted in increased abundance and size of cladocerans.

With an increase in the abundance of large zooplankton, one would expect an increased loss rate, and consequently a decreased biomass, in the phytoplankton. In most of the above cited experiments, that is what happened. The positive effect of planktivorous fish on phytoplankton biomass was initially documented by Hrbacek *et al.* (1961), but the deliberate alteration of fish populations to control phytoplankton by

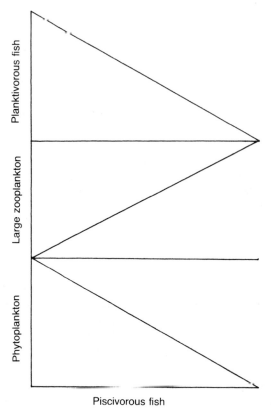

Figure 8.1. Hypothetical relationships between the biomass of piscivorous fish, planktivorous fish and phytoplankton

preserving large, grazing efficient zooplankton, was originally termed 'biomanipulation' by Shapiro *et al.* (1975). Later the process was referred to as 'cascading trophic interactions' by Carpenter *et al.* (1985). It is also known as the 'top-down effect', as opposed to the 'bottom-up effect' (i.e. through nutrients), of controlling phytoplankton. The essence of the process can be illustrated with three relationships: indirect between piscivorous fish and planktivorous fish; direct between piscivorous fish biomass and the biomass of large zooplankton; and a resulting inverse relationship between piscivorous fish biomass and phytoplankton biomass (Figure 8.1). As an example, Shapiro *et al.* (1975) showed that phytoplankton biomass in enclosures were ultimately (50 days) reduced to low levels by the grazing of *Daphnia* regardless of whether the water had been enriched as long as perch were absent (Figure 8.2). The trophic cascade, however, does not always carry down to the phytoplankton, especially in highly enriched waters (Post and McQueen, 1987), and the

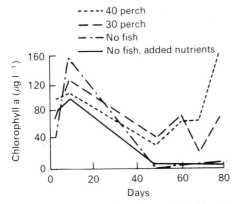

Figure 8.2. The control of algal biomass by *Daphnia* with and without the presence of planktivorous fish (Shapiro *et al.*, 1975)

reason is not well understood. It could be due to the reduced filtering capabilities encountered by *Daphnia* feeding on filamentous blue-greens, because blue-greens are promoted by physical–chemical conditions associated with enrichment, or by stimulation from increased nutrient recycling (Post and McQueen, 1987), or by the inability of the process to reduce phosphorus to controlling levels in highly enriched lakes (Benndorf, 1990). Additional examples and the feasibility of biomanipulation for lake restoration will be discussed in Chapter 10.

8.4 TEMPERATURE AND OXYGEN

The lethal temperature (LT_{50}) for most crustacean zooplankton is rather high, between 28 and 35°C, and acclimation temperature seems to have little influence in adjusting the lethal temperature (Welch and Wojtalik, 1968). For example, *Eurytemora affinis*, an estuarine copepod, had a lethal temperature between 28 and 32°C whether the acclimation temperature was 15, 20 or 25°C (Heinle, 1969a). The acclimation time in that experiment was two months or about one generation time. The rather absolute nature of the lethal temperature for zooplankton will be illustrated again in Chapter 10 with the discussion of invertebrate survival during passage through power-plant cooling systems. There are stenothermal species, however, such as *Limnocalanus* discussed below.

Considering their apparent high thermal tolerance, compared to typical maximum temperatures in lakes, there would seem to be some capacity for heat assimilation with respect to zooplankton in most temperate lakes. One, therefore, might expect that moderate increases in temperature from the discharge of heated water would increase zooplankton production. That is because metabolic activity increases with

TABLE 8.4.   Zooplankton (herbivores) production and turnover rate (P:B) over a trophic series of lakes and with heat added

Lake	Trophic state	Production (kcal m^{-3} day^{-1})	P:B
Naroch	Mesotrophic	0.12	0.06
Miastro	Mesotrophic-eutrophic	0.30	0.08
Batorin	Eutrophic	0.71	0.18
Mikorzynskie	Eutrophic, unheated	0.62	0.11
Lichenskie	Eutrophic, heated	1.38	0.24

Source: Patalas (1970)

temperature resulting in increased growth so long as the organism's optimum is not surpassed.

Such an increase in production occurred in a Polish lake (Lichenskie) that received heated water from a power plant. The temperature of that lake reached an average during the warm part of the summer of 27.1°C compared to 21.6°C in an unheated control lake (Mikorzynskie). Zooplankton production and turnover rate (P:B) doubled in the heated lake compared to the unheated lake (Table 8.4). Both lakes were eutrophic and had similar biomass and species composition, although *Daphnia* were less important in the heated lake. So the effect of an average difference of 5–6°C resulted in a doubling in herbivorous zooplankton production. That was in spite of the dominance of the phytoplankton being *Melosira granulata* and *Microcystis aeruginosa*, which are not easily consumed (see Chapter 6). Although zooplankton production and turnover rate increased through the trophic series, with even a greater increase when increased temperature was combined with a eutrophic state, there still may have been a shift in size and a decrease in grazing efficiency as discussed above. *Daphnia* did decrease in importance in the heated lake and copepods were the principal producers in both lakes. Such increases in production could result even if the detritus/ bacteria pathway were increasingly utilized by smaller zooplankton throughout the trophic series shown in Table 8.4. However, phytoplankton production also increased with added heat from 3.7 to 7.3 g O_2 m^{-2} day^{-1}, which suggests that a change in energy pathway was not associated with the temperature effect. However, the quality of the lakes may have declined with eutrophication (e.g. the dominance of *Microcystis*) and heat, in spite of the increased production. Moreover, other species such as many fish are less tolerant of temperature increases than are zooplankton and would take precedence when judging whether or not eutrophication and/or heated water input were benefits.

Zooplankton is adversely affected by low dissolved oxygen. The large copepod, *Limnocalanus*, was nearly eliminated in Lake Erie as eutrophication increased. The reason was thought to be related more to planktivory than to enrichment *per se*. Gannon and Beeton (1971) suggested that decreased hypolimnetic DO forced the ordinarily cold-loving organism from the hypolimnion into the epilimnion where DO was plentiful, but was more susceptible to grazing by increasing populations of yellow perch. Reversing the process has been suggested as a way to enhance the preservation of large-bodied zooplankton subjected to excessive planktivory. By oxygenating anoxic, relatively dark hypolimnia, refuges could be created into which zooplankton could migrate during the day when they are most susceptible to sight-feeding planktivores. This has been shown to work in enclosure experiments (McQueen and Post, 1988).

Cladocerans living in low-DO waters often become red in colour as a result of producing haemoglobin (Fox, 1950). Red *Daphnia* survived longer, fed on *Chlorella* at a higher rate and produced more eggs than did pale organisms under low DO. Red patches at the surface of highly enriched environments, such as sewage lagoons, are often the result of swarms of cladocerans with high haemoglobin content.

9
MACROPHYTES

The higher plants in aquatic environments are usually rooted with rigid cell walls and asexual (e.g. fragmentation, tubers) and sexual (flowers and seeds) reproduction. Asexual, or vegetative, reproduction is usually highly effective in expanding a plant's distribution within a water system. Plants emphasized here are the submersed species, which are usually rooted and occupy the littoral zones of lakes and often the slow-moving or stagnant reaches of rivers. Emergent species dominate in the very shallow edges of the littoral and in marshy wetland areas, but they usually do not present the nuisance problems provoked by some submersed species.

Of interest are the effects of macrophytes on water quality, such as nutrient recycling through senescense, and controls on growth and distribution of especially the nuisance submersed species, such as water milfoil and hydrilla. Much has been learned about these aspects in the past 20 years. Rooted submersed macrophytes are known to depend largely on the sediments for their nutrition. They release very little nutrient to the surrounding water via excretion, but some contribute substantially to internal nutrient loading through senescense and decay. Although sediment texture, organic and nutrient content have been shown to affect the distribution and growth of rooted macrophytes, sediment characteristics have not been effective in predicting such attributes. However, given an adequate substratum, the maximum depth of colonization can be predicted.

9.1 HABITATS

9.1.1 Running water

Most rooted macrophytes are not able to withstand too great a current and an adequate sediment deposit must exist for their rooting. Light is generally satisfactory, because rivers are usually shallow, but turbidity may still limit in some instances. Butcher has listed five types of waters, based on current and sediment type, that influence nutrient availability and in which the associated turbidity affects light penetration (Butcher, 1933; Hynes, 1960). The types with representative plants are as follows.

1. Torrential on rock or shingle – this habitat contains mostly mosses (*Fontinalis*); larger macrophytes are not stable in the high velocities.
2. Non-silted on shingle – a 'sluggish' stream; milfoil (*Myriophyllum*) and water crowfoot (*Ranunculus*).
3. Partly silted on gravel and sand – this habitat, because of bottom type and turbidity, has, for example, *Ranunculus*, pondweeds (*Potamogeton*), and arrowweed (*Sagittaria*).
4. Silted on silt – the sedimented bottom and increased turbidity favours pondweeds (*Potamogeton*), water lilies (*Nuphar*), and the aquarium plant (*Elodea*).
5. Littoral – very little current with mud substrate. Emergent vegetation is common, for example, reeds and bulrushes.

9.1.2 Standing water

The distribution of macrophyte development in standing water is dependent on slope in the littoral region. The more gradual the slope, the greater the potential to collect sediment, as well as the more area available to light penetration and thus optimal growth. Spence (1967) has observed in Scottish lochs that macrophyte diversity and abundance were favoured by brown muds and an alkalinity $>50 \, \text{mg} \, \text{l}^{-1}$ as $CaCO_3$. Vegetation was usually absent from rocky or sandy beaches.

The characteristic zones in the littoral region are as follows.

1. Emergent vegetation zone: grasses, rushes, and sedges, which depend on lake sediments for nutrients but otherwise carry on photosynthesis in the atmosphere, which supplies CO_2.
2. Floating leaf plant zone: water lilies, some pondweeds, and duckweed (*Lemna*), most of which obtain their nutrients primarily from sediments but carry on some photosynthesis in direct contact with the atmosphere.
3. Submerged vegetation zone: pondweeds, hornwort (*Ceratophyllum*), naiads (*Najas*), milfoil, arrowweed, and stonewort (*Chara*), which obtain nutrients from both soil and water and are dependent on CO_2 in water for photosynthesis.

9.2 SIGNIFICANCE OF MACROPHYTES

Rooted macrophytes are significant in lakes in at least four ways. With respect primarily to the submersed growth forms in the littoral zone, macrophytes have the following effects: contribute productivity themselves and provide a substratum for periphyton and insects to attach; contribute to lake ageing by recycling nutrients and accumulating sediment; serve to protect juvenile fishes from predation; and create nuisance biomass levels, especially if they are exotic species, that interfere with water recreation.

Because the colonizable area for submersed macrophytes is limited by light penetration to the lake bottom, macrophytes normally supply a relatively small fraction of primary productivity in large deep lakes. That is, littoral production is much less than pelagic production. However, Wetzel and Hough (1973) argue that most of the lakes in the world are small and shallow and consequently have a large fraction of their productivity supplied by macrophytes and attached periphyton. It follows, therefore, that macrophytes must provide a substantial portion of primary productivity in freshwater although there are few cases where they actually dominate productivity (Wetzel, 1975).

Macrophytes also recycle nutrients from the sediment that contributes to phytoplankton production, which in turn contributes additional sediment (together with the organic matter from macrophytes themselves) that is trapped by macrophytes further extending the littoral zone. In that manner, macrophytes accelerate lake ageing. Carpenter (1981) quantitatively analysed this process for Lake Wingra, Wisconsin and concluded that ageing was more sensitive to internal phosphorus recycling via rooted macrophytes (milfoil) than to external sediment and nutrient inputs.

Submersed macrophytes provide a surface for fish egg incubation, e.g. in the yellow perch. They also protect juvenile fish from predation. That function has a positive effect in the case of northern pike (*Esox lucious*), which are very cannibalistic. Without substantial macrophyte cover their survival is low. On the other hand, if macrophyte stands are too dense and widespread, protection from predation, especially in the centrachids (sunfishes) and yellow perch (*Perca flavens*), leads to stunting. Therefore, numbers increase but not biomass (Nichols and Shaw, 1986). Reducing the area colonized by macrophytes (and concentrating the fish for a summer) by drawing down the water level led to decreased numbers and increased size of black crappie and, consequently, increased growth of large-mouth bass in Long Lake (Kitsap), WA (Gross, 1983). Thus, there seems to be an optimum density and distribution of macrophytes with respect to fish production and individual size. Critical limits for surface coverage of macrophytes have been suggested, but would no doubt vary from lake to lake depending on fish species and macrophyte density (i.e. g or stems m^{-2}).

Excessive macrophyte abundance can represent a nuisance to water recreation, such as swimming, fishing, boating and water skiing. Increased density and distribution of macrophytes are often perceived by lake-shore residents and water users to occur over a period of 10 to 20 years. Such increases may be caused by sediment accretion due to increased nutrient input and recycling as suggested by Carpenter (1981) and involve native species. However, many problems with submersed macrophytes are due to exotic species, such as Eurasian water milfoil (*Myriophyllum spicatum*), curlyleaf pondweed (*Potamogeton crispus*), elodea (*Elodea canadensis* and *E. densa* or *Egeria densa*), and hydrilla (*Hydrilla verticillata*). Milfoil, *E. densa*, hydrilla and curlyleaf pondweed are exotic to the US and *E. canadensis* is exotic to Europe. It is of interest to review briefly the case of milfoil in the US.

Milfoil had been present in the Chesapeake Bay region since 1881, but did not reach nuisance proportions until the 1950s and 1960s, possibly related to hurricanes accompanied by increased salinity and input of sediment (Nichols and Shaw, 1986). Subsequently, it spread to the mid-western US and the northwest where it reached nuisance proportions in the 1960s and 1970s, respectively. In some areas it has declined inexplicably to 10–20% of earlier, explosive levels (Elser, 1967; Wile, 1975; Carpenter and Adams, 1977; 1980b). In those areas it has begun to fluctuate from year to year in a fashion similar to native species (Nichols and Shaw, 1986).

Much of the success of milfoil is related to an effective method of vegetative reproduction. The plant easily fragments with parts floating for long distances before settling to the sediment to colonize a new area. Although the plant is exotic to the US, environmental reasons have been suggested to account for its success. Coffey and McNabb (1974) observed that milfoil grew better under the ice at low light intensities than other species. It therefore had a larger biomass at the beginning of the growing season than other species, which it could shade. However, milfoil rose to dominance in parts of Lake Washington, especially Union Bay, in spite of increasing clarity due to decreased phytoplankton biomass (recovery from eutrophy to oligotrophy) and no ice cover. While this case does not support the low-light explanation, its invasion into many rather turbid Tennessee Valley reservoirs where macrophytes were previously scarce is consistent with the idea. Its ability to reach the surface in water as deep as 3 m and then extend horizontally creating a tangled mass of stems and leaves on the surface allows it to compete effectively for light among other species and in turbid environments.

Milfoil may also be favoured by water hardness and pH (Hutchinson, 1970b). Results from Swedish lakes showed that it seemed to prefer higher pH and Ca content compared to two other species of *Myriophyllum* (Figure 9.1). Hutchinson hypothesized that its success may be related to its ability to use HCO_3. Spence (1967) also noted the presence of *M. spicatum* in lakes with HCO_3 $\geqslant 60 \, mg \, l^{-1}$. However, Nichols and Shaw

Figure 9.1. Distribution of three species of *Myriophyllum* in Sweden as a function of pH and water hardness (modified from Hutchinson, 1970b)

(1986) suggest that explanation may be over-simplistic even though they cite evidence that it tolerates salinity up to 2% and grows vigorously at salinities to 1%. Its explosion and dominance in Lake Washington and Lake Sammamish, WA, indicates that it can be successful in soft-water lakes as well. Nevertheless, eutrophication, which tends to increase pH and shift the carbon equilibrium to more HCO_3 and less CO_2, due to greater demand on CO_2 by phytoplankton, may contribute to the explosive success of milfoil. On the other hand it has risen to dominance in oligotrophic systems as well (Newroth, 1975).

Alternatives to control macrophytes will be discussed in Chapter 10.

9.3   EFFECTS OF LIGHT

Submersed macrophytes are adapted to high light intensities. This is understandable considering that they can grow to great abundance in water less than 1 m deep. For example, *E. canadensis* was shown to have an optimum between 75 and 100% of full sunlight, based on *in situ* measurements of ^{14}C assimilation (Figure 9.2). Note that rates were very low for plants exposed to 12.5% of sunlight. Nichols and Shaw (1986) cite studies showing the optimum for elodea to occur between 15% and full sunlight, as well as it being found deeper than other species in some instances. Thus, elodea is tolerant of a wide range of light, but of special interest is its utilization of high light intensities. In contrast, phyto-

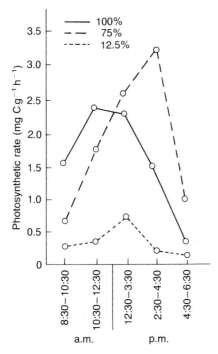

Figure 9.2. Photosynthetic rates of *Elodea canadensis* suspended in pickle jars at the indicated percentages of incident light in a shallow pond (modified from Hartman and Brown, 1967)

plankton are generally considered to be saturated at between 30 and 50% of full sunlight (see Chapter 6).

The apparent benefit of high light intensities to submersed macrophytes is consistent with regression models developed by Canfield *et al.* (1985) for 108 lakes and Chambers and Kalff (1985) for 90 lakes in which the maximum depth of plant colonization (MDC) was a function of the Secchi disc depth, which disappears at $\approx 10-15\%$ of surface intensity. The MDC can vary depending on the species present. From the following models, the actual depth of colonization is slightly more than the Secchi depth: $\log MDC = 0.62 \log SD + 0.26$ (Canfield et al., 1985); $Z_c^{0.5} = 1.33 \log SD + 1.4$ (Chambers and Kalff, 1985).

The great sensitivity of colonization depth to relatively high light intensities can have marked year-to-year effects on plant abundance and distribution in reservoirs subjected to variations in turbidity. For example, the year-to-year variation in plant abundance in Pickwick Reservoir, Alabama was tied closely to light and rainfall during the spring growth period (Figure 9.3). The two years in which the plants (mostly

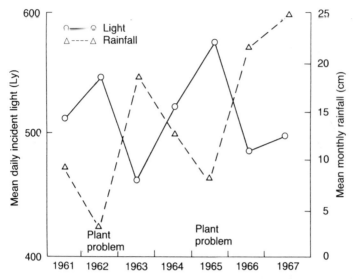

Figure 9.3. Mean daily incident light and mean monthly rainfall recorded at Florence, Alabama, during the 'critical' period of plant growth in Pickwick Reservoir from 15 April to 31 May during 1961–7 (Peltier and Welch, 1970)

Najas) reached nuisance proportions had sunny springs with low rainfall. Nuisance proportions refer to a coverage of 2400 ha, of the 20 000 ha reservoir, with submersed macrophytes that reached the surface, allowing mats of periphyton to develop. The two sunny, dry springs at Pickwick Reservoir with more incident light and less rainfall resulted in greater water clarity, because less rainfall produced less runoff and less turbidity. As a result, the depth of 10% surface intensity ranged from 1 to 2 m, which would represent a large difference in reservoir area colonized. Robel (1961) has also shown the effect of turbidity on the abundance of *Potamogeton pectinatus* (sago pondweed) in a Utah marsh (Figure 9.4).

The lake area colonized can similarly vary greatly with the concentration of phytoplankton and in turn with the level of enrichment. Secchi disc depth is a function of both phytoplankton and inorganic sediment (as in Pickwick Reservoir). The potential inverse relationship between phytoplankton and macrophytes must be considered when choosing restoration alternatives for lakes. For example, for Green Lake, Seattle (WA), it was shown that halving the summer average TP content would decrease the chlorophyll *a* content threefold and increase the area colonizable by macrophytes by threefold as well. The effect of water clarity on area colonizable by macrophytes is an important consideration when manipulating the nutrient, and hence phytoplankton, concentration over

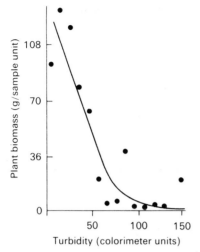

Figure 9.4. Relation between *P. pectinatus* biomass and turbidity in a Utah marsh (Robel, 1961)

the long term. However, sediment type and nutritional content are also important as will be shown.

9.4 EFFECTS OF TEMPERATURE

Temperature acts as a control on light-saturated rates of photosynthesis and could be expected to approximate the Q_{10} rule, over a portion of its temperature range at least, as was shown with plankton and periphytic algae. Field observations of *P. pectinatus* have shown the maximum growth rate to occur in the spring, coincident with greatest increase in light while temperature remained rather low (10°C). Hodgson and Otto (1963) state that the pondweed can develop at temperatures from 10–28°C. The interpretation must be that the greatest rate of growth occurred in young plants, but that this rate would increase as temperature increased up to its thermal optimum. This is because light-saturated rates of photosynthesis increase with temperature. Thus, if the river had been heated during the spring when light increased, growth would have been even greater than observed.

A temperature increase from heated-water discharges would probably select species with temperature optima matching the new conditions, as was the case with algae. Hynes (1960) observed that *Vallisneria spiralis* (wild celery) is usually a warm-water aquarium plant in the UK and has been found in the natural environment only where heated effluents occur.

TABLE 9.1.   Growth (stem elongation) of *P. pectinatus* in aquaria with continuous flow

	NO_3-N (mg l^{-1})	PO_4-P (mg l^{-1})	22-day growth (mm)	
			Sediment	Sand
Full nutrient	1.71	0.26	359	310
Half nutrient	0.89	0.11	303	264
Tap water	0.44	0.03	351	288
Tap water with sealed roots	0.58	0.02	303	84

Source: Peltier and Welch (1969)

However, nuisance macrophyte growth related to heated-water discharges has not been a recognized problem.

9.5   EFFECTS OF NUTRIENTS

Although distribution and growth of submersed macrophytes is dependent on year-to-year and seasonal variations in water clarity, the more long-term increases (and decreases) are usually attributed to increases in enrichment. While N, P and C (and possibly K) are the important macronutrients for submersed macrophytes, as they are for algae, experimental work shows that macrophytes may not be affected by relatively high concentrations of N and P (500 and 50 μgl^{-1}) in the surrounding water as are plankton algae. For example, experiments with *P. pectinatus* (sago pondweed) conducted in flow-through aquaria to simulate river growth conditions, showed no significant difference in growth over a tenfold gradient in SRP, which should have been the limiting nutrient (Table 9.1). Growth appeared to be enhanced slightly if plants were rooted in river sediment. This suggests that growth was either saturated at low water concentrations ($<20\mu$gl^{-1}, which was the lowest concentration tested) or that water concentrations were relatively unimportant and the plants were more dependent on sediments for nutrition. Plants with roots in sand sealed from overlying water showed poor growth indicating the roots to be more important in nutrient uptake than shoots.

The dominant effect of sediment nutrition and minor effect of high nutrient concentrations in the water was also shown for milfoil (Table 9.2). There was some limitation at very low concentrations (5 μgl^{-1} SRP) with roots in sand, but growth in full sediment (undiluted with sand) was nearly five times greater than growth in full sand and at very high water concentrations of nutrients. Likewise, *Najas* growth was found to be proportional to the fraction of lake sediment (mixed with sand), irres-

TABLE 9.2. Growth (biomass dry weight) of *Myriophyllum spicatum* under different treatments of sediment type and water nutrient concentration

Water nutrient (mg l^{-1})	Sand/sediment (%/%)			
	100/0	*67/33*	*33/67*	*0/100*
0.5 P; 5.0 N	1.43	–	–	–
0.05 P; 0.5 N	1.26	–	–	–
0.005 P; 0.05 N	0.68	3.78	4.51	5.63

Source: modified after Mulligan and Barnowski (1969)

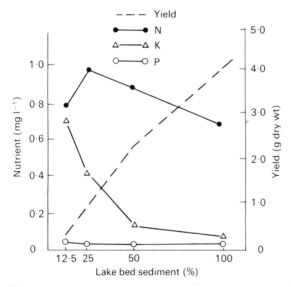

Figure 9.5. Effect of increasing fractions of lake-bed sediment (mixed with sand) on the yield of *Najas* and the measured concentrations of nutrients in experimental containers (Martin *et al.*, 1969)

pective of water concentrations, although there appeared to be some biomass dependent depletion of K (Figure 9.5). Gessner (1959) cited similar results with elodea and milfoil and Chambers *et al.* (1989) have shown that *P. crispus* grew better in sediments enriched by sewage effluent regardless of the surrounding water nutrient content.

The more definitive experiments evaluating the relative role of sediments versus surrounding water as sources of nutrition, are those using ^{32}P. Carignan and Kalff (1980) showed that most of the P taken up by milfoil, *Heteranthera* and *Vallisneria* came from the sediments; 95–103% in oligotrophic, 86–96% in mesotrophic and 70–74% in eutrophic lakes. Gabrielson *et al.* (1984) found that 85% of the foliar demand of *E. densa* for P originated from the sediment and also found, as have others (see Nichols and Shaw, 1986), that there was no net excretion of P to the surrounding water. Other researchers to demonstrate that sediments are the principal source of the plant's nutrition when surrounding water concentrations are relatively low, are: Bristow and Whitcombe (1971) for P with *M. brasiliense*, *M. spicatum* and *E. densa*; Nichols and Keeney (1976a; 1976b) for N with milfoil; Bole and Allan (1978) for P with milfoil and hydrilla; and Best and Mantai (1978) for N and P with *M. exalbescens*. Schultz and Malueg (1971) did show, however, that elodea, *P. amplifolias* and *V. americana* took up more P from the water than from sediments. The preponderance of results, however, demonstrate that the roots of several species of submersed macrophytes supply most of the plant's nutritional needs from the sediments. Foliar uptake may supply most needs if surrounding waters are relatively high in dissolved nutrients.

While the above experimental results show that sediments are a more important source for plant nutrition than the surrounding water, they do not explain the factor(s) in sediments responsible for improved growth. If sediment were the principal source of P in P-limited lakes, then one might expect a relationship between sediment P content (e.g. pore-water SRP + hydroxide exchangeable P) and plant growth, within and among lakes. However, such a relationship has not been found, which suggests that the availability of P in most lake sediments is large and some other factor(s) in sediment is more critical to growth and distribution. In support of this, critical tissue concentrations have been determined experimentally for several species by relating tissue content to growth (yield) in plants exposed to complete media of elements (Figure 9.6). Tissue analyses of naturally occurring populations have shown very few to be nutrient-limited (Gerloff and Krombholz, 1966). Exceptions were in Lake Wingra, Wisconsin where Gerloff (1975) showed K to limit milfoil growth (yield) and Schmitt and Adams (1981) found that net photosynthesis of milfoil was reduced at tissue levels of P (0.28%) much higher than those (0.13%) that limit short-term yield.

Organic content has been considered as a critical factor and favours plant growth when at intermediate levels, e.g. 10–25% of dry matter (Nichols and Shaw, 1986). In support of this, Misra (1938) observed a species selection by sediments of different humus content in Lake Windemere, UK following work by Pearsall (1920; 1929) that plant distribution was controlled by the substratum. Three types of sediment were identified in the lake and each had a characteristic plant assemblage (Table 9.3). Laboratory experiments showed that *P. perfoliatus* grew best

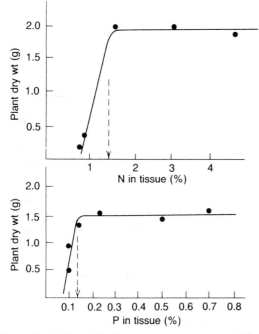

Figure 9.6. The critical tissue concentrations of N and P (concentrations allowing maximum growth, dashed line) in algae-free cultures of several species of macrophytes were consistently around 1.3% N and 0.13% P. The above results are with *Vallisneria americana* (modified from Gerloff and Krombholz, 1966)

TABLE 9.3.   Association of plant and sediment types in Lake Windermere, England, and 1 month's growth of *P. perfoliatus*

Mud type	Dominant plant species	Humus (%)	Dry weight (mg)
Inorganic course brown silt	*Isoetes* (quill wort)	8.04	467
Moderately organic black flocculent mud	*Potamogeton perfoliatus*	12.26	778
Highly organic brown mud	*Sparganium minimum* and *Potamogeton alpinus*	24.00	298

Source: Misra (1938) with permission of Blackwell Scientific Publications Ltd

on the sediment type (intermediate humus content) in which it was found in the lake. Subsequent experiments showed that each dominant species preferred the sediment type in which it naturally occurred. Spencer (1990) recently found that *P. pectinatus* biomass and tuber production increased with increased humus addition to sediment and suggested that plants may prefer intermediate sediment density, not too flocculant and not too dense (sandy). The role of humus was one of reducing sediment density. On the other hand, additions of labile organic matter such as glucose (Spencer, 1990) or plant remains (Barko, 1983) have shown, respectively, either neutral or negative effects on plant growth. Apparently too much organic matter may in some cases create conditions of organic acid accumulation, low pH, increased metal availability and inhibiting gases that together are inhibitory to plant growth. At intermediate levels, however, organic matter may create a sediment density most favourable for growth because of greater nutrient availability (Barko and Smart, 1986).

The increased growth and distribution of rooted submersed macro-phytes resulting from eutrophication is thus apparently an indirect effect of lake-water enrichment. Enrichment of the lake water will result in increased phytoplankton production, which will in turn result in increased sedimentation of organic matter. Enrichment of the water will probably not contribute directly to increases and decreases in macrophyte biomass. If there is ample shallow area in the lake with a gradual enough slope to accumulate the increased production of phytoplankton-derived low density organic sediment, conditions should improve for macrophytes to root and maintain themselves. The importance of slope in this regard has been demonstrated by Duarte and Kalff (1986), who showed that for 17 lakes in which turbidity and light did not limit growth (SD > 2 m), 72% of the variation in macrophyte biomass was explained by slope. The relationship was steeply inverse below a critical slope of 2.24% and if organic content were included, the variance explained increased to 88%. They suggest that at greater slope, high-water content, low density sediment tends to be unstable.

As enrichment of lake water increases, the maximum depth of colonization should decrease due to reduced clarity of the water restricting the availability of light. Thus, slope, that controls the distribution of suitable sediment, and available light are apparently the principal factors controlling the growth and distribution of submersed macrophytes. The increase in macrophyte biomass with eutrophication, until light begins to limit, is probably related to the increased deposition of organic matter that creates a low-density sediment more suitable for rooting. Wetzel and Hough (1973) have suggested a generalized sequence among the various types of primary producers and increased eutrophication (Figure 9.7). Phytoplankton and periphyton (attached algae) dominate as enrichment begins to increase, followed by submersed macrophytes. Periphyton are probably the first indicator of increased enrichment because its attachment near nutrient sources allow it to respond to increased nutrient concentrations, while dilution would prevent pelagic phytoplankton from

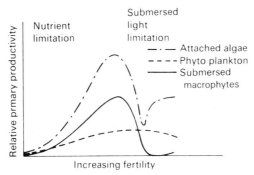

Figure 9.7. A generalized scheme for the changes in submersed macrophytes, attached algae and phytoplankton in lentic environments (modified from Wetzel and Hough, 1973)

experiencing such increased concentrations. This has been shown in ultraoligotrophic Lakes Tahoe (Loeb, 1986) and Chelan (Jacopy *et al.* 1990). The subsequent greater increase in periphyton than phytoplankton as enrichment proceeds, as depicted in Figure 9.7, is probably related to the increased surface area for attachment provided by macrophytes. The dominance of periphyton and macrophytes in this example suggests that the lake is relatively shallow with a sufficiently small bottom slope to accumulate suitable sediment for macrophyte growth over much of its area. In deeper lakes with greater littoral bottom slope, a greater time separation may occur in the sequence of phytoplankton/periphyton and macrophytes in order to develop suitable sediment. The sediment and macrophyte development process, as Carpenter (1981) has suggested, includes a feedback loop from macrophytes to phytoplankton through P recycling.

Shading of macrophytes by periphyton may also account for the greater rate of increase by periphyton and, as Phillips *et al.* (1978) have demonstrated for the Norfolk Broads, UK, it is the most important reason for the decline in submersed macrophytes depicted in Figure 9.7.

9.6 NUTRIENT RECYCLING

Because rooted macrophytes obtain most of their nutrient demand from the sediment, the fate of their tissue P and N content may be important to the nutrient economy of a lake if there is substantial coverage by plants. As indicated above, most research has shown that plants do not excrete significant amounts of dissolved P or N into the surrounding water (Nichols and Shaw, 1986). However, they can contribute significant amounts of nutrients through senescense and decay, which takes place throughout the summer growing season in some species. Milfoil continually sloughs leaves during the summer; the plants can senesce and break apart as they remain in the water column with much of the plant material near the water surface. A similar pattern has been observed for *P. praelongus* with most of its die-back complete by late July.

Landers (1982) showed that enclosures with milfoil in an Indiana reservoir contained more SRP than enclosures without milfoil or in the open water. Enclosures with milfoil also contained over four times more chlorophyll *a* than those without. Although this source of internal loading of P is considered important, realistic estimates of whole-lake contributions of P from macrophytes exist only for milfoil in Lake Wingra, Wisconsin. Smith and Adams (1986) determined shoot and root uptake of ^{32}P by plants in laboratory experiments and tissue content of P was determined in plants in the lake throughout the year. By combining laboratory and lake data, they constructed the following budget for P in the lake's weedbeds: in $gP\,m^{-2}\,year^{-1}$, uptake by roots (from sediment) 2.2 plus uptake by shoots (from water) 0.8 = 3.0; loss to water from healthy shoots 0.0 plus loss to water by senenscense/decay 2.8 minus uptake from water 0.8 = gain to water (internal loading) 2.0 or in $mgP\ m^{-2}\,day^{-1}$, whole year on daily basis = 5.5 and summer on daily basis = 17.0.

These results are from 1977, after milfoil had started its decline to ultimately ≈10% of its previous level in the 1960s and 1970s. Carpenter (1980a) estimated a contribution of $3.0\,g\,m^{-2}\,year^{-1}$ from weed beds in 1975 when the plants were more abundant. As an indication of the significance of milfoil senescense as a source of P, the summer weedbed release of $17\,mg\,m^{-2}\,day^{-1}$ is intermediate in the range of measured release rates from sediments in anoxic hypolimnia: $1.5-34\,mg\,m^{-2}\,day^{-1}$ (Nurnberg, 1984).

In contrast, invasion of a Norwegian lake (Steinfjord) by *E. canadensis* in 1976 resulted in increased phytoplankton on only a few occasions in the 1980s following summer die-back (Rorslett *et al.*, 1986). While the plant's establishment raised the lake P storage (water + plants) from 2 to 6 tonnes, there was no significant change in water P content and the lake remained mesotrophic. Large year-to-year fluctuations in *E. densa* in Long Lake (Kitsap), WA have also produced opposite effects on water P content (Welch and Kelly, 1990). A virtual crash in the plant in 1985 and 1986, resulting in a 75% reduction in overall lake macrophyte biomass, produced the marked increases in lake water P content. The plant's normal presence as a thick carpet was hypothesized to protect the sediment–water interface from wind mixing and, thus, reduce internal loading rather than increase it through senescence and decay.

10
LAKE AND RESERVOIR RESTORATION

The control of eutrophication by restoring degraded lakes and reservoirs is a relatively new science. While some activity began in the 1960s, organization of knowledge into an assortment of techniques to improve the quality of standing bodies of freshwater did not occur until the 1970s. The US Environmental Protection Agency (USEPA) began funding lake restoration activities in 1976 and instituted a Clean Lakes Program in 1980. As a result, activity has profoundly increased. Dunst *et al.* (1974) reported a total of 81 restoration projects in the US, employing one or more techniques, that were either started or finished prior to 1974. Between 1976 and 1987, USEPA had funded 362 projects through the Clean Lakes Program (USEPA, 1987). Additional projects were also funded solely through state 'Clean Lake Programs'.

The need for increased restoration activities was evident from several surveys. The North American Lake Management Society reported in 1983 that 12 000 lakes in 38 states were adversely affected by macrophytes and algae. The American Water Works Association (AWWA, 1987) reported that 61% of surface drinking water supplies in the US and Canada had algae-related (primarily planktonic blue-greens) taste and odour problems that affected a total of 38 million people. In response to the need, books and manuals have been published on techniques for restoring lakes and reservoirs (Cooke *et al.*, 1986; Moore and Thornton, 1988; Cooke and Carlson, 1989).

The water quality problems caused by eutrophication were briefly mentioned in Chapter 7, but will be reiterated here in more detail. Increased enrichment of lakes and reservoirs can have the following direct and indirect effects: increased biomass of planktonic algae, with usually a dominance of blue-greens, several of which produce unsightly surface scums; taste and odour problems, caused largely by excretory products of blue-green algae, that can taint drinking water supplies and fish flesh; oxygen depletion, caused by decomposition of sinking algae, that may render the hypolimnion anoxic or low enough in DO to limit the production and even survival of cold-water fish; reduced transparency of the water; increased incidence of swimmer's itch, as a result of increased production of algae and detritus eating snails, the intermediate host for the parasite; increased trihalomethane precursors resulting from the increased organic matter; and possible increased incidence of toxic blue-green algal blooms.

Lake quality can degrade from either increased abundance of algae or macrophytes, or both. Increased algal abundance is a direct result of increased concentration of limiting nutrient, usually P, and would be expected to respond to reductions in nutrient concentration. However, increased macrophyte abundance is an indirect effect of enrichment, as indicated in Chapter 9, and therefore would not be expected to respond quickly to nutrient controls. Therefore, most external and in-lake controls on P are expected to reduce biomass and change species composition of plankton algae. In many cases, however, controls have not resulted in significant reductions in P concentrations, so reductions in algae and changes in species composition could not be expected. Restoration results have ranged from dramatic success to little or no change in water quality. In-lake controls on P are designed to reduce internal loading.

10.1 PRE-RESTORATION DATA

To estimate the degree of success of a restoration measure(s) to control lake P content, there are some important data needs and some alter-native ways of using those data to estimate lake response and, therefore, benefits versus costs of treatment. Most important are the budgets for water and P. Without a reasonably accurate budget on sources and losses of water volume, a sufficiently accurate budget on P becomes an impossible task. Point and non-point sources of P are described in Chapter 5. If sources of P via groundwater are important, then more direct estimates of groundwater than simply calculation of the unknown residual in the budget may be necessary (Winter, 1981). From measured inputs and outputs of P, a residual accounting for exchange across the sediment–water interface can be calculated. A negative residual is an estimate of net internal loading (see Chapter 4). N loading is usually not computed, because of the following: there are very few cases where N control has been attemped or was successful; there is one more gain (N fixation) and one more loss (denitrification) than in the P budget, so it is more difficult to construct; and P is usually easier to control, because of more sinks and fewer sources than N (see Chapter 4).

The next step is to develop a P model that is calibrated and verified for the lake in question. There are several alternatives and a Vollenweider-type, mass-balance model is often chosen (see Chapter 7). However, a dynamic P model may be more useful than a steady-state one. With a dynamic model, seasonal effects of inputs from external and/or internal sources can be separated, which will allow alternatives for selection of average P values to relate to algal biomass (Ahlgren *et al.*, 1988; Welch *et al.*, 1986) (see Chapter 7). Algal biomass, as chlorophyll *a*, and subsequently Secchi transparency, are usually predicted from predicted P concentration based on choices of several regression models available (see Chapter 7 and Ahlgren *et al.*, 1988). However, development of predictive equations based on data from the lake in question would be preferred (Smith and Shapiro, 1981). Predictions of species composition (e.g. percentage of blue-green algae) are not yet highly reliable (see Chapter 7).

Sediment cores can be used to estimate the net sedimentation rate of P. This can be compared to the current net annual retention of P, as well as to past rates of sedimentation or retention. The sum of the annual P net sedimentation rate and the measured outflow rate of P can represent an additional estimate of the current external loading rate. Sedimentation rate can be determined either by stable lead measurements, assuming that the lead increase was due to the use of leaded gasoline, starting about 1930, or by lead-210 dating of the core (Schell, 1976). The latter gives recent rates as opposed to post-1930 average rates that are obtained from using stable lead. Because of the decrease in use of leaded gasoline starting in about 1972, stable lead may also give an estimate of recent sedimentation rates. Net sedimentation rate of P can also be determined by calibrating the Vollenweider model, i.e. solving for the unknown σ (see Chapter 7).

There are nearly 20 different techniques that have been used to restore and/or improve/preserve the quality of lakes and reservoirs. These can be divided into external and in-lake controls. External controls are largely designed to control P and include diversion of point sources of stormwater or wastewater (sewage effluent), advanced treatment of wastewater (alum, lime or ferric chloride) and detention/retention, infiltration or polishing (with alum) of stormwater. In-lake treatments include P inactivation (with alum or Riplox), dilution/flushing, dredging, artificial circulation, hypolimnetic aeration, hypolimnetic withdrawl, N addition, biomanipulation (planktivore predation/removal), mechanical harvesting (including rotavating), bottom covering, biological harvesting (grass carp) and draw-down (lake-level lowering). Several have been predictable and cost effective, while others have been less so. However, with most there has been insufficient controlled, whole-lake demonstration, with adequate data collection, with which to establish a completely reliable understanding of cost effectiveness.

10.2 EXTERNAL CONTROLS ON P

10.2.1 Diversion and wastewater treatment

These two techniques have been employed most to restore and/or pre-serve the water quality in lakes and reservoirs. Diversion of treated sewage usually involves the construction of interceptor sewers and other piping systems through which treated sewage is transported away from the degraded lake to a body of water with greater assimilative capacity. Advanced wastewater treatment is usually synonymous with P removal with either alum (aluminium sulphate), lime (calcium carbonate) or ferric chloride. A large volume sludge is created with advanced treatment that requires disposal, usually on land.

Recovery following external P reduction has not always been prompt or complete. Cullen and Forsberg (1988) have reviewed the response in 46 lakes resulting from decreases in external P loading by either diversion of point sources or advanced treatment. The responses to P input reduction varied from 'sufficient to change the trophic state' (13) to 'no response' (6), with the remaining 27 responding by 'either P and algae reduction with no trophic state change' or 'P reduction with no change in algae content'. The failure of lakes to recover promptly (in proportion to flushing rate) following external P control is usually due to high rates of recycling of nutrients from sediments, termed internal loading. Shallow lakes are the most difficult to improve (Ryding and Forsberg, 1976). In fact, internal loading was observed to increase on the average following external P load reduction in a study of the recovery of 18 European lakes (Sas et al., 1989). However, in most cases lake P concentration will decline in spite of internal loading, which has also been found to eventually decline (Ahlgren, 1977; Welch et al., 1986; Sas et al., 1989). Unfortunately, there has been little success at predicting the persistence of internal loading.

Diversion of secondary treated sewage from ten treatment plants that discharged effluent to Lake Washington is one of the most celebrated cases in the world. The lake had begun to recover promptly even before the three-year construction project, which diverted 88% of the P load, was complete (Edmondson, 1970, 1972; Edmondson and Lehman, 1981). The diversion from the lake was accomplished over the period 1964 to 1967. TP declined from a mean annual $64\,\mu g\,l^{-1}$ prior to diversion to an equilibrium value of about $21\,\mu g\,l^{-1}$ in 1972, five years after diversion was complete (Figure 10.1). However, it had declined to about $25\,\mu g\,l^{-1}$ by 1969. The lake should have reached 10% of its total decrease in 2.2 years, based on a first-order decline [$\ln 10/(\rho + \rho^{0.5})$], where $\rho = 0.4\,year^{-1}$]. Thus, Lake Washington responded precisely as the Vollenweider model predicted (see Chapter 7).

Figure 10.1. Relative values in surface water (upper 10 m) of phosphate-phosphorus and nitrate-nitrogen (January to March), total P (whole year), and summer (July and August) chlorophyll *a* in Lake Washington. The 1963 mean values, plotted as 100%, were in $\mu g l^{-1}$; total P, 65.7; phosphate-P, 55.3; nitrate-N, 425; and chlorophyll *a*, 34.8 *1933 data (Edmondson, 1978)

Algal biomass as chlorophyll *a* declined from a four-year summer mean of $36 \mu g l^{-1}$ before diversion to a seven-year mean of $6 \mu g l^{-1}$ after, in direct proportion to the decline in P (Figure 10.1). This was some of the first direct evidence of the singular importance of P to algal control. However, the fraction of blue-greens in the algal biomass did not decline until the mid-1970s, which was promptly followed by a dramatic increase in *Daphnia* in 1976 (Edmondson and Litt, 1982). Transparency had improved from 1 m, as a summer average, prior to diversion, to 3.1 m in the mid-1970s, but the advent of *Daphnia* reduced the chlorophyll *a* by one-half and approximately doubled the transparency. Thus, the lake condition in the late 1970s of about $17 \mu g l^{-1}$ TP, $3 \mu g l^{-1}$ chlorophyll *a* and a Secchi transparency of nearly 7 m was the result of both chemical and biological recovery.

Unfortunately, the Lake Washington case is not typical as indicated by the analysis of Cullen and Forsberg above. The principal reasons that Lake Washington recovered so promptly and completely are its relatively great depth (64 m maximum, 37 m mean), fast renewal rate (0.4 year^{-1}), its oxic hypolimnion and its relatively short history of enrichment. The lake's large hypolimnion and short period of enrichment (first signs were observed in the early 1950s, Edmondson *et al.*, 1956) prevented the hypolimnion from reaching an anoxic state. Thus, internal loading was

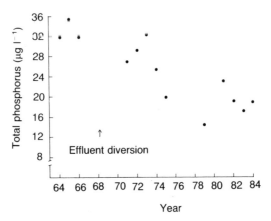

Figure 10.2. Whole-lake TP in Lake Sammamish, WA before and after waste-water diversion (Welch *et al.*, 1986)

insignificant and demonstrates the advantage of treating a lake before reaching an advanced state of eutrophy.

In contrast to Lake Washington, the response of Lake Sammamish, which lies about 12 km from Lake Washington, is half as deep (18 m mean depth) and has about the same renewal rate ($0.55\,\text{year}^{-1}$), was substantially delayed following diversion of sewage and dairy wastewater in 1968 (Welch *et al.*, 1980; 1986). There was a slight decrease in whole-lake TP, from 33 to $27\,\mu\text{g}\,\text{l}^{-1}$, following diversion of ≈35% of its P load, but not until seven years after diversion did TP approach the ultimate equilibrium level of $18\,\mu\text{g}\,\text{l}^{-1}$ (Figure 10.2). The latter value ($\pm 2\,\mu\text{g}\,\text{l}^{-1}$) has persisted during the 1980s. Change in chlorophyll *a* and Secchi transparency paralleled that of TP. Chlorophyll *a* decreased by 50% and transparency, which had remained at a summer mean of 3.3 m during the early 1970s after diversion, increased to a mean of 4.9 m.

The delay in recovery of Lake Sammamish was due to internal loading from anoxic sediments. Although the lake had not exceeded a state of mesotrophy during its enrichment, it nonetheless had an oxygen deficit rate typical of eutrophy due to its relatively small hypolimnion. The ultimate improvement in the lake was related to a reduction in the sediment release rate of P, which was in turn related to a decline in oxygen deficit rate (Figure 10.3).

A similar long-term response to nutrient diversion has occurred in Lake Norrviken, Sweden. However, in contrast to Lakes Sammamish and Washington, Lake Norrviken was hypereutrophic prior to diversion having received industrial and domestic wastewater for nearly 100 years (Ahlgren, I. 1977). A large fraction of incoming P was diverted (87%) and the lake P content decreased predictably (see Chapter 7) as high-P lake

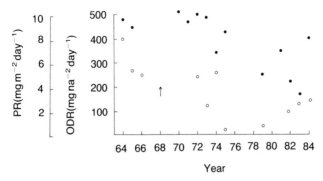

Figure 10.3. Sediment P release (PR) and oxygen deficit rate (ODR) in Lake Sammamish before and after wastewater diversion (Welch *et al.*, 1986). ○, PR; ●, ODR

Figure 10.4. Lake mean concentrations of total phosphorus in Lake Norrviken. □—□, Theoretical dilution curve without sedimentation. Arrows, three water retention times after onset of restoration (modified from Ahlgren, 1988)

water was replaced with the new low-P inflow water (Ahlgren, I. 1977; 1978; 1979). Initially, from 1969 to 1975, average annual lake TP content decreased from 450 to $150 \mu g l^{-1}$, which was still well within the hyper-eutrophic state. Summer TP dropped from $263 \mu g l^{-1}$ before to values of generally $100 \mu g l^{-1}$ or less after 1974. Internal loading showed a decrease through 1977, but increased again as indicated in Figure 10.4. Chlorophyll *a* steadily decreased from summer concentrations in excess of $100 \mu g l^{-1}$ to $36 \mu g l^{-1}$ in 1980. Inorganic N also decreased in the lake from levels $>1000 \mu g l^{-1}$ to usually $100 \mu g l^{-1}$ or less and the decrease in algal bio-mass is more closely correlated with the N decrease, which is considered the more probable cause (Ahlgren, I. 1978). Transparency approximately

TABLE 10.1.   Nutrient loading to the Madison lakes

| Lake | Loading ($g\,m^{-2}\,year^{-1}$) | |
	N	P
Mendota	2.2	0.07
Monona	8.8	0.90
Waubesa	47.0	7.0
Kegonsa	6.8	3.9
Koshkonong	9.8	4.3

Source: Lawton (1961)

doubled but the depth of visibility was still only 1.2 m (Sas et al., 1989). The phytoplankton in summer is no longer a monospecific crop of Oscillatoria (Ahlgren, G. 1976).

Sewage effluent from the city of Madison, Wisconsin, USA was diverted in 1958 from Lakes Waubesa and Kegonsa, the last two lakes in the Madison chain of lakes. Effluent contributed 88% of the P to Lake Waubesa and the four- to sevenfold increase in loading to the lower lakes is shown in Table 10.1. Following diversion, winter SRP decreased from $>500\,\mu g\,l^{-1}$ to $\approx100\,\mu g\,l^{-1}$ as predicted from a simple hydraulic washout model (Sonzogni and Lee, 1974). Although there was little effect on algal biomass, due to the continued high P, dominance changed from 99% Microcystis to a more mixed assemblage (Fitzgerald, 1964). Through the early 1970s, Stewart (1976) found no significant change in either oxygen or transparency that was detectable above normal year-to-year fluctuations. Since 1975, SRP during spring in the lower lakes has decreased to near undetectable levels from values between 50 and $90\,\mu g\,l^{-1}$ (Lathrop, 1988; 1990). This further decrease is attributed to interception of sewage effluents from communities upstream from Madison that drained to Lake Mendota, resulting in lower P in the inflow to the lower lakes, as well as less spring runoff since 1977. Transparency and chlorophyll a have improved accordingly (Lathrop, 1988; 1990).

The most studied case of lake response to advanced wastewater treatment (AWT) is that of Shagawa Lake, Minnesota, USA. Treatment of sewage effluent with lime beginning in 1973 removed 85% of the incoming P to the lake (Larsen et al., 1975). The lake responded quickly with average annual TP dropping from $51\,\mu g\,l^{-1}$ to $30\,\mu g\,l^{-1}$ and spring chlorophyll a also dropping by 50%. However, the lake remained eutrophic during summer due to internal loading from an anoxic hypolimnion and the partial entrainment of that high-P water during summer storm events (see Chapter 4). Although summer levels of TP also declined since treatment, they were still too high to allow much of a

Figure 10.5. Comparison of predicted (—) and observed (●) P concentrations in Shagawa Lake, MN before (upper line) and after (bottom line) institution of phosphorus removal from sewage effluent (modified from Larsen, *et al.*, 1979)

decrease in summer chlorophyll *a*. This is shown in Figure 10.5 by the decreasing trend in TP between 1973 and 1976 compared to the predicted pretreatment level (Larsen *et al.*, 1979). Shagawa Lake is expected to continue to recover, in spite of continued internal loading, as has been observed in other lakes, but the rate of recovery will be slower than if internal loading did not exist (Sas *et al.*, 1989).

Recent results from Shagawa Lake show that the summer increase in TP has decreased from $35-50 \mu g l^{-1}$ to $20-30 \mu g l^{-1}$, but that internal loading was still occurring 16 years after treatment (Wilson and Musick, 1989). The persistence of internal loading was indicated by outflow TP still exceeding inflow TP with internal loading accounting for $\approx 30\%$ of the annual loading. Peak chlorophyll *a* concentrations during summer were less in 1988 than in the early 1970s; $30-45 \mu g l^{-1}$ versus $50-60 \mu g l^{-1}$, respectively.

10.2.2   Watershed treatment

After any sewage effluents have been diverted or given advanced treatment, the next most important sources of enrichment causing eutrophication are stormwater runoff and failing individual septic systems. Although the P content of stormwater runoff is not as high, by a factor of ten or more, as that of sewage effluent and the fraction that is SRP is usually much less, it is nonetheless an important source of enrichment to lakes. There are several techniques to remove P from stormwater, but their effectiveness varies greatly and there are few cases of lake improvement from treatment of stormwater. Detention of stormwater in ponds results in effective settlement of P if time is sufficient, but it is mostly the particulate fraction. Directing stormwater through a grassy swale can remove a significant fraction of P if contact time between water and the soil/plant surface is sufficient. If areas of high percolation capacity are available, infiltration of stormwater is a possibility. Infiltration can result in a long and effective contact between water and soil. The use of wetlands for P retention is also a possibility. Detention ponds with polishing (removal of P with alum or iron) are a more effective, but costly, alternative.

The removal efficiency of P in detention basins varies from 35 to 60% over a detention time range of 2 to 15 days (Walker, 1987b; Benndorf and Putz, 1987). The longer the detention, the greater the area and/or volume necessary to achieve the desired removal. However, constraints on available area in urbanized watersheds can limit detention basin size so that adequate removal efficiency is not possible to protect a lake or reservoir sufficiently. As a result, water from 'pre-dam' detention basins, at flows up to $5 \, m^3 \, s^{-1}$, has been treated with ferric chloride to precipitate P (Bernhardt, 1981). Wahnbach Reservoir and Lake Schlachten in the Federal Republic of Germany are examples of eutrophic water bodies that have recovered to a state of high quality following such treatment (Sas *et al.*, 1989).

Leachate from inadequately percolating or failing septic tank (on-site) drainfields may be another important source of enrichment, although if the drainfield is working properly, P is effectively sorbed in the soil and significant amounts may not reach the lake. P retention is especially effective if there is an unsaturated zone above the groundwater level. As groundwater level raises, the effectiveness of drainfields decreases.

10.2.3   Dilution/flushing

The nutrient content of lakes can be reduced, or diluted, by adding low-nutrient water, the availability of which is the greatest limitation to this technique. Where such low-nutrient sources were used, marked improvement in lake quality resulted; e.g. Green Lake and Moses Lake in Washington state, USA (Cooke *et al.*, 1986; Welch and Patmont, 1980). If low-nutrient water is not available, algal biomass may be reduced by increasing the flushing rate to a level that significantly increases the loss rate of cells, relative to their growth rate. A critical level in that regard seems to be $\approx 10\%$ day^{-1} which, for anything but a small lake, would require a large daily supply of water. Where increased flushing rate with high nutrient water was used, the results were unsatisfactory; e.g. Rotsee near Lucerne, Switzerland, where flushing rate was increased from 0.1 to 0.7% day^{-1} in 1921–1922 (Stadelman, 1980; see Cooke *et al.*, 1986).

The most extensive dilution project is Moses Lake, where low-nutrient Columbia River water has been routed through the 2400 ha lake at various intervals during spring-summer since 1977. On a time-weighted average, dilution water inflows have entered the lake at 8.2 m^{3}s^{-1} during the subsequent 12 years, replacing water in the arm of the lake proximal to the inflow at an average rate of 5.8% day^{-1}, but in the whole lake at only 0.27% day^{-1}. TP and chlorophyll *a* were reduced by >50% and transparency doubled in the proximal arm during the first years of treatment. Similar improvements were also evident in other parts of the lake accessed by wind-enhanced transport of the water. Algal biomass control was due to decreased NO_3 in the inflow, even though little change in that variable was noted in the lake (Welch *et al.*, 1984).

Further improvements in the quality of Moses Lake have resulted from diversion of sewage effluent in 1984 and a progressive decrease in P concentration in 50% of the lake's undiluted inflow water. P has become the most limiting nutrient and further control hinges on the rather large and variable ($\pm 100\%$) internal load of P. Moreover, while dilution effectively reduces in-lake P content, the benefit diminishes as water input increases, due to enhanced internal loading of P, apparently resulting from the reduced lake P causing an increased diffusive gradient between sediment and overlying water. While simple dilution should provide a two-thirds reduction in lake TP, the effect of internal loading results in only a one-third reduction (Welch *et al.*, 1989a; Welch and Jones, 1990).

10.3   IN-LAKE CONTROLS ON P

Techniques that are most appropriate at controlling internal loading of P are P inactivation, Riplox, hypolimnetic aeration, hypolimnetic withdrawal, and dredging. Three of these five techniques are aimed primarily at controlling the iron redox mechanism (see Chapter 4). Alum $[Al_2(SO_4)_3 \cdot (H_2O)_{18}]$, when added to lake water, forms an hydroxide floc that rather quickly settles to the lake bottom creating a barrier to further releases of sediment P. The continued effectiveness of Al is due to its solubility being independent of redox, which in contrast controls Fe. Hypolimnetic aeration produces an oxic hypolimnion, which should increase the control of P by Fe and prevent sediment release. The Riplox technique is designed to reduce interstitial SRP by oxidizing the organic matter in sediments. Oxidation is accomplished by denitrification through the addition of $Ca(NO_3)_2$ under alkaline conditions. Fe is added if naturally low or if S is high. The idea is to keep Fe in the oxidized ferric state and P complexed to the ferric hyroxy complex.

Alum has been used extensively in the USA with about 30 projects being supported by USEPA and additional projects receiving sole support from individual states. Hypolimnetic aeration has also been rather popular, but while its record is good for increasing hypolimnetic oxygen content without destratifying the lake, it has been much less effective than alum in controlling sediment-P release. Riplox is a relatively new technique, having been used in only three lakes, two in Sweden and one in Minnesota, USA (Cooke et al., 1986).

Representative examples of treatment effectiveness for alum, Riplox and hypolimnetic aeration are given in Table 10.2. Of these three techniques to retard sediment-P release, alum was most effective and consistent and these results are typical of other treated lakes (Cooke et al., 1986). The first three lakes listed are thermally stratified in summer, which is the case for most lakes that have been treated. Long Lake is shallow (mean depth 2 m) and unstratified and there were doubts that treatment of a shallow lake would be successful, because three previous shallow-lake treatments were unsuccessful (Welch et al., 1988b). However, the Long Lake treatment was highly effective for four years (Table 10.2). Moreover, lake TP has averaged 40% less than the pretreatment level for 8 of 9 years following the application. Two of the previous three shallow-lake failures were due to continued high external P loading. Treatment of stratified lakes has been, without exception, highly effective and longevity, where evaluated, has exceeded ten years. However, whether stratified or unstratified, unless external P loading is controlled first, an alum treatment may fail.

Riplox has a greater longevity potential than alum. While the alum floc layer is covered with new sediment and dispersed downward diminishing its effectiveness, Riplox removes the organic matter and the cause for Fe reduction. Sediment oxygen demand in Lake Lillesjon has continued to decline even after ten years and hypolimnetic P has remained low (Ripl, 1986). Unfortunately, external P loading in the other two lakes treated (Long Lake, MN and Lake Trekanten, Sweden) remained great enough to offset any improvement in water quality even though sediment-P release was controlled (see Long Lake, Table 10.2). Internal loading recurred in Trekanten, apparently because newly deposited algal organic matter reduced sulphate to sulphide, which tied up iron allowing P

TABLE 10.2. Representative results from in-lake restoration techniques to reduce internal loading from eutrophic lake sediments. Values are summer means and are averaged if more than one year's data were available (from Welch, 1988)

Treatment	Lake	Whole-lake TP ($\mu g l^{-1}$) Pre	Post	Reduction (%)	Internal Loading (kg) Pre	Post	Reduction (%)	Reference
Alum	Dollar, OH	240	52	78	–	–	–	Cooke et al. (1986)
	W. Twin, OH	97	32	67	50	19	62	Cooke et al. (1978)
	Annabessacook, ME	72	34	53	1800	625	65	Dominie (1980)
	Long, WA	65	30	54	165	0	100	Jacoby et al. (1982); Welch et al. (1982)
Riplox	Lillesjon, Sweden	3000	50/100[a]	97	–	–	–	Ripl (1986)
	Long, Mn	87	85[b]	2	480	96	80	Noonan (1986)
Hypolimnetic aeration	Sodra Horken, Sweden	400	25/50[a]	91	–	–	–	Bjork (1985)
	Kolbotvatn, Norway	900	600[a]	33	–	–	–	Holton (1981)
Hypolimnetic withdrawal	Ballinger, WA	–	–	–	–	–	86	USEPA, (1987)
	Mauensee, Switzerland	–	–	–	–	–	98	Gachter (1976)
Dredging	Reither See, Austria	41	21	49	–	–	–	Pechlaner (1978)
	Lilly, WI	40	18	55	–	–	–	Dunst (1981)
	Trummen, Sweden	1000	100	90	–	–	–	Bjork (1985)

[a] Hypolimnion only; [b] epilimnion only

release. Excess iron was not added to Trekanten. A comparison of alum and Riplox treatments both showed well over 90% reduction in P release from Green Lake (Seattle, USA) sediment in laboratory incubated cores even after 64 days (unpublished data).

Hypolimnetic aeration has been least effective and most inconsistent of the three techniques to control sediment-P release (Table 10.2). The technique was effective in controlling hypolimnetic P in Sodra Horken, but the unimpressive results from Kolbotvatn are more typical of other treated lakes (Cooke et al., 1986). The technique does not seem to have any long-term effect. Sediment-P release is usually partially controlled when aeration is in progress, but rather quickly resumes once aeration is curtailed.

The advantage of alum is that the hydroxide floc forms a physical–chemical barrier that traps dissolved P that would diffuse from sediments to overlying water even if the water is anoxic. Furthermore, it may also be a barrier to other mechanisms of release, such as microbial mineralization, excretion from microbial cells and blue-green algal migration. On the other hand, Riplox and hypolimnetic aeration will work only if oxic conditions exist. However, Riplox does remove organic matter, which is the cause for anoxia in sediments, so it has a long-term potential.

Hypolimnetic withdrawal has been successful is several lakes (Table 10.2). Nurenburg (1987) reviewed data from 17 cases, and showed that hypolimnetic TP decreased in 10 of 11 lakes where data were available as well as in the epilimnion in 8 of 11. Epilimnetic decrease in TP was correlated with the magnitude of TP export from the lake. However, the technique has not been successful in improving water quality in lakes that are meromictic in which the monomolimnion remained anoxic (Pechlaner, 1978) or where external loading remained high, e.g. Lake Ballinger.

Lake Trummen represents the classic case for dredging, which can provide near permanent control on P internal loading by simply removing the P-rich layer in the sediments (Table 10.2). The P-rich layer extended to 0.5 m, so 1 m of sediment was removed. Lilly Lake was dredged to greater depths, ranging from 1.8 to 6 m, because the principal objective was to control macrophyte growth and increase fish habitat. However, the improved transparency, to some extent related to P control, allowed macrophytes to colonize to a greater depth even though macrophyte biomass was 70 to 85% less (Dunst, 1981).

Comparative costs for the above mentioned in-lake treatments are shown in Table 10.3. Only application costs are considered. Retreatment over a 20-year period is assumed necessary only for alum in an unstratified lake. Clearly, alum is probably the most cost-effective of the five treatments for controlling sediment-P release. Riplox should be equally effective and more long-lasting, although the practical evidence is not yet

TABLE 10.3. Cost and possible frequencies of in-lake treatments for a hypothetical 100 ha lake over 20 years

Control of	Treatment	Whole lake	Hypolimnion only	Number of treatments	Total cost ($ × 10^{-3})
Phosphorus	Alum	X (unstratified)		4	185
	Riplox	X		1	45
			X	1	350[a]
			X	1	175
	Hypolimnetic aeration		X	6 months/yr	230
	Hypolimnetic withdrawal		X	6 months/yr	210[b]
	Dredging	X		1	1475
Biomass	Complete circulation	X		6 months/yr	185[a]
	Biomanipulation				
	Fish removal				
	Netting[b]	30% (littoral)		2	45[a]
	Poisoning[c]	X		2	25
	Piscivore stocking[c]	X		4	130
	Macrophyte harvesting	30% (littoral)		Annual	230
	Macrophyte grazing[d]	30% (littoral)		2	40[a]

Costs are median values (or single examples[a]) in 1988 dollars converted from Cooke et al. (1986) except as indicated (from Welch 1988)

[b] Wagner, personal communication; [c] Mongello, personal communication (stocking 0.5 kg trout at 50 kg ha^{-1}); [d] Devils L., or Bonar, personal communication. X = whole lake or hypolimnion only

convincing, but the cost is over fourfold greater While hypolimnetic aeration is less cost-effective than alum for P control, there may be other reasons for aeration, such as creating an aerobic environment. Dredging is clearly the most costly, however, in return there is the added benefit of having a deeper lake with possibly less noticeable problems with macrophytes.

10.4   IN-LAKE CONTROLS ON BIOMASS

Techniques to control plant biomass may be possible in cases where external sources of nutrients cannot be controlled. The causes for nuisance blooms of blue-green algae have become better understood in recent years and some of the important factors can be affected by artificial circulation. One of the most frequently used techniques to improve lake quality is artificial circulation (whole-lake aeration), but the goals for that technique are several. As a result, algal biomass has been reduced in some cases but increased in others (Pastorak et al., 1982). Control of algal biomass has been attempted by manipulating grazer food chains with some success, but mostly in small-scale experiments (Andersson et al., 1978; Benndorf et al., 1984; Shapiro et al., 1975; Lynch and Shapiro, 1981; Shapiro, 1979; Shapiro and Wright, 1984; Wagner, 1986). More whole-lake experiments are needed before restructuring fish communities to achieve algal control can be considered reliable. A high effective and popular (in the USA) biological technique is the use of triploid grass carp (*Ctenopharyngodonu idella*) to control aquatic macrophytes. However, there are unsolved problems, such as appropriate stocking rates, nutrient recycling and interactions with other fish populations (Cooke et al., 1986).

The understanding of blue-green algal blooms, in relation to their buoyancy regulation mechanism (Reynolds et al., 1987) and changes in CO_2/pH (King, 1970; Shapiro, 1973; Paerl and Ustach, 1982; Shapiro, 1984) has increased in recent years and directly affects the feasibility of controlling such blooms with artificial circulation. Such blue-greens as *Aphanizomenon*, *Microcystis* and *Anabaena* can conceivably rise through a 10 m water column in only a few hours if turgor pressure in their gas vacuoles is high, which occurs if CO_2 or light limitation curtail the cellular accumulation of carbohydrate. Previously accumulated carbohydrate can be assimilated into proteins at depth, where N and P are available for growth (and carbohydrate dilution) and photosynthesis is limited by light. This results in gas vesicle production and turgor rise in the vacuoles. Cell phytosynthate (carbohydrate) accumulates when cells enter the surface water with greater light and CO_2 availability, but with N and P limitation. Either vacuole collapse or increased carbohydrate ballast causes the cells to sink. Such a buoyancy response to CO_2 limitation and N addition was shown for *Oscillatoria* (Klemer, 1973;

Klemer et al., 1982), for Anabaena with N and P addition (Booker and Walsby, 1981) and with pH adjustment (CO_2 limitation) for Aphanizomenon and Anabaena (Paerl and Ustach, 1982).

The dominance by blue-greens may occur under conditions of strong stratification or infrequent diel mixing, because their buoyancy mechanism allows them to exploit nutrients and light from a water column that becomes increasingly turbid as enrichment proceeds. Diatoms and green algae would be less favoured under such relatively static, light limiting conditions, even though they generally grow faster than blue-greens (Reynolds et al., 1987). Increased circulation could induce the algal community to revert to a dominance by diatoms and greens in spite of high nutrient concentrations. This could occur because increased circulation would lower pH and raise CO_2, causing gas vacuole collapse/carbohydrate dilution and increased light availability to mixed plankton cells. Experiments in 7-m deep enclosures in a stratified lake have shown that rapid deep mixing was more successful than slow deep mixing in lowering pH, raising CO_2 and decreasing blue-green dominance (Shapiro et al., 1982). Observations following artificial circulation have shown that a shift occurred from blue-greens to greens in about one-half the cases examined (Pastorak et al., 1982). In the majority of cases where a shift occurred, pH also decreased. However, blue-greens increased in the other half of the cases where there was not a shift. Thus, while there is some evidence that changes in CO_2/pH as a result of circulation will cause a shift, there is also a moderate possibility a shift will not occur and blue-green abundance may even increase. However, as Shapiro et al. (1982) showed, whether a shift occurs depends on mixing rate. To achieve adequate mixing, a critical air flow rate of $9.2 \, m^3 \, km^{-2} \, min^{-1}$ has been recommended (Lorenzen and Fast, 1977), although results suggest that the critical air flow may also depend on other factors, such as velocity of mixing, rate of CO_2 introduction, buffering capacity of the water and rate of nutrient input to the photic zone (Shapiro 1984).

Control of algal biomass may also be attained by increasing the grazing rate by zooplankton, and, thus, the loss rate of algae. While the effect of planktivorous fish on algal abundance has been known for some time (Hrbacek et al., 1961), attempts to refine biomanipulation as a restoration technique have intensified during the past 10 to 15 years (Shapiro et al., 1975; Shapiro, 1979). Grazing loss rates of algae can be increased by preserving the large-bodied zooplankton, especially Daphnia. This may be accomplished by: complete removal of fish by poisoning and replacement with a favourable planktivore: piscivore ratio; improving that ratio by planting piscivores; or installing hypolimnetic aeration to provide a refuge for Daphnia from predation. While there has been some success with the technique in experimental enclosures (e.g. Andersson et al., 1978; Lynch and Shapiro, 1981; McQueen et al., 1986), results from whole-lake treatments are scarce (Stensen *et al.*, 1978; Shapiro and Wright, 1984; Benndorf *et al.*, 1984, Wagner, 1986). In most instances, the presence of planktivores resulted in removal of the large-bodied zooplankton. And without planktivores, algal biomass was often reduced, which of course is the desired effect (Figure 8.2). This was even true of blue-greens except where *Aphanizomenon* formed flakes (Lynch and Shapiro, 1981). Some experiments, however, have shown that large-bodied zooplankton are preserved, but without the desired reductions in algal biomass (McQueen and Post, 1988).

Although experiments with biomanipulation have often been effective, there are too few cases from which to develop definite procedures. The complete removal of fish from Round Lake, Minnesota, USA, followed by restocking with a reduced planktivore:piscivore ratio (165:1 to 2.2:1), was successful for two years (Shapiro and Wright, 1984). *Daphnia* quickly became dominant causing a reduction in chlorophyll *a* of >50% and an increase in transparency from 2 m to >4 m. Planktivores may be effectively removed by netting, as was done in a small New Jersey pond. There, the planktivore:piscivore ratio was changed from 6:1 to 2.4:1. An increase in the existing daphnid species doubled the overall zooplankton size, but large-bodied daphnids did not appear, presumably because planktivore density was still too high, although planktivores remained controlled for four years (Wagner, 1986; Wagner, personal communication). Piscivores (rainbow trout) were introduced to a small quarry in the Federal Republic of Germany at a rate of 117 kg ha^{-1}, matching the abundance of planktivores. The planktivores were decimated after one year. The planktivore reduction was apparently responsible for the post-stocking dominance (95%) of the zooplankton biomass by large-bodied daphnids (Benndorf *et al.*, 1984).

The use of triploid (sterile) grass carp to control macrophytes has become very popular and is legal in 22 states in the USA (Bonar, personal communication). Stocking densities have ranged from 27 to 185 ha^{-1} and in most cases 90% or more plant removal effectiveness was achieved within two or three years (Cooke *et al.*, 1986). The use of grass carp will probably increase because control is relatively quick and lasts for several years. However, grass carp can potentially increase nutrient recycling, which could cause algal blooms. They could also have adverse effects on other fish populations through removal of habitat, competition for food and introduction of disease. Until these potential problems are resolved, the technique cannot be embraced with complete confidence.

There are several techniques designed to physically remove rooted macrophytes, including harvesting (mowing), rotavating, bottom covering and dredging (Cooke *et al.*, 1986). Harvesting is the most common mechanical technique and although relatively slow compared to, e.g. herbicides, harvesting has several advantages: nutrients are removed at a rate that could significantly reduce the lake-P content; risk of toxicity to animals and public objection to a toxicant are avoided; and some plant habitat will remain. Neither herbicides nor harvesting are very long-lasting, however, Although the beneficial effects of nutrient removal by harvesting are logical (Wile, 1975; Carpenter and Adams, 1977; Welch *et al.*, 1979), they have not been adequately demonstrated.

Example costs for in-lake treatments are compared in Table 10.3. Mechanical harvesting and complete circulation are relatively more expensive than the biological techniques of macrophyte and algal control. Also, fewer treatments may be necessary for the biological than physical techniques, although frequency, as well as costs, are not firmly established. While costs for grass carp may be relatively consistent, because the sources for triploid individuals are limited, sources are not so limited for piscivores and their costs may vary widely. Greater use of biological techniques can be expected in the future because of their relatively low cost, minimal side effects and effectiveness in waters where nutrients cannot be controlled.

Part Three

Effects of Wastes in Running Water

Running waters, or lotic environments, contrast sharply in many respects to their lentic counterparts. The fauna and flora are distinct from that in lentic systems. Productivity in running water is dependent primarily on attached algae, which is part of the periphyton. That is because plankton algae adapted to lentic waters are usually intolerant of the short residence time and turbulence that characterize running water. However, a modest population abundance of plankton algae can be found in reasonably clear, slow-moving rivers. The dominance of periphytic (attached) producers in running water explains their reported low abundance of algae (in the water) per unit phosphorus (Soballe and Kimmel, 1987). The effect of residence time on dominant producer organisms is illustrated in Chapter 11.

The ecological effects of pollution have been described more for running than standing water (e.g. Hynes, 1960; Klein, 1962). The majority of the effort to control domestic and industrial wastes has been directed to protecting running water environments, e.g. restoring oxygen to acceptable levels in the Thames (UK), Willamette (USA) and Cuyahoga (USA) Rivers. That organic and toxic wastes would have a greater effect in lotic than in lentic waters is largely because populations of organisms downstream from the waste source are exposed to high concentrations of waste with little or no opportunity for avoidance. The plug-flow conditions of wastewater transported downstream from a point source would explain the greater exposure than if the same volume of waste were dispersed into a lake from a point source. This is consistent with pollution-caused fish kills being observed more often in steams than lakes. For these reasons, effects of organic and toxic waste are discussed under running water, primarily in Chapters 12 and 13. Eutrophication in running water is usually not included in discussions on effects of pollution and there has been little effort at setting criteria or controlling nuisance periphytic algal biomass in streams. Eutrophication is often mistakenly thought to be a problem in lakes only. The effect of stream enrichment and difficulty in setting nutrient criteria are treated extensively in this part.

As in lakes, there is substantial variability in organism populations in streams due to natural factors, such as light, temperature, current velocity, depth and substratum character. The effect of these factors must be considered in separating the effects of wastes. Establishing acceptable experimental controls to define variations in populations due to natural factors is often difficult. While upstream sections from the source may represent acceptable controls, one or more of the above-mentioned factors are often substantially different between upstream and down-stream sections. This problem will be discussed primarily in Chapter 12.

11
PERIPHYTON

Periphyton is the slimy material attached to the surface of rocks but also other substrata (e.g. wood, sediment, etc.) in streams and in the littoral zone of lakes. It is composed primarily of algae, but can be dominated by bacteria and fungi if there is a source of dissolved organic matter and can include invertebrates (e.g. protozoans, insects, etc.). The periphyton mat is usually rather thin, but very slippery in fast-flowing streams with moderate enrichment. In that case, the community is usually dominated by diatoms, such as *Navicula*, *Gomphonema*, *Diatoma*, *Cymbella*, *Cocconeis* and *Synedra*. Most of these genera were not mentioned in the discussion on plankton because they are peculiar to the periphyton.

In streams enriched with inorganic nutrients (N and P), long waving strands of filamentous green algae can develop, especially if current velocity is relatively low. Typical representatives of the filamentous green algae are *Cladophora* and *Stigeoclonium*. Blue-green algae, such as *Phormidium* and *Oscillatoria* may also be important, but green algae are usually more abundant in running water. These mats can represent an aesthetic and water quality nuisance as great as that from planktonic blue-green algae in lakes. Once biomass reaches a level that is unstable under the existing current velocity, the mats slough and the dislodged biomass may create a secondary BOD downstream. Diatoms can also develop a thick brownish mat and can even appear filamentous in character under enriched conditions.

Filamentous bacteria, such as *Sphaerotilus*, may develop to nuisance levels in streams receiving organic enrichment. The filaments of bacteria are greyish or colourless, but they may become covered with diatoms (e.g. *Gomphoneis*) and be mistaken for a diatom mat and, conversely, mats of filamentous-appearing diatoms may be mistaken for *Sphaerotilus*. Fungi and ciliated protozonas, such as *Stentor*, *Carchesium* and *Vorticella* may also be abundant in the periphyton mat in organically enriched streams. Insects, such as chironomid larvae (midges), also occur in the thick mat and are considered part of the periphyton, whether it is composed primarily of algae or bacteria. Even in unenriched streams, other insects, such as mayfly nymphs, blackfly (*Simulium*) larvae and even rotifers may be abundant in the mat.

While the emphasis in this chapter is on the character and cause of nuisance levels of periphyton in running water, filamentous green algae can reach very visible nuisance levels in standing water as well. As an example of nuisance conditions in lakes, Wezernak *et al.* (1974) mapped the extent of mats of the filamentous green alga *Cladophora* (blanket weed) along the Lake Ontario shoreline using remote sensing. A nearshore strip 350 m wide was 66% covered with *Cladophora* from Niagara to Rochester, New York, amounting to about $15\,700\,\text{kg}\,\text{km}^{-1}$. Waves break the strands loose from their substrata and deposit windrows of decaying plant material on the beach in such situations. Such nuisance conditions may be much more obvious to lake shore users than those caused by phytoplankton in the open water. In this chapter, results from studies of filamentous green algae in lakes and estuaries will be integrated with those from streams to understand better the causes for nuisance levels.

11.1 SIGNIFICANCE TO PRODUCTIVITY

The turbulence and short residence time in streams is intolerable to planktonic organisms and, therefore, attached algae are the dominant primary producers. However, plankton may represent a significant contribution in deep, slow rivers where periphyton may be limited by light and substrate instability. Even plankton productivity may be limited by light in such environments if turbidity is too high. Rickert *et al.* (1977) illustrated the distribution of algal biomass per unit photic zone with detention time, showing the shift from periphyton to plankton dominance in running water as detention time increases (Figure 11.1). However, they omitted filamentous green algae, which are often important in moderate velocity (and even high velocity), shallow rivers.

The river continuum concept of Vannote *et al.* (1980) is useful in considering community shifts from shallow, fast-flowing streams (usually low-order tributaries) to slow-flowing rivers (usually high-order). Heterotrophic fungi and bacteria usually dominate the shaded, fast-flowing, shallow first- to third-order streams in the forest, which is the principal source of primary production and organic matter to the stream, i.e. allochthonous sources. The organic matter from leaves, twigs and other large particulate material, along with attached bacteria and fungi, is degraded by the detritivore insects, as well as by the fungi and bacteria. As a result, respiration exceeds primary production in such environments ($P/R < 1$). Farther downstream in the fourth- to sixth-order range, shade

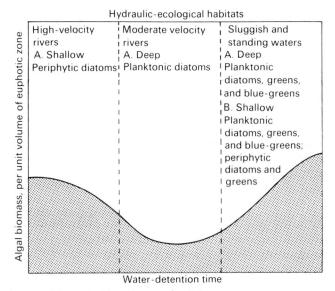

Figure 11.1. Conceptual diagram relating algal biomass and dominant algal types to water-detention time and light penetration. Shallow implies that the euphotic zone extends to the bottom; deep implies the euphotic zone does not reach the bottom (Rickert *et al.*, 1977)

by the forest canopy and the terrestrial source of organic matter decrease, or that source becomes well diluted by the increased stream volume. In such reaches, in-stream primary production by periphytic algae becomes dominant, grazing invertebrates increase and the stream is autotrophic with $P/R > 1$. Still farther downstream (seventh order), the P/R ratio may again drop below 1 if turbidity and depth create light-limited conditions.

Periphyton comprises a relatively small fraction of the primary production in deep lakes, whose area of shallow littoral is relatively small. The contribution to production depends on the percentage of area where shallowness is sufficient to allow enough light to reach the bottom. If macrophytes are abundant, the zone of periphyton growth can be larger due to the large lighted substratum supplied by the rooted plants. Periphyton production per unit area can be great in oligotrophic lakes or shallow lakes with large littoral areas (Loeb *et al.*, 1983) and in the intertidal areas of estuaries. In such cases, the contribution to total production is significant. While rooted plants, logs and rocks are suitable substrata, shifting sand beaches are not stable substrata and, therefore, support little periphyton.

Wetzel (1964) compared the three sources of primary production in a

TABLE 11.1.   Comparison of the three sources
of productivity in Borax Lake, California

Source	Annual mean $(mg\ C\ m^{-2}\ day^{-1})$	Total annual mean $(kg\ C\ lake^{-1}\ yr^{-1})$
Phytoplankton	249.3	101.0
Periphyton	731.5	75.5
Macrophytes	76.5	1.4
Total	1057.3	177.9

Source: Wetzel (1964)

large (44 ha) shallow saline lake in California (Table 11.1). In the shallow area, periphyton production per unit area exceeded phytoplankton and macrophyte productivity. But overall, phytoplankton was greatest because of the larger pelagic area. This is probably typical of most shallow lakes. Although not the largest in most cases, the periphyton contribution may still be significant. There are other examples showing that the contribution of periphyton to production in lentic waters can range from 1% in an oligotrophic lake without macrophytes to 62% (the dominant producer) in a shallow, rapidly flushed lake where phytoplankton production is controlled by washout (Wetzel, 1975, p. 414).

Streams have often been considered less productive than lakes. However, McConnell and Sigler (1959) studied a 32-ha area in the upper canyon of the Logan River, Utah, USA which is a beautiful, fast-moving, productive trout stream. They measured periphyton productivity at $1020\ mg\ C\ m^{-2}\ day^{-1}$ for that year. Considering that $\approx 200\ mg\ C\ m^{-2}\ day^{-1}$ annually is the lower limit for the meso-eutrophic threshold of a lake (Rodhe, 1969), this stream exceeded that rate by 5.0 times in the upper canyon and 20 times in the lower river. Thus, relatively shallow (less than the photic-zone depth), swift streams can produce large quantities of organic matter per unit area and still not take on the nuisance conditions of many eutrophic waters. However, blue-green algae dominated in the canyon with an average chlorophyll a of $300\ mg\ m^{-2}$ and Cladophora was prevalent downstream. Effects of enrichment in streams will be discussed later.

11.2   METHODS OF MEASUREMENT

Periphytic biomass can be collected by scraping an area from natural or artificial substrata and analysing for wet weight, dry weight, ash-free dry weight (AFDW), pigment content, or cell and/or species counts. The available time and specific purpose usually dictate the extent of analysis.

The most convenient determinations for biomass are chlorophyll *a* and AFDW. The ratio of AFDW to chlorophyll *a* is often used as a heterotrophic–autotrophic index for water quality.

Accumulation or accrual of biomass on artificial substrata, such as Plexiglas slides, glass microscope slides, wood shingles and concrete blocks, can be conveniently collected and analysed for any of the above indices. The general procedure for using artificial substrata is to stagger incubation times for pairs of slides in order to develop a curve of biomass accrual over periods of 2, 4, 6 and 8 weeks. The purpose is to determine the time to maximum accrual rate and maximum biomass. The time for maximum accrual depends on growth and loss rates. Once determined for a particular stream and season, the incubation period of maximum accrual is often used exclusively to minimize sample analysis.

The disadvantages of artificial substrata are the following: the species that normally occurs on the stream bottom may not be selected, diatoms are usually the first to colonize and more time is required for filamentous green algae, thus, maximum biomass may be underestimated; and productivity as indicated by net accrual rate is probably over-estimated because accrual starts from a bare area in contrast to that proceeding on natural stream substrata over the same time period.

Advantages of artificial substrata are as follows: they provide standardized, readily comparable substrata of known area on which organisms at each station have an equal chance for attachment and growth; a precise, comparable rate of accrual can be determined; data collection is easy; and analysis of material can provide sensitive indices of water quality (e.g. AFDW/chlorophyll *a*) and effects are integrated over time.

Productivity and respiration rates by periphyton can also be determined by light/dark 'bell jars' or a light/dark flow-through box using O_2 evolution/uptake, or ^{14}C uptake (Vollenweider, 1969b) or the O_2 curve method of Odum (1956). These methods are more time-consuming, but in some cases may be preferred due to their greater sensitivity and desire to understand the metabolic dynamics of the stream.

11.3 FACTORS AFFECTING GROWTH OF PERIPHYTIC ALGAE

11.3.1 Temperature

The same principles pertaining to the effect of temperature on phytoplankton growth (Chapter 6) also apply to periphyton, but some added examples may be useful. McIntire and Phinney (1965) showed that the short-term metabolism of periphyton in an artificial stream was greatly affected by temperature. The following changes in respiration rates were in response to abrupt temperature changes over a period of 5 h: a change from 6.5 to 16.5°C produced a change in metabolism from 41 to 132 mg

Figure 11.2. Relation between respiration rate and temperature in stream periphyton, with rates indicated for an interval of 10°C (Phinney and McIntire, 1965)

$O_2\,m^{-2}\,h^{-1}$ and change from 17.5 to 9.4°C a change in metabolism from 105 to 63 mg $O^2\,m^{-2}\,h^{-1}$.

The photosynthetic rate increased accordingly in response to a temperature rise over an 8-h period at a light intensity of 20 000 lux (1966 footcandles, fC): a change from 11.9 to 20°C produced a change in photosynthesis from 335 to 447 mg $O_2\,m^{-2}\,h^{-1}$.

There was no effect of temperature on photosynthesis below 11 000 lux (1000 fC), which is ≈28% of photosynthetically active radiation (PAR) in average full sunlight (14% of total). In this case, the Q_{10} for photosynthesis was ≈1.7 and that for respiration was ≈2.0 (Figure 11.2). The respiratory Q_{10} can be calculated as follows:

$$Q_{10} = \left(\frac{K_2}{K_1}\right)^{10/t_1 - t_2} = \left(\frac{110\,\text{mg}\,O_2\,m^{-2}\,hr^{-1}}{56\,\text{mg}\,O_2\,m^{-2}\,hr^{-1}}\right)^{10/10} = 1.96$$

Periphyton can thus be expected to respond rather quickly, at $Q_{10} \approx 2$, over a similar range in temperature. However, Q_{10} may not be the most appropriate model to represent temperature response (see Chapter 6). At low light intensities an increase in temperature would cause a disproportionate increase in respiration over that of photosynthesis and net production would decrease in the short term.

What can be expected in the long term from a temperature increase? As with the plankton, periphyton community composition would probably change gradually towards a dominance by blue-green algae and lowered diversity. Coutant (1966) showed such a trend in some of the first results on the effects from a heated (power-plant) discharge. The study was done in the Delaware River where temperature at times reached 40°C (104°F). When the temperature exceeded about 30°C (86°F), blue-green algae were favoured.

Figure 11.3. Results of periphyton collection from artificial substrata over a 7-day period ending 4 June 1959, in the Delaware River at Martin's Creek Plant (after Trembly, 1960; Coutant, 1966)

The shifts in community composition and diversity in response to heated water are shown in Figures 11.3 and 11.4. The increased temperature would also likely result in greater light-saturated rates of photosynthesis and net production. In support of that, Coutant (1966) noted that periphytic biomass was 8-times greater and chlorophyll *a* was 2.5-times greater in a heated channel (38–40°C) compared to ambient levels in the Columbia River, where the maximum daily mean reached 21°C.

11.3.2   Light

The periphytic algal community can also be expected to respond to light in a fashion similar to the phytoplankton (see Chapter 6). However, depth of the photic zone and depth of mixing are not critical factors in shallow, fast-flowing streams. Rather, there is the question of too much light and whether periphyton can adapt to nearly full sunlight conditions. Adaptation in the periphytic algae does occur and the direct effect of light is modified by high and low light-adapted communities ('sun' and 'shade').

Figure 11.4. Comparison of microorganisms in warm water and normal Delaware River water at Martin's Creek Plant (after Trembly, 1960; Coutant, 1966)

This is important because shaded conditions along streams are common and the extent to which adaptation can occur determines how much of a possible increase in enrichment could effectively be utilized. Moreover, shade is often removed and the effect of enrichment in low/high light streams is of interest.

McIntire and Phinney (1965) showed very interesting physiological and compositional shifts in their artificial stream community in response to changing light intensities (Figure 11.5). The principal effect of shade adaptation was a slightly higher rate of photosynthesis, with increasing light, at low-light intensity, but a lower light-saturated rate for the shade-adapted community. Sun and shade adaptation was performed at $\approx 7\%$ and 3% of full sunlight (14 and 6% PAR).

The light-adapted periphyton appeared to saturate ($\approx I_c$) at $\approx 800\,\text{fC}$ (Figure 11.5), which is $\approx 11\%$ of full sunlight or $\approx 22\%$ of full PAR (McIntire and Phinney, 1965). Furthermore, the periphyton showed no inhibition up to $\approx 27\%$ full sunlight (54% PAR), suggesting tolerance over a rather wide range of intensity. Jasper and Bothwell (1986) showed that a periphytic diatom community exposed to natural light was saturated at $500\,\mu\text{E}\,\text{m}^{-2}\,\text{s}^{-1}$ from April to September and at $150\,\mu\text{E}\,\text{m}^{-2}\,\text{s}^{-1}$ from January to March; these levels correspond to 25 and 7.5% of full

Figure 11.5. Effect of light on shade-adapted (200 fC) and light-adapted (550 fC) communities composed of diatoms (D) blue-green algae (B/G), and green algae (G). The light-adapted community consisted of 46% D, 42% B/G, and 12% G. The shade-adapted community consisted of 67% D, 26% B/G, and 7% G (McIntire and Phinney, 1965)

TABLE 11.2.   Biomass accumulated in artificial stream communities held in light (550 fC) and shade (200 fC) conditions (values are maxima)

| Duration | Biomass (mg slide^{-1}) | |
	Light	Shade
180 h	140	5.4
2 wk	120	89
6.5 mo	593	–
9 mo	–	565

Source: McIntire and Phinney (1965)

sunlight. Furthermore, that community was not inhibited during summer unless intensities exceeded that of full sunlight by factors of 2 to 3. Thus, stream periphyton appear to be highly adapted to wide ranges in light intensity. However, as shall be discussed later, mat thickness may be subject to light limitation.

In spite of the overall slower growth in the McIntire and Phinney shade-adapted periphyton, biomass accrual attained levels similar to that at high light but required more time (Table 11.2). Biomass in the lighted stream reached an apparent saturated biomass level in two-thirds the time necessary for the shaded stream.

The interaction of low light and nutrient enrichment on the McIntire and Phinney periphyton is worth mentioning at this time although effects

Figure 11.6. Light- and shade-adapted periphyton community response to increased light over a gradient of CO_2 content (McIntire and Phinney, 1965)

of enrichment will be discussed more fully later. The effect of shade adaption was that nutrient enrichment did not increase the photosynthetic rate, but there was a positive response to CO_2 increase in the light-adapted community (Figure 11.6). The explanation for this phenomenon seems to be that if photosynthesis-caused CO_2 removal is great enough and the CO_2 level (or any other nutrient) becomes low enough to limit growth, the light-adapted community will remove more CO_2 and cause self-limitation. As a result, increases in CO_2 will stimulate further increases in photosythesis. Because the pH change was only 1.45, it was discounted as directly influential in that case. However, as will be shown later, situations may exist in which nutrient content is low enough to provoke limitation even in shade-adapted communities so that increased enrichment would have a stimulatory effect.

The interaction of light and temperature can have a dominant influence on the seasonal occurrence of any given species. Alteration of either factor can shift species composition due to tolerance, as was shown with limits to both factors. For example, there is interest in the extent to which light and temperature cause seasonal changes in the abundance of *Cladophora glomerata*, a nuisance filamentous green algae in Lake Erie.

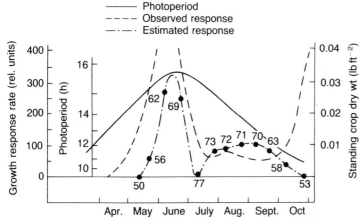

Figure 11.7. Observed biomass of *Cladophora* in Lake Erie and estimated levels based on experimental light and temperature response. Observed values for temperature given in degrees Fahrenheit (Storr and Sweeney, 1971)

For temperature, Storr and Sweeney (1971) found that its optimum growth occurred at 18°C, while growth stopped at 25°C. Predictions of growth rate from natural photoperiod and temperature alone, based on laboratory growth related to photoperiod and temperature, compared rather closely with observed biomass (Figure 11.7). In spite of sub-optimum temperature and photoperiod in October, which resulted in low predicted biomass, the observed biomass increased, probably in response to increased nutrient content following lake overturn. A similar increase in periphyton biomass in the autumn following lake turnover and SRP increase was also observed in Lake Sammamish (Porath, 1976). This was in spite of decreasing light conditions. Thus, a close fit between periphyton growth and the physical factors light and temperature may hold for a rather stable nutrient regime, but where limiting conditions occur, fluctuations in nutrient levels can dominate growth response.

11.3.3 Current velocity

Current velocity is often the most important factor controlling the biomass of stream periphyton. The frictional shear stress created on the periphyton mat at high velocity may substantially erode or scour the attached organisms from the substratum (Horner and Welch, 1981). Thus, the biomass of periphyton observed represents a balance between growth and loss due to scouring. As a result, the frequency of flood events has been shown to be an important determinant of periphytic biomass in streams (Tett *et al.*, 1978; Biggs, 1988; Biggs and Close, 1989). Several

Figure 11.8. Periphyton biomass loss rates before and immediately (15 min) after velocity increase and addition of solids (to $25\,\mu g\,l^{-1}$ TISS (total inorganic suspended solids)) in channels that receive $25\,\mu g\,SRP\,l^{-1}$. Velocities of 60 and $20\text{--}60\,cm\,s^{-1}$ refer to channels with and without addition of solids (from Horner *et al.*, 1990)

researchers have observed that $\smallsmile 50\,cm\,s^{-1}$ represents a reasonable estimate of a maximum tolerable velocity beyond which loss rates substantially reduce biomass (see Horner and Welch, 1981). Field observation in western Washington streams tend to agree with such a limit.

Observations in artificial channels, on the other hand, have shown that periphyton loss rates were very low even at velocities of $70\,cm\,s^{-1}$ (Horner *et al.*, 1983). Adaptation to high velocities ($>80\,cm\,s^{-1}$) have been observed by others (Traaen and Lindstrom, 1983). That suggests some other factors interacting with velocity to enhance scouring in natural streams. That other factor may be suspended matter, because the addition of sediment (glacial flour) to those same artificial channels resulted in enhanced scouring (Horner *et al.*, 1990) (Figure 11.8). Thus, a synergistic effect of velocity and suspended sediment is usually a normal occurrence in natural streams since sediment transport increases with velocity. In addition, sudden increases from relatively low to high velocity (e.g. 20 to $60\text{--}70\,cm\,s^{-1}$) will result in increased scouring, even without suspended sediment, if the mat had developed under stable flow (Horner *et al.*, 1990). Also, high relative changes in velocity are required to slough low biomass, but high biomass is susceptible to low relative changes (Biggs and Close, 1989). Limitation by light may also be a consequence of increased suspended sediment, but the enhanced scouring effect is prob-

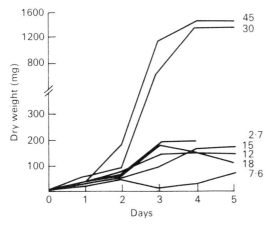

Figure 11.9. Effect of current velocity on *Sphaerotilus* growth. In all cases sucrose was added at $5\,mg\,l^{-1}$; and velocity is indicated in $cm\,s^{-1}$ (modified from Phaup and Gannon, 1967)

ably more important in shallow streams where light is usually super-saturated.

Velocity increases over the range below the $50{-}60\,cm\,s^{-1}$ level have been shown to enhance nutrient uptake and growth of periphyton. The stimulatory effect of velocity is shown by the growth of *Sphaerotilus* in artificial channels (Figure 11.9). Velocities above $\approx15\,cm\,s^{-1}$ greatly increased accrual rates. Such a critical level has been shown for filamentous green algae in artificial channels (Horner *et al.*, 1983; 1990) and was suggested earlier by Whitford (1960). As in Figure 11.9, there seemed to be little difference in accrual at velocities $>15\,cm\,s^{-1}$ in the artificial channels.

An explanation for the stimulatory effect of velocity was given by Schumacher and Whitford (1965), following observation that ^{32}P uptake by *Spirogyra* increased with velocity increases over a range of $0{-}4\,cm\,s^{-1}$ (Table 11.3). They concluded that the velocity effect was one of maximizing turbulent diffusion. The turbulence across the mat surface by current velocity maintained a maximum gradient of nutrient concentration between the ambient water and the surface of periphytic algal cells. Logically, as mat thickness increases, diffusion of nutrients (and light penetration as well) into the mat decreases at any particular velocity. Uptake of SRP per unit biomass of filamentous green algae (Horner *et al.*, 1990) and respiration rate per unit biomass of heterotrophytic slime bacteria (Quinn and McFarlane, 1989) have both been shown to decrease with increased mat thickness. Thus, a velocity increase should tend to increase diffusion and enhance greater mat thickness and, as a result, greater areal biomass.

TABLE 11.3.   Effect of velocity over a range of low rates on the uptake of ^{32}P by *Spirogyra*

Velocity (cm s^{-1})	Uptake (counts min^{-1} g^{-1} dry weight)
0	78 424
1	92 748
2	98 602
4	158 861

Source: Schumacher and Whitford (1965)

The taxonomic composition of periphytic communities is also related to velocity. McIntire (1966) showed that a felt-like community, composed largely of diatoms, developed in unenriched artificial channels at a velocity of 38 cm s^{-1}. However, at a velocity of 9 cm s^{-1}, long oscillating filaments of the green algae *Stigeoclonium*, *Oedogonium* and *Tribonema* developed. That phenomenon can be observed in natural streams as well; mats of filamentous green algae become abundant at low summer flow-rates where at higher flows, brown mats of diatoms occurred. As will be shown subsequently, taxonomic composition is also affected by enrichment and as Stanford and Ward (1988) and Biggs (1985) have suggested, the low-flow proliferation of filamentous green algae during summer low flow may be partly related to infiltration of higher-nutrient groundwater.

11.3.4   Inorganic nutrients

That nutrient concentration can limit periphyton accrual in running water, in spite of diffusion enhancement by velocity, has been shown experimentally *in situ* and in artificial channels alike. Stockner and Shortreed (1978) increased periphytic diatom accrual in stream-side channels (with ambient light) tenfold by increasing SRP from <1 to 9 μg l^{-1}. Perrin *et al.* (1987) increased periphyton biomass from <10 to >100 mg chlorophyll *a* m^{-2} by increasing SRP from a background level of 1 to 20 μg l^{-1} in another oligotrophic British Columbia stream. A lesser increase was observed at SRP addition to 15 μg l^{-1}. They also showed that the filamentous green algae *Ulothrix* and *Spirogyra* replaced diatoms in the enriched areas during the late-summer, low-flow period, indicating that dominance by filamentous green algae during low flow may be related to increased nutrients as well as low velocity. Elwood *et al.* (1981) also increased periphytic accrual and produced mats of filamentous green and blue-green algae by raising in stream SRP content above the background of 4 μg l^{-1}. Finally, proliferations of filamentous green and blue-green algae (exceeding 500 mg chlorophyll *a* m^{-2}) have been produced in artificial channels by raising SRP from ≈ 2 to $15\,\mu\mathrm{g\,l^{-1}}$ and higher (Horner et al., 1983).

Clearly, eutrophication of running water can produce large masses of periphyton, especially filamentous green algae, that can represent a nuisance condition to water quality and recreational water use (Biggs, 1985; Freeman, 1986; Biggs and Price, 1987). Mats and/or filamentous strands can break loose and clog water supply intakes, contribute to oxygen demand, alter the substratum and habitat for bottom fauna, interfere with angling and generally degrade the aesthetic environment. However, these detrimental effects have not been quantitatively related to biomass level. From 19 cases of stream enrichment reported in the literature and results of a multi-stream survey in which the percentage stream area covered by filamentous forms increased with total biomass, a lower threshold for possible nuisance conditions of $100-150\,\mathrm{mg}$ chlorophyll $a\,\mathrm{m^{-2}}$ was suggested (Horner et al., 1983; Welch et al., 1988c).

The filamentous green alga, Cladophora, known as blanketweed, has special significance with respect to nuisance conditions. It is especially obnoxious because of the long streamers that can develop, tear loose after collecting sediment and drift downstream. Some of the highest periphyton biomass levels, $1200\,\mathrm{mg}$ chlorophyll $a\,\mathrm{m^{-2}}$, have been determined for Cladophora growth in streams (Welch et al., 1990). Cladophora and other filamentous greens can also proliferate and cause nuisances in the littoral zone of lakes proximal to sources of wastewater or agricultural runoff, as is the case in Lakes Erie and Huron (see Preface). Although Cladophora could accumulate, and at greater depths probably due to greater light penetration in areas not affected by wastewater, experimental results in situ showed that enrichment with P was the principal causative factor producing nuisance proliferations (Neil and Owen, 1964). The experiment consisted of adding N, P and organic fertilizer, and combined N, P and K, to nearshore areas in Lake Huron with no existing Cladophora growth. All experimental areas were seeded with rocks covered with the alga. The experimental period was from 24 June to 25 September and final results are illustrated in Figure 11.10. Inorganic nutrient enrichment produced the greatest proliferation of biomass with P alone producing nearly as much as N, P and K together. N was obviously not the nutrient limiting the growth of Cladophora. The rate of $2.7\,\mathrm{kg\,day^{-1}}$ of P input represents a daily concentration renewal in the affected area (based on crude assumptions of depth/area) of $\approx 250\,\mu\mathrm{g\,l^{-1}}$ and a wasteload from ≈ 1000 people. In other work on Cladophora problems in the Great Lakes, the biomass near waste inputs were found to be longer, greener, denser and have higher cellular P content (Lin, 1971). Higher densities of littoral periphyton (i.e. $100-150\,\mathrm{mg}$ chlorophyll $a\,\mathrm{m^{-2}}$) have also been measured adjacent to developed areas of the

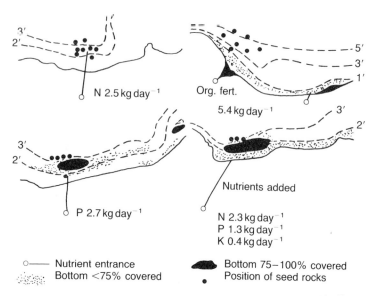

N 2.5 kg day^{-1}

Org. fert.

5.4 kg day^{-1}

P 2.7 kg day^{-1}

Nutrients added

N 2.3 kg day^{-1}
P 1.3 kg day^{-1}
K 0.4 kg day^{-1}

o—— Nutrient entrance   Bottom 75–100% covered
Bottom <75% covered   • Position of seed rocks

Figure 11.10. Effect of inorganic nutrients on *Cladophora* growth in Lake Huron. Experimental period was 24 June to 25 September (Neil and Owen, 1964)

Lake Tahoe watershed (Loeb, 1986) and in the vicinity of nutrient-rich tributaries to Lake Chelan (Jacoby *et al.*, 1990) relative to periphyton biomass (<10–20 μg chlorophyll a m^{-2}) near undistributed areas of the lakes' watersheds.

The problem of determining the critical concentration of P, beyond which nuisance biomass levels can be expected, has received considerable attention. Although there are 'critical' concentrations or guidelines for P that can be related to the trophic state and expected phytoplankton biomass in lakes, no such relationships exist for streams. Fundamentally, it is not surprising that development of guidelines for streams has been much slower. The basis for guidelines for P in lakes is analogous to conditions in continuous culture systems where the steady-state biomass of algae per unit volume of water is proportional to the inflowing nutrient concentration (or initial concentration prior to growth). Biomass per unit volume increases as the soluble nutrient per volume is converted into biomass. Thus, direct correlations between total limiting nutrient (e.g. TP) and algal biomass (e.g. chlorophyll *a*) such as those developed by Dillon and Rigler (1974b) and Jones and Bachmann (1976) are logically expected. On the other hand, periphyton biomass in running water (or even in littoral areas of lakes) accrues on a substratum surface and extracts its nutrients from the overlying water continually supplied from upstream (or groundwater sources). Therefore, the limitation for maximum biomass per area of periphyton is not limited to the mass of nutrient per volume. Rather, in that sense, one would expect that total biomass developed over a stream reach may be related to the total mass of limiting nutrient originating upstream. However, in the case of biomass per unit area at one point, it can be easily demonstrated that mass of nutrient from upstream (i.e. loading or concentration × flow rate) is many orders of magnitude greater in a very short time than is needed to produce periphytic algae per unit area at the maximum observed rate. Therefore, the pertinent unit for nutrient uptake, as well as biomass accrual, is nutrient concentration, not supply rate. Growth rate, and ultimately biomass, of periphyton is controlled by the concentration of soluble, limiting nutrient and that unit is the basis for modelling (McIntire et al., 1975; Newbold et al., 1981; Horner et al., 1983). However, determination of available nutrient concentrations in streams is complicated by algal uptake, and thus removal from the overlying water, and contributions from diffuse groundwater sources that are difficult to measure.

If biomass is in effect saturated above some nutrient concentration, then it follows that biomass should not be enhanced if nutrient concentration is raised above the growth-saturating level, which could represent the 'critical' level for running water. That is consistent with the general failure to develop meaningful causal relations between periphytic biomass and nutrient concentrations in streams; high biomass is often coincident with low soluble (and also total) nutrient concentrations, because the periphyton mat has absorbed it from the water (Jones et al., 1984; Welch et al., 1988c). That problem was avoided to some extent by Biggs and Close (1989), who correlated ($r^2 = 0.53$) mean annual periphyton biomass with average SRP content over a 13-month period.

Growth-rate saturating concentrations have been found to be very low for thin films of diatoms in artificial channels ($1-4\,\mu g\,l^{-1}$ SRP) (Bothwell, 1985; 1988) as well as for *Cladophora* ($5\,\mu g\,l^{-1}$ SRP) (Freeman, 1986). Such low growth-saturating levels make control of biomass by manipulation of in-stream nutrient concentration virtually impossible and cast doubt on experimental observation of biomass increase in response to increased SRP concentrations above background levels, which were similar to the above growth-saturating levels (see earlier results of enrichment).

The apparent explanation for increased biomass accrual from enrichment of waters with low, but rate-saturating concentrations, is due to mat-thickness limitation. In order to develop a thick mat, and hence a high periphyton biomass, increased nutrient concentration is apparently necessary to offset the diffusion limitation created by increased mat thickness. In this regard, Bothwell (1989) has shown that while growth rate is saturated at $1-2\,\mu g\,l^{-1}$ SRP, maximum biomass of diatoms ($150\,\text{mg}$ chlorophyll $a\,\text{m}^{-2}$) after some accrual time period, continues to

Figure 11.11. Maximum periphyton biomass on natural rocks in artificial channels during three separate experimental periods (symbols) of about three weeks each, at an average light intensity of $194\,\mu\mathrm{E\,m^{-2}s^{-1}}$, an average temperature of 19°C and current velocities of $20\,\mathrm{cm\,s^{-1}}$ ($\blacksquare$, $\blacktriangle$) and $40\,\mathrm{cm\,s^{-1}}$ ($\bullet$). The line is calculated from the steady-state model with $K_s = 5\,\mu\mathrm{g\,l^{-1}}$ SRP and $K_1 = 1.46 + 0.276$ ($r^2 = 0.85$) (Walton, 1990)

increase up to $25\,\mu\mathrm{g\,l^{-1}}$. A maximum biomass of diatoms and blue-green algae (*Phormidium*) of almost 1000 mg chlorophyll $a\,\mathrm{m^{-2}}$ in three weeks required an in-channel SRP of nearly $20\,\mu\mathrm{g\,l^{-1}}$ (Walton, 1990) (Figure 11.11). In the same channel system, a mat dominated by a filamentous green alga (*Mougeotia*) required SRP of $\approx 7\,\mu\mathrm{g\,l^{-1}}$ to reach a maximum 2-week biomass of 350 mg chlorophyll $a\,\mathrm{m^{-2}}$ (Horner *et al.*, 1990). Thus, the nutrient concentration that limits maximum biomass, because of mat-diffusion limitation, would seem to be the most appropriate for management of stream quality. Such a maximum-biomass limiting value for filamentous forms may, therefore, range from about 7 to $20\,\mu\mathrm{g\,l^{-1}}$ SRP, probably determined as an annual mean.

Maximum biomass limitation by SRP is illustrated in Figure 11.12 by means of a kinetic-based model (Horner *et al.*, 1983; Welch *et al.*, 1989). The model was calibrated against experimental data from artificial channels in which SRP and velocity were varied and therefore represent potential stream biomass without loss due to grazing. The model formulation is as follows:

$$B = B_{\max} - \frac{K_2 V^\theta}{K_1 \mu L (K_f + K_{f0})}\left[1 - e^{-[K_1 \mu L (K_f + K_{f0})]t} \right]$$

where B is periphyton biomass expressed as mg chlorophyll $a\,\mathrm{m^{-2}}$, $B_{\max}$ is the maximum possible, μ is the Michaelis–Menton uptake rate with $\mu_{\max} = 0.22e^{t/10}$ (Eppley, 1972), L is a light reduction factor, assumed to be 1.0 for stream conditions (0.755 in laboratory due to 10% of full sunlight), while K_1, K_2 and θ are empirical constants with values of 1.2 (or a function of SRP), 0.3 and 0.45. V is current velocity in $\mathrm{cm\,s^{-1}}$, K_{f0} is

Figure 11.12. Predicted and observed biomass as a function of mean in-stream SRP in the lower Spokane River, WA. Observed biomass was from natural (crosses) and artificial (closed squares) substrates. The upper and lower bounds on predicted biomass represent uncertainties due to variation in temperature, velocity and accumulation period. Open circles represent mean growing season SRP and periphyton biomass on natural substrata for six other streams (Welch *et al.*, 1989b)

non-turbulent mass transfer at $0.0094\,\mathrm{cm\,s^{-1}}$, and K_f is turbulent mass transfer equal to $(DV/\pi 1)^{0.5}$ with $D = 1.5 \times 10^{-5}\,\mathrm{cm^2\,s^{-1}}$ and t is time in days. B_{max} attained on Plexiglass slides was $560\,\mathrm{mg\,m^{-2}}$ chlorophyll *a* (Horner *et al.*, 1983) and $1000\,\mathrm{mg\,m^{-2}}$ on natural rocks (Walton, 1990).

The comparison of expected versus observed biomass in a natural stream showed two main points: expected maximum biomass saturation is $\approx 10\,\mu\mathrm{g\,l^{-1}}$ SRP; and most observed stream biomass (and filamentous green algae were dominant), whether on natural or artificial substrata, was much less than that expected. One explanation for the much lower than expected stream biomass is a loss due to grazing by macroinvertebrates.

11.4 GRAZING

The removal of periphyton biomass by macroinvertebrate grazing may be the most important control on periphyton in streams of moderate velocity even if enrichment occurs. Several researchers have demonstrated that snails and caddis flies are capable of holding periphyton to 5–50% of the ungrazed biomass (Lamberti and Resh, 1983; Jacoby, 1985, 1987; Lamberti *et al.*, 1987; McCormick and Stevenson, 1989, Walton, 1990). A large-cased caddis fly, *Dicosmoecus gilvipes* (2–4 cm) has been used experimentally by several investigators and found to be a highly effective grazer, producing some of the highest removal rates observed, even under enriched conditions. The organism reduced periphyton (diatoms) biomass by 80% *in situ* at natural densities ($41 \pm 8\,\mathrm{m^{-2}}$) in an unenriched

Cascade Mountain foothill stream (Jacoby, 1987) In laboratory channels, biomass (composed of diatoms and filamentous forms) was reduced to 5% of the unenriched level at a density of 200 organisms m^{-2} (Lamberti et al., 1987) while at 100 m^{-2}, the caddis fly reduced biomass of primarily a filamentous blue-green and diatom to 7% of the ungrazed level in unenriched and 12% in enriched (SRP 25 μg l^{-1}) treatments (Walton, 1990). Small mayflies, such as Nixe rosea (Jacoby, 1987) and Centroptilum elsa (Lamberti et al., 1987), consumed periphyton but were not effective in reducing periphyton biomass of either diatoms or filamentous forms. Snails, on the other hand, have been shown to be effective grazers (Jacoby, 1985), but less so than Dicosmoecus (Lamberti et al., 1987).

The growing evidence on the effectiveness of grazers in reducing periphytic biomass of both diatoms and filamentous green and blue-green algae, may help to explain the frequently observed poor relationship between biomass and nutrient concentration. Considering the relatively low, biomass-saturating-SRP concentration (in P-limited streams), the occurrence of periphytic biomass levels of the order of 1000 mg chlorophyll a m^{-2} may be more related to poor habitat conditions for macroinvertebrate grazers than to enrichment, regardless of how nutrient availability is expressed.

Results from seven New Zealand streams receiving point-source enrichment suggest that grazing was a more important determinant of periphyton biomass than were nutrients (Welch et al., 1992). All streams received treated sewage effluent; five from oxidation ponds, one from a trickling filter plant and one from a combination of primary and secondary treatment of sewage, dairy and slaughterhouse wastewaters. SRP concentrations only doubled or tripled from low upstream levels (5 μg l^{-1}) in two streams, but increased to well over 100 μg l^{-1} in all others. From model predictions (see above), most observed biomass levels were much below that expected from SRP, temperature and velocity. Rather, the data suggested a grazer-threshold effect, grazer densities $>300 (0.1$ m$)^{-2}$ being associated with periphyton biomass <50 mg chlorophyll a m^{-2}. The highest periphyton biomasses ($\approx$1200 mg chlorophyll a m^{-2}, Cladophora dominated) occurred where grazer densities were very low. In one case a thick mat of primarily Cladophora developed on flat bedrock throughout the stream in a virtual absence of grazers due to the limited habitat.

Evaluating the effect on enrichment on stream periphytic algae should probably begin with an assessment of substratum. Rubble bottom streams with relatively little fine sediment clogging up interstices may be able to assimilate greater inputs of enrichment than streams with relatively little clean and secluded surface area as macroinvertebrate habitat. This may point to management strategies of preventing flow variation and sediment input to preserve a grazer-rich substratum. Such an approach may be more effective than putting the effort into controlling nutrient inputs, especially since biomass-saturation levels of SRP are so low ($<20\,\mu g\,l^{-1}$) and SRP levels in some cases are $>1000\mu g\,l^{-1}$.

11.5 ORGANIC NUTRIENTS

Dissolved organic nutrients favour bacterial slimes dominated by filamentous bacteria (*Sphaerotilus*) and fungi. Bacterial slimes are aggregates of *Sphaerotilus*, *Zooglea* and fungi. *Sphaerotilus* may have a different appearance morphologically under varying environmental conditions. For example, *S. natans* may appear similar to *Cladothrix dichotoma* and in turn to *Leptothrix ochracea*. All these organisms are obligate organotrophs. The can create nuisance slime conditions, usually in running water, downstream from a source of dissolved organic substances. They have created nuisance problems near untreated waste discharges from pulp mills, dairy plants, landfills (leachate) and slaughterhoues as well as domestic sewage.

The culture-determined nutrient requirements for *Sphaerotilus* are monosaccharide and disaccharide sugars at a concentration of 20 to $50\,mg\,l^{-1}$ as well as organic acids, amino acids and inorganic N and P (Phaup and Gannon, 1967). The presence or absence of *Sphaerotilus*, however, seems definitely related to the presence of sugars and low molecular weight BOD (Curtis and Harrington, 1971; Quinn and McFarlane, 1989).

As with algae, the response of heterotrophic slimes has also been studied in experimental channels. Phaup and Gannon (1967) determined the effects of sucrose and velocity on biomass accrual and community type in experimental, 210-m long, outdoor channels. The periphyton that developed on strings held in the running river water without additional enrichment included *S. natans*, the algae *Melosira*, *Navicula*, *Cosmarium*, *Euglena*, the protozoans *Tetrahymena*, *Colpedium*, amoeba, and some insects: chironomids (midges) and simulids (blackflies). Sucrose enrichment increased biomass accumulation on the string substrata. Growth was maximum within 30 h with $5\,mg\,l^{-1}$ sucrose and increased with current velocity from 18 to $45\,cm\,s^{-1}$, at a temperature of 20 to 28°C as noted earlier (Figure 11.9). The river water contained $500\,\mu g\,l^{-1}$ N and $1\,\mu g\,l^{-1}$ SRP. The maximum biomass accrued was not obtained at temperatures below 17°C, because *S. natans* was replaced by another filamentous bacterium that was much less prolific at higher temperature. This finding has a bearing on the winter phenomenon in which *S. natans* often covers more stream distance than in summer. The cause for this is related to temperature and metabolic rate. At high temperature, utilization of nutrient and growth are also high, resulting in the quick depletion of nutrients downstream. Consequently, the minimum required organic content occurs over a shorter distance than in winter when temperatures are low.

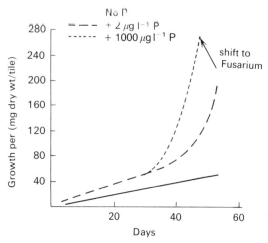

Figure 11.13. Effect of phosphate on slime community growth. Channels contained $10\,\text{mg}\,l^{-1}$ sucrose and $1.9\,\text{mg}\,l^{-1}$ NH$_4$Cl (Ormerod *et al.*, 1966)

In a similar experimental channel study, Ormerod *et al.* (1966) showed that a succession of species within the bacterial slime community was affected by a combination of pulp-mill wastewater and inorganic nutrients. Bacterial slimes replaced algae when spent sulphite liquor (SSL) was added, but algae was only partly replaced with sewage addition, because apparently the additional inorganic nutrients favoured algae. Two streams were studied with the same concentration of SSL ($0.1\,\text{ml}\,l^{-1}$). The following organisms dominated under two different stream conditions: softwater stream, *Fusarium aquaeductum*; hard-water stream, *Sphaerotilus natans*. Under variable temperature with velocity of $5\,\text{cms}^{-1}$, the following change in the periphyton occurred: SSL addition at $0.1\,\text{ml}\,l^{-1}$, change from algae to *Sphaerotilus*; SSL + sewage at $0.5-1.0\,\text{ml}\,l^{-1}$, algae increased with some *Fusarium*. At a constant temperature of $12°C$ and a water velocity of $15\,\text{cms}^{-1}$, the following community changes were observed (Figure 11.13): sucrose addition at $10\,\text{mg}\,l^{-1}$, change from algae to *Sphaerotilus* and *Zooglea*; sucrose ($10\,\text{mg}\,l^{-1}$) + SRP at $2\,\mu\text{g}\,l^{-1}$ and up to $1000\,\mu\text{g}\,l^{-1}$, change from algae to *Sphaerotilus* and then to *Fusarium*.

Although an increase in temperature resulted in an increase in growth rate (Figure 11.14), the pattern of ultimate dominance did not change – only the rate of change. Regardless of temperature, the addition of P favoured dominance by *Fusarium*. This is most interesting because it shows that although heterotrophic microorganisms are favoured by dissolved organic matter, the species actually dominating can be strongly influenced by inorganic nutrients. The addition of other nutrients, such as

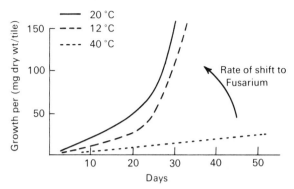

Figure 11.14. Effect of temperature on slime community growth. Channels contained $10\,mg\,l^{-1}$ sucrose, $0.19\,mg\;NH_4Cl\,l^{-1}$ and $0.34\,mg\;K_2HPO_4\,l^{-1}$ (Ormerod *et al.*, 1966)

N, K, Na, Mg, Ca, Fe and vitamin B_{12}, had no effect upon succession.

Nuisance biomass levels of *Sphaerotilus* were also produced in a natural stream (Berry Creek, Oregan, USA) by adding sucrose at $1-4\,mg\,l^{-1}$ (Warren *et al.*, 1964). Effects of those nuisance biomass levels on fish and invertebrates will be discussed later.

Although 'outbreaks' of nuisance biomass levels of heterotrophic slimes have been clearly associated with the discharge of organic waste-water from a variety of activities (Gray and Hunter, 1985), establishing a critical concentration of organic matter has been rather difficult. The usual measures of organic matter, such as BOD_5, COD and DOC (dissolved organic carbon), have not always been useful. Curtis and Harrington (1971) stressed that heavy slime outbreaks have been noted where very low BOD concentrations occurred, as well as no slime development in spite of existing high BODs. They suggested that the former could have been due to the presence of rapidly degradable substances, such as sugars, whereas the latter may have been due to inhibitors.

Quinn and McFarlane (1989) have recently provided some practical insight to the problem. They showed that low molecular weight compounds ($M_r < 1000$) determined by COD and BOD_5 in water passed through an ultra-filtration membrane were the key factors determining respiration rates of heterotrophic slimes in artificial channels. They further showed that restricting BOD_5 to $5\,mg\,l^{-1}$ eliminated nuisance biomass levels downstream of sources if the wastewater were from domestic sewage or a slaughterhouse. However, dairy wastewater contained a higher proportion of low M_r organics and if that source represented half the BOD load entering the river, the critical BOD_5 needed to eliminate nuisance biomass was $2-3\,mg\,l^{-1}$ (Quinn and McFarlane, 1989).

Figure 11.15. Biomass of periphyton in two artificial streams, one receiving inorganic and the other organic (yeast extract) N. Diatoms and green algae were dominant at the start of the experiment in both streams, bacteria peaked early in stream with organic N, but at the end, green algae dominated the inorganic stream and blue-greens dominated the organic stream (Ehrlich and Slack, 1969)

The concentration of total dissolved organic matter is, therefore, of less importance than its composition. If the DOC is diverse, with few compounds of low M_r, nuisance conditions may not develop. For example, Cummins *et al.* (1972) added leaf leachate, a natural source of allochthonous organic enrichment, to an artificial stream at concentrations up to $30 \, mg \, l^{-1}$ DOC, which is 10-times the natural level and similar to that received by heavily polluted streams. In 10 days, 85% of the DOC was utilized and no change in the diversity of macroinvertebrates resulted. Moreover, a diverse fungal flora developed without dominance by *Sphaerotilus*, which probably would have occurred if the $30 \, mg \, l^{-1}$ had been DOC as sucrose or low-M_r organics.

Other indications that a diverse DOC source will not lead to a dominance by nuisance heterotrophic slimes was apparent in experiments by Ehrlich and Slack (1969). Figure 11.15 shows the accumulation of periphyton biomass in two artificial streams; one received inorganic N and the other a yeast extract of $50 \, mg \, l^{-1}$. The rate of increase of algal biomass was greatest with the addition of inorganic N because it was readily available and mineralization was unnecessary. With the yeast extract, an intermediate step was required for bacterial mineralization, which then led to algae. Although blue-green algae dominated in the stream receiving organic N, filamentous heterotrophic slime bacteria like *Sphaerotilus* did not.

Based on the above evidence from artificial-stream studies, one can conclude that the principal cause for nuisance levels of heterotrophic slime growth (mainly *Sphaerotilus*) is the presence of low-M_r, highly degradable, dissolved organic matter, particularly sugars.

1. Wastes containing sugar in the range of $1-30\,\text{mg}\,\text{l}^{-1}$ should lead to nuisance heterotrophic slimes dominated by *Sphaerotilus*. BOD_5 concentrations causing nuisance slimes could be as low as $2-3\,\text{mg}\,\text{l}^{-1}$ if a high proportion of the wastewater is low M_r organics.

2. Sugars and/or low M_r organics plus inorganic nutrient (mainly P) enrichment should lead first to *Sphaerotilus* then to a domination by fungi.

3. Diverse DOC compounds even up to $30\,\text{mg}\,\text{l}^{-1}$ (such as natural leaf leachate) should lead to a variety of nonfilamentous (and therefore non-nuisance) heterotrophs and algae.

11.6 PERIPHYTON COMMUNITY CHANGE AS AN INDEX OF WASTE TYPE

Although organic wastewater, such as sewage, can result in an increase in periphyton biomass with mono-species dominance (e.g. *Sphaerotilus* or *Cladophora*), certain species are typically characteristic of various levels of organic matter, resulting in a downstream community change or succession during natural degradation. Fjerdingstad (1964) has described such a successional response of the periphyton community to a gradient of sewage enrichment (Table 11.4). This successional response is similar to the Saprobien System first described by Kolkwitz and Marsson (1908) and has often been characterized as representing the process of 'stream purification'. However, the chemical concentrations are only relative. For example, according to the previous discussion on BOD and filamentous bacteria, problems could still be present in the recovery zone even if BOD is $<10\,\text{mg}\,\text{l}^{-1}$.

The length or presence of these zones in a stream depends upon flow rate (dilution), waste concentration and temperature. These factors, in turn, determine the dispersed concentration, its contact time and, therefore, the uptake rate of organic matter (i.e. BOD). A high flow rate and low temperature results in a low concentration and low uptake rate, so that a lesser effect (but possibly still a nuisance biomass) is usually stretched over a longer stream distance. With a low flow and high temperature, the uptake rate is greater so the BOD is depleted faster and the adverse effect occurs over a shorter stream distance.

Sladeckova and Sladecek (1963) have described the succession of periphyton in a series of Czechoslovakian reservoirs polluted with sugar-beet waste and sewage. The combined effect of these wastes showed

TABLE 11.4. Water quality characteristics and the associated response of the microorganism community to organic waste

Zone	Chemical	Biological
Oligosaprobity (clean water)	BOD < 3 mg l^{-1} O$_2$ high Mineralization of organic matter is complete	Diatoms diverse Filamentous green algae present Filamentous bacteria scarce Ciliated protozoans scarce
Polysaprobity (septic)	H$_2$S high O$_2$ low NH$_3$ high	Algae present but not abundant Protozoa absent Bacteria abundant–faecal, saprobic
α Mesosaprobity (polluted)	Amino acids high H$_2$S low–none O$_2$ < 50% saturated BOD > 10 mg l^{-1}	Algae scarce–some tolerant forms[a] Filamentous bacteria abundant[b] Ciliated protozoans abundant[c] Few species–biomass great
β Mesosaprobity (recovery)	NO$_3$ > NO$_2$ > NH$_3$ O$_2$ > 50% saturated BOD < 10 mg l^{-1}	Diatoms not diverse–biomass great[d] Ciliated protozoans only present[e] Blue-green algae abundant[f] Filamentous green algae abundant[g]
Oligosaprobity (clean water)	Stream recovered or 'purified'	

[a] *Gomphonema, Nitzschia, Oscillatoria, Phormidium, Stigeoclonium* often dominate. [b] *Sphaerotilus, Zooglea, Beggiatoa.* [c] *Colpidium, Glaucoma, Paramecium, Carchesium, Vorticella.* [d] *Melosira, Gomphonema, Nitzschia, Cocconeis.* [e] *Stentor.* [f] *Phormidium, Oscillatoria.* [g] *Cladophora, Stigeoclonium, Ulothrix.* Source: Fjerdingstad (1964); Sládečková and Sládeček (1963)

typical zones, which are illustrated (Figure 11.16). These zones are analogous to those in streams shown in Table 11.4. Three representative periphyton communities in the reservoirs, receiving various amounts and types of waste effluents (sugar beet and sewage), were collected on glass slides. The communities indicate different degrees of water quality change: community 51 = relatively 'clean' water conditions (oligosaprobity); community 52 = 'recovery' conditions (beta-mesosaprobity); community 53 = 'polluted' conditions (alpha-mesosaprobity).

Note the lack of diatoms and preponderance of ciliated protozonas and filamentous bacteria in the 'polluted' water community (community 53). According to this system of evaluating water-quality change, the various species are grouped according to their position with respect to distance or travel time from a waste outfall and, therefore, to the amount of 'self-purification' that has occurred. The occurrence of these groups of species or communities is considered indicative of various degrees of organic 'pollution'.

Another useful method of evaluating the effects of organic wastewater is the heterotrophic or autotrophic index, HI or AI, respectively. AI is

Figure 11.16. Illustrated pattern of periphyton response to waste in three Czechoslovakian reservoirs. See text for explanation (from Sládečková and Sládeček, 1963)

more appropriate from the standpoint of unit values and has been the more accepted terminology (Collins and Weber, 1978). The index is determined as the ratio of AFDW:Chl*a* in the periphyton collected on artificial substrata and indicates the degree of effect of organic waste. AI increases in proportion to the concentration of organic matter (or BOD), because heterotrophs occupy a greater portion of the biomass as organic waste increases.

Figure 11.17. Changes in periphyton biomass (Chlorophyll *a*, AFDW) before and after diversion of slaughterhouse wastewater from South Branch, Canterbury, New Zealand, Diversion was complete before August. Dashed line is recommended upper limit of AI for clean water (see text). •, Control; ▲, affected section (Biggs, 1989)

Collins and Weber (1978) suggested a value of 400 as the upper limit indicating clean water conditions. An interesting example of the effectiveness of the index was shown by Biggs (1989) for a spring-fed New Zealand river. Upstream and downstream stations were monitored before and after the diversion of slaughterhouse wastewater. Figure 11.17 illustrates the quick recovery (to below 400) of the periphyton community following the beginning of volume reduction of the discharge in May 1986 and complete diversion in August 1986. Prior to diversion, the AI downstream was >10-fold higher than the upstream control, but shortly after diversion values were similar. Biggs also found AI to be highly correlated with BOD_5 ($r^2 = 0.86$). Heterotrophs were dominated by *Sphaerotilus* and *Zooglea*.

11.7 EFFECTS OF TOXICANTS

Toxicants may affect periphyton both positively and negatively, but periphyton is not usually investigated as an indicator of toxicity. Total biomass would normally be expected to decline downstream from the input of an inhibitor, but the tolerant species would become domi-

TABLE 11.5.   The effects of Cu waste on stream periphyton collected on artificial slides

Upstream station	Periphyton density $1000\,mm^{-2}\,(3\,wk)^{-1}$
Cu^{2+} low	*Stigeoclonium, Nitzschia, Gomphonema,*
	Chaemosiphon, and *Cocconeis* were most abundant
Downstream from waste	Density $150–200\,mm^{-2}\,(3\,wk)^{-1}$
Cu^{2+} $1.0\,mg\,l^{-1}$	*Chlorococcum, Achnanthes*

Source: Hynes (1960)

nant and possibly increase in abundance. The increase in some species may result from nutrients released from dead animals. Grazers may be eliminated, which should allow a rapid biomass increase in algal periphyton as a result of no grazing once the toxicant has dispersed. Insecticides would probably produce such an effect, because their negative effect on algae would be minimal. For example, about a week after spraying with DDT to control forest insects, forest streams have become intensely green. The positive effect of insecticide on periphytic biomass has also been shown in artificial channel experiments (Eichenberger and Schlatter, 1978).

Metal toxicants have a more detrimental effect on periphyton. The addition of Zn^{2+} to artificial streams showed an increased reduction in the number of algal species over a concentration range of $0–9\,mg\,l^{-1}$ (Williams and Mount, 1965). Noteworthy in that experiment was the elimination of *Cladophora* at the lowest concentration tested, $1\,mg\,l^{-1}$. In this instance water was taken from a holding pond that also supplied a high concentration of particulate food; consequently, *Sphaerotilus* and fungal slime organisms increased as Zn^{2+} increased and algal species decreased. Without the dead organisms, little increase in decomposer biomass would have occurred.

Cu^{2+} is another heavy metal to which algae are very sensitive. $CuSO_4$ is a common algicide. Hynes (1960, p. 81) cited work by Butcher showing the effects of Cu^{2+} in industrial waste in an English stream (Table 11.5). Recovery of species diversity and biomass ($33\,000\,mm^{-2}$) occurred 8 km downstream from the point of discharge.

Diatom diversity was shown to be a sensitive indicator of toxic effects of Zu and Cu in a Montana stream (Prickly Pear Creek) which received leachate from abandoned mine tailing ponds (Kinney, personal communication; Miller *et al.*, 1983). Figure 11.18 shows the marked reduction in diatom taxa at the two downstream stations where Zu and Cu concentrations were elevated most above US EPA criteria determined for fish. Further downstream, as metal concentrations dropped, taxa increased to near the level observed at the upstream control station.

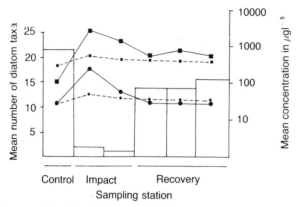

Figure 11.18. Mean diatom taxa related to stream concentrations of total Zn and Cu and EPA acute criteria for those metals in Prickly Pear Creek, Montana in 1980 (data from Miller *et al.*, 1983). ■—■, Zn concentration; ■---■, Zn criteria; ●—●, Cu concentration; ●---●, Cu criteria

11.8 NUISANCE

Under proper conditions of light and temperature, which usually sets limits on nutrient-saturated growth rate, enrichment with either inorganic or organic nutrients promotes the dominance of certain prolific periphyton species and thus results in an increase in production and possibly a nuisance biomass. Whether such a biomass is considered to be an undesirable nuisance condition may depend on how the users of the stream are affected. Some of the adverse effects that may be caused by high periphyton biomass are as follows.

1. A secondary BOD is created that can deplete O_2 downstream as the filaments break off, float away and decompose.

2. Water intakes may be clogged with floating clumps of filaments.

3. Undesirable taste and odours can be created if the affected stream is used for a water supply.

4. Dense mats can cover the bottom, restricting intragravel water flow, which can inhibit fish reproduction.

5. Dense mats and long stringy filaments can interfere with fishing, recreation and aesthetic enjoyment.

6. Dense mats and filamentous clumps reduce the habitat for benthic animals and can cause direct physical damage.

The relative importance of these and other effects were determined for 103 cases of heterotrophic slime outbreaks in Ireland (Gray and Hunter, 1985) (Table 11.6). Although these effects on stream uses seem to prove

Table 11.6. The relative importance of various adverse effects of slime growths in Irish rivers ($n = 103$)

Effect	%
Appearance and amenity	86.4
Smell and deoxygenation	46.6
Damage to fish	37.9
Sloughed flocs	24.3
None	12.6
Smell only	11.7
Deoxygenation only	9.7
Habitat destroyed	3.9
Spawning ground smothered	2.9
Possible public health risk	2.9
Blockage of flow	1.0
Agricultural supplies affected	1.0
Public supplies affected	1.0
Slime growth on fauna	1.0

From Gray and Hunter, (1985)

that excessive biomass levels do in fact represent a nuisance, the increased production could be interpreted as a possible benefit under some circumstance if it leads to increased production of fish food (invertebrates) and fish. Examples of benefits were illustrated in Berry Creek, Oregon, where sucrose added at concentrations from $1-4\,mg\,l^{-1}$ produced mats of *Sphaerotilus* and dense populations of midges (Warren *et al.*, 1964). The diversity of both periphytic microscopic and macroscopic organisms decreased, while mats of heterotrophic slimes, which may not be aesthetically pleasing, and fish production nonetheless increased. Table 11.7 shows the results of annual biomass levels of benthos, fish-food consumption and fish production in energy units in enriched and unenriched sections of the stream.

The production of cut-throat trout appears to have been enhanced in this case from the enrichment, in spite of other adverse effects of the mats of *Sphaerotilus* on macroinvertebrate diversity. The insect biomass was composed mostly of chironomids. From these results one could argue that the benefit of increased fish production met the goal for sound water-quality management; the ecosystem was used but not abused. However, the following points must be considered in similar cases.

1. The knowledge base necessary for 'fine tuning' the discharge of wastewater to allow 'benefits' without 'detriments' for a variety of ecosystems receiving a variety of wastes.

TABLE 11.7.   Insect biomass, fish-food consumption, and fish (cut-throat trout) production in artificially enriched (1–4 mg l^{-1} sucrose) and unenriched sections of Berry Creek, Oregon

	Annual values (kcal m^{-2})				
	Unenriched (U) dark	Unenriched (U) light	Enriched (E) dark	Enriched (E) light	E:U
Insect biomass					
1960–61	2.19	5.03	20.4	12.2	4.5
Fish-food consumption					
1961	8.38	6.06	20.15	15.64	
1962	9.46	8.09	19.07	21.55	
1963	7.45	8.36	13.05	9.00	
Mean	8.43	7.50	17.42	15.40	2.1
Fish production					
1961	−0.21	0.01	2.13	2.20	
1962	0.49	−0.07	3.70	4.80	
1963	0.58	0.99	2.51	1.65	
Mean	0.29	0.31	2.78	2.88	6.3

Values are totals for a portion of each year.   *Source:* modified from Warren *et al.* (1964)

2. The costs and problems of monitoring stream conditions to provide data for fine tuning.

3. The instability risks of intense management.

Considering the last point, even if a fine-tuned system is achieved, it may be highly susceptible to environmental changes. For example, trout are known to grow very fast in highly productive areas, but with associated temperature, pH, NH_3 and O_2 conditions that may become marginal at times. For Berry Creek, slight changes in weather patterns, such as cloudiness and/or high temperature, could cause total elimination of the highly prized, fat trout. Moreover, the trout were planted as adults and whether or not reproduction was successful in the enriched portions of the stream was not determined. For example, Smith *et al.* (1965) showed interference from *Sphaerotilus* on reproduction by wall-eye pike. Thus, total cost may be less if waste inputs were minimized rather than attempting the objective of use without abuse.

12
BENTHIC MACROINVERTEBRATES

The benthic macroinvertebrates of running waters include many orders of organisms that represent an interesting, usually unappreciated part of the aquatic environment. Benthic fauna in standing waters, which will be briefly discussed in this chapter, are relatively unnoticed as well. The mere existence of large populations of strange-looking nymphs, naiads and larvae of insects within the rubble and gravelly substrata of streams is fascinating. That so few people know of their existence is a pity. They have an important function in the aquatic food web and, as such, the structural composition of the macroinvertebrates will reflect changes in the quality or quantity of energy input.

The macroinvertebrates include grazers, detrital feeders and predators. They process and utilize the energy entering streams from either autochthonous periphyton production or from allochthonous sources, such as leaves, needles and other particulate matter from the forest, or organic wastes from humans (e.g. sewage) or other animals in the watershed. Macroinvertebrate species can be separated into feeding types, such as large-particle detritivores (shredders), small-particle detritivores (collectors and selectors), grazers (periphyton scrapers) and predators. Furthermore, the representation of the community among these groups is distinctive, as streams change naturally from upstream, low-order and usually heterotrophic, to autotrophic in mid-regions and ultimately back to heterotrophic, high-order lower downstream (Cummins, 1974). In lakes or the sea they are largely dependent upon the autochthonous or allochthonous production that is sedimented to the bottom. A diverse community of macroinvertebrates is instrumental in stream purification – the processsing of organic matter from either human or natural sources that leads ultimately to CO_2, water and heat. As noted in the preface, the efficiency with which that process proceeds depends to a large extent upon the diversity of the community – the more individual niches that are occupied, the quicker and more completely will be the conversion process. Overloading streams with organic enrichment or the introduction of toxic substances will tend to decrease the diversity and thus the efficiency of 'purification'.

The macroinvertebrates of fresh water are dominated by the insects and their diversity is greatest in running water. The orders Ephemeroptera (mayflies), Plecoptera (stone flies), Trichoptera (caddis flies), Diptera (true flies) and Odonata (dragonflies and damselflies) usually make up the majority of the biomass. Because of the abundance of species – 50 to 100 is not unusual in a running water environment – their identification to the species level is difficult. Other important groups in fresh water are the Mollusca (snails and clams), the Annelida (worms and leeches) and the Crustacea (scuds, sow bugs and crayfish). In the marine environment the latter groups represent the community to the practical exclusion of the insects. In brackish water environments there is considerable overlap and some of the more tolerant freshwater species of insects can be relatively important. See Appendix A for a classification system for freshwater macroinvertebrates.

The principal purpose in this chapter is to treat the macroinvertebrates as indicators of the kind and relative magnitude of waste entering aquatic systems. As such, they offer advantages not available from any other group. They are sedentary and relatively easy to sample compared to the highly motile fishes, and because their longevity is greater, fluctuations in biomass and species composition are less pronounced than in plankton. Therefore, the macroinvertebrate fauna can readily reflect changes in the input load or type of waste even when sampled as infrequently as monthly or bimonthly. Semiannual sampling is often adequate to detect significant changes.

The life history of the macroinvertebrates includes either three or four stages in the case of the insects; egg, naiad (or nymph) and adult, or egg, larvae, pupa and adult. Although completion of more than one life cycle in a year is not uncommon, particularly with the midges, many require one or more years for completion (Usinger, 1956). The Oligochaeta, Mollusca and Crustacae have two- or three-stage life histories and have one or more generations per year. Most of the insect's life is spent in the immature stage, with the adult terrestrial portion usually devoted to the purpose of reproduction. Therefore, bottom substratum samples largely contain the naiads and larvae of insects, and major changes occur in the species composition through the spring and summer because emergence time varies among the populations with photoperiod and temperature. Natural community changes can be identified apart from the effect of wastes by previous study or with a representative control area. The mere presence of a variety of long-lived species that remains relatively stable in diversity and biomass from one month to another means that water quality has remained relatively unchanged. Any alteration in water quality, even though brief, can readily be reflected in that community. If decimated in the interim, the only replenishment source for the remainder of that life cycle is from tributaries or upstream. Repopulation can be relatively fast as a result of drift. As indicators of water quality, together with their inherent interspecies variability in tolerance to the variety of wastes that enter aquatic ecosystems, the macroinvertebrates are extremely valuable.

12.1 SAMPLING FOR BENTHIC MACROINVERTEBRATES

12.1.1 Quantitative estimates

The existing standing crop and species representation of stream invertebrates can be sampled with several types of units, including the following.

1. Surber square foot sampler is most often used in shallow areas.
2. Standard area enclosure (heavy metal frame) with a screen is often used in riffle areas of deeper streams.
3. Hess sampler is a cylindrical unit used in deep or shallow riffles and is more efficient than the above two units.
4. Peterson and Ekman grabs are used either in deep pools in rivers or in lakes and the Ekman is restricted to small-particle substrates.

None of these samplers is representative for all habitats (riffles, pools, shorelines, etc.). The size of the area sampled must be determined by the density of organisms so that a minimum sample size is possible with minimum sorting effort. Selectivity of certain samplers for different organisms and size must be considered. For example, the Surber selects for those organisms that wash from rocks easily and are large enough to be caught with a net. Small and newly hatched larvae and naiads are easily lost.

Artificial substrata have been used with considerable success. Multiplate samplers, which are similar to slides for periphyton, are one type (Hester and Dendy, 1962). Wire baskets filled with natural rubble substrata and buried at each station have been successful (Anderson and Mason, 1968; Mason *et al.*, 1970). Some species are probably missed by these methods, particularly less mobile ones. However, comparisons have shown good agreement with dredge collections (Anderson and Mason, 1968). Also, more individuals are usually collected with the artificial units. Of distinct advantage is the comparison among stations by using the same type substratum. Nevertheless, some analysis of the naturally occurring community structure should be included to determine what portion of the community is sampled with artificial substrata.

12.1.2   Qualitative estimates

Hand nets can be used in weedy areas, along mudbanks and in debris or in places where it is difficult to use any of the above quantitative samplers. A visual search on protruding substrata such as logs, large rocks, sticks etc. is often useful.

Although these procedures give occurrence and therefore contribute to a species list, they cannot give quantitative data on relative abundance and thus are not as useful for communicating findings and statistical analysis.

12.2   NATURAL FACTORS EFFECTING COMMUNITY CHANGE

Substratum and current are important factors in selecting kinds and quantity of macroinvertebrates. Macon (1974), with respect to the effect of current on macroinvertebrate community type, grouped velocities as very swift ($>100 \, \text{cm s}^{-1}$), swift ($50–100 \, \text{cm s}^{-1}$), moderate ($25–50 \, \text{cm s}^{-1}$), slight ($10–25 \, \text{cm s}^{-1}$), and very slight ($<10 \, \text{cm s}^{-1}$). Current erodes substrata composed of either rocks, stones or gravel, which interact to determine the kinds of organisms in the community (Hynes, 1960). Current determines sediment transport. The greater the current, the larger the substratum type (gravel to boulders) and the freer the interstices within the substratum of fine sediments; as current decreases, deposition increases and the interstices tend to fill up. Physical characteristics of the two types of environments, running and standing water, and associated community types are compared.

12.2.1   Running water

Eroding substrata is the rule in running water, particularly in the upper reaches of streams (low order). Species are usually numerous and of many groups, including worms, leeches, snails, clams, insects, and some crustaceans (scuds), but insects usually dominate the community. In such an environment the substratum is usually clean, interstices are free of sediment and animals are adapted in characteristic ways: limpets adhere to smooth stone surfaces, insects have sharp claws (stone flies, caddis flies, and mayflies) to cling to stone surfaces. Some are dorsoventrally flattened so that the whole body fits firmly against stones (Heptageniidae). Many species, particularly stone flies, do not tolerate sediment because it fills interstices within the substratum, which hinders attachment for herbivorous (or predaceous) feeding and restricts hiding space. Dense algal growth on rocks tends to clog interstices, thus producing a surface unfavourable for flattened forms but favourable for midges and small mayflies. Snails and leeches attach to relatively clean rocks. Many of

TABLE 12.1.   Fauna abundance in English streams versus substratum type; fauna abundance in an unpolluted African river in summer

Substratum type	Abundance (No. m^{-2})	
English streams		
Loose stones	3316	
Stones embedded in the bottom	4600	
Small stones with fine gravel	3375	
Blanket-weed on stones	44 383	
Loose moss	79 782	
Thick moss	441 941	
Pondweed on stone	243 979	
African river	Coarse sediment	Stones
Eroding substrata	6710	4730
Stable depositing	12 590	7570
Unstable depositing	4450	6660

Source: English streams: Hynes (1960), Table 2 data from Percival and Whitehead (1929); African river: Chutter (1969), Tables II and VI

these forms can tolerate velocities from $100-200\,cm\,s^{-1}$ (Macon, 1974, p. 136).

An example of how the type of eroding substrata affects density, particularly as the substratum becomes more fixed, is shown in some results from English streams (Table 12.1). The number of taxa varied from 11 to 25 but showed no special trend with substratum type. However, obvious species selection did occur within these types of eroding substrata. For example, *Rhithrogena* (mayfly) preferred stones free of vegetation, whereas *Hydropsyche* (net spinning caddis) was mostly found on periphyton-covered stones. Midges comprised from 40 to 54% of the fauna on the four vegetation-covered substratum types, but only 5 to 17% on the other three types. Diversity (ratio of species number to individuals) decreased considerably as density increased.

Depositing substrata, where interstices are filled because of reduced current, is typical in lower reaches of streams where sediment accumulates. This can also occur from watershed or in-channel erosion or sediment containing waste. In such an environment burrowers become dominant, e.g. worms which live under the sediment surface, build tubes and feed selectively on detritus, as well as insects that burrow (midges and ephemerid mayflies). Without a clean hard substratum to adhere to and living space among sediment-free rubble, clingers disappear. Midges, most of which are tube builders and net spinners, become abundant. Also detrital-feeding clams occur. However, abundance of individuals may not be greatly reduced in depositing substrata compared to eroding substrata.

Stability of substrata may have more influence on organism density than sedimentation (Table 12.1). For example, the hydraulic effects of urban stormwater runoff may have a greater effect on macroinvertebrate abundance than do pollutants in the stormwater including sediment. The increased area of impervious surface in the watershed (streets, houses, etc.) results in much higher peak flows than the stream had experienced resulting in substantial substrata movement.

Typical representatives of the depositing and eroding substrata environments were clearly illustrated by Hynes (1960, pp. 30–31).

12.2.2   Standing water

Lakes are represented in general by two types of substrata, which relate to trophic state. An oligotrophic lake without an accumulation of sediment in the littoral region has clean rocks and fauna similar in some respects to that of eroded substrata in streams, e.g. stone flies and case-building caddis, but few of the burrowing types. Where fine sediment prevails at depth, the community is composed of largely midges, clams and worms. Eutrophic lakes and reservoirs with macrovegetation and sedimented littoral regions harbour a varied fauna. In the depositing substrata community in such an environment, whether at depth or in shallows, midges usually dominate and some species of clams and worms occur, and densities are much higher than in oligotrophic lakes. A varied fauna occurs in and on the vegetation, including insects represented by stone flies, mayflies, dragonflies, beetles and midges.

12.2.3   Experimental controls

To monitor and detect effects in receiving waters resulting from waste discharges, a 'control' must be selected to account for the natural variations caused by current, substrata, elevation and temperature. Control stations must be located in areas unaffected by the waste but comparable in substratum and current to the test (waste-affected) stations. Otherwise, a significant difference in mean densities, species abundance or diversity between test and control cannot be interpreted as clearly an effect of only the waste. If control stations are physically comparable with test stations, then interpretation of effect can be rather straightforward. If controls are not comparable, conclusions must be qualified.

If control stations are unavailable, i.e. current, substratum and elevation at stations unaffected by waste are not similar to test stations (or if there is no unaffected upstream section), a control period may be required before waste is introduced to receiving waters. Such a period must be sufficiently long to include normal variability. One year is usually a minimum, but two years of data are much preferred to allow for variations in temperature and flow.

Because elevation difference usually causes emergence differences within insect groups, the interpretation of the presence or absence of such a group must not be mistaken for an elevation-emergence effect.

12.3 OXYGEN AS A FACTOR AFFECTING COMMUNITY CHANGE

Certain species and even whole groups of macroinvertebrates show various levels of tolerance to DO (dissolved oxygen). The groups in which most species are usually intolerant of low DO are the stone flies, mayflies and caddis flies (Gaufin and Tarzwell, 1956). Thus, a diverse representation among these orders indicates clean water, especially with respect to organic waste.

However, even within these groups many species can survive for a limited period of time at concentrations as low as $1 \, \text{mg} \, l^{-1}$ (Macon, 1974). Resistance to low DO can be partly due to morphological adaptations for maintaining a current over the animal's gills, such as with most mayflies (Hynes, 1960).

At the other extreme, there are a few organisms whose special physiological and morphological capabilities allow them to tolerate or resist low DO for a considerable period. These include the air breathers: the moth fly (*Psychoda*), mosquito (*Culex*), rat-tailed maggot (*Eristalis*), pulmonate snails (*Physa* and *Limnaea*), beetle, and adult water boatman and back swimmers. Some tubificid worms (*Tubifex* and *Limnodrus*) and chironomid midges (*Chironomus*) have haemoglobin, which can increase in oxygen-poor conditions. A high glycogen content and reduced activity allow these organisms to withstand prolonged oxygen minima (even anoxia) by meeting reduced metabolism with an aerobic glycolysis. Haemoglobin has a greater affinity for O_2 and probably loads up more easily under lower O_2 tension than haemocyanin. The success of tubificids, particularly *Tubifex tubifex* and *Limnodrilus hoffmeisteri*, in environments that become anaerobic for extended periods, partly results from their rapid reproductive rate. They can quickly repopulate the area once the anaerobic period has passed. The same can be said for the tolerant chironomids (Brinkhurst, 1965).

Those species with an intermediate tolerance to low DO include the crustaceans, *Asellus* and *Gammarus*, blackflies (*Simulium*), some leeches (Hirudinea), and craneflies (*Tipula*), and *Sphaerium* and *Pisidium* clams may be abundant in low-oxygen environments. Some mayflies are adapted to low tensions with the consumption of O_2 remaining rather constant down to low levels. Figure 12.1 shows curves for *Cloean* and *Leptophlebia* (Fox *et al.*, 1935; Macon, 1974). Thus, these two species can show normal activity down to very low O_2 levels, whereas one *Baetis* species failed to survive at low concentrations.

As indicated earlier, some mayflies tolerate sedimented conditions

Figure 12.1. Respiration rate (O_2 consumption) versus O_2 concentration in four ephemeropterans. (Fox *et al.*, 1935)

Figure 12.2. Respiration rates of aquatic insects determined at 20°C. Insects indicated in figure in descending order from left to right are *Tipula* (crane fly), *Caloteryx* (dragonfly), *Limnephilus* (caddis fly), *Pteronarcys* (stone fly), *Hetaerina* (dragonfly), *Paragnetina* (stone fly), *Macronemum* (caddis fly), *Ephemera* (mayfly), *Potamanthus* (mayfly), *Baetisca* (mayfly), *Leptophlebia* (mayfly) (modified from Olson and Rueger, 1968)

under which they can maintain adequate circulation of water over gills. *Cloean* (above) has such a morphological advantage. Because some caddis flies (*Hydropsyche*) can undulate and thereby increase circulation, they are found in areas of intermediate DO. Dragonflies and damselflies (Oonata) are often found in silty conditions. They have highly developed rectal and caudal gills, respectively, as well as other structures and thus are able to tolerate low DO for short periods (Gaufin and Tarzwell, 1956).

The demand for DO has been found to vary greatly within the so-called intolerant groups, the stone flies, mayflies and caddis flies (Figure 12.2). Thus the concept of group tolerance is only very generally valid. The species variation among the five mayflies shown in Figure 12.1 also illustrates the difficulty in generalizing about group tolerance. On the other hand, oxygen consumption rates of the individual species, according

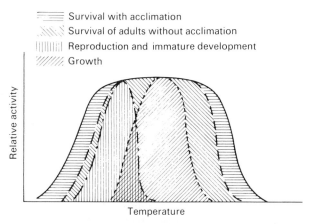

Figure 12.3. Generalized temperature tolerance graph for the important life stages of an invertebrate

to Olson and Rueger (1968), correspond well with their observed distribution in organically polluted streams. In order to use the occurrence or absence of aquatic macroinvertebrates as an indication of oxygen availability, at least the genera, if not the species, level must be identified because of the variability among species. This point will be taken up later.

Experimental evidence indicates that no single critical level of DO exists for macroinvertebrates in streams (Macon, 1974). The incipient limiting levels vary even within a family from rather low to quite high concentrations and vary even more with current velocity. For organisms adapted to high velocities (*Rhithrogena*), the lethal limit for DO increased greatly at low velocity (from 3 to $6 \, \text{mg} \, \text{l}^{-1}$); but for *Ephemerella*, which is morphologically capable of increased circulation at low DO and/or velocity (Ambuhl, 1959), the limit remained rather low and constant $(1-2 \, \text{mg} \, \text{l}^{-1})$ over a wide range of velocity.

12.4 TEMPERATURE

The range of tolerance for each critical stage in an organism's lifetime can be quite different for each species. Thus, an alteration in environmental temperature can result in changed community composition even if the ultimate lethal temperature in summer is not exceeded for any species. That is, a species can be eliminated from the community without mortality of adults. The replacement may result from an altered ability to utilize food resources or to reproduce and thereby compete for living space. A general tolerance pattern for a hypothetical species is shown in Figure 12.3.

TABLE 12.2.   Zooplankton mortality compared to the maximum temperature and the ΔT

	Mortality (%)	Maximum	ΔT (°C)
Chalk Point Plant, Chesapeake Bay[a]	90	37	6.4
British plant on Thames[b]	0	24.4	7
Paradise Plant, KY[c]	100	35.5	9

[a] Heinle (1969b); [b] Markowski (1959); [c] Welch (1969b)

Although the range for survival of adults is quite large, the optimum range for growth usually occurs at a lower temperature. The reproductive stage and the early free-living stage in the development of the young are therefore critical to population survival. Gradual shifts in species dominance can thus occur as the seasonal pattern of daily mean temperature changes from normal. This has been shown for quite small increases in temperature. Tolerance is apparently great for relatively wide diurnal fluctuations.

Although most of these principles have been demonstrated more thoroughly for fish, they probably also hold for most macroinvertebrates. In the determination of effect, the important factor associated with temperature change is time. A very high temperature can often be tolerated for a short time, whereas long-term survival can be guaranteed only at usually much lower temperature. Thus, when stating tolerance limits for any life stage of a given species the time of exposure must be included. Other factors that are almost as important include time and temperature for acclimation, food availability, DO, level of activity, age or developmental stage and presence of inhibitory substances.

12.4.1   Temperature criteria – rate of rise

Aquatic organisms that are entrained and pass through power plants may experience a relatively large and rapid rise in temperature (i.e. without cooling towers/ponds); the entrained water can undergo the increase in temperature in a matter of minutes. Some evidence suggests that invertebrates can tolerate a relatively high $\Delta T/t$ (increment rise above ambient per unit time). Direct observations of the condition of entrained zooplankton show that rate of rise is not detrimental unless the final temperature reached equals or exceeds the acute lethal level. This is indicated by results summarized in Table 12.2. The lethal temperature for zooplankton seems to be the same whether subjected to abrupt or gradual temperature increase (Heinle, 1969a; Welch, 1969b). There was little difference observed in the 24-h TL_{50} (temperature at 50% survival) for

Eurytemora affinis whether previously acclimated for months at 15, 20 or 25°C. A similar result was found with *Mysis relicta* (Smith, 1970), although the lethal temperature for that cold-water species was 16°C.

12.4.2   Temperature criteria – increment rise above ambient

The total increment rise above ambient temperature (ΔT) can be important because reproduction in benthic animals in temperate areas is triggered primarily when an appropriate temperature level is reached rather than by a daily rate of rise. For the most part, temperature level seems to be more important than photoperiod. Moreover, wide diurnal fluctuations in temperature can apparently be tolerated, whereas small changes in daily means cannot. As an example, a 7-year study in a British pond showed that the dominant mayfly emerged when the temperature in the spring reached 10–11°C, which occurred within a 2-week period in April in spite of what seemed to be a wide variation in weather conditions from year to year. The slow response of water bodies to climate change evidently makes water temperature a safer cue for aquatic animals than for terrestrial animals.

The seasonal changes (low temperature in winter and high in summer) are thus very important. The life cycles of aquatic animals are geared to seasonal temperature changes as well as to light. Crustaceans, such as the crayfish, must go through a low-temperature stage in which they do not moult but rather put their energy into reproductive cell development. Crayfish held in a cooling pond that was warm all year did not stop moulting and growing and did not reproduce until the winter temperature was reduced. The same response has been suggested for insects that responded with lower adult longevity and poorer emergence if held at constant but sublethal temperatures (Nebeker, 1971a; 1971b). Also, emergence may occur as much as five months early in heated stream sections. This would be detrimental to adults in temperate climates (Nebeker, 1971b). A similar increment rise above ambient may not have as important an effect on life cycles in tropical climates, because natural temperature changes are relatively small.

12.4.3   Temperature criteria – daily maximum

During the period of high temperature, most damage results if the maximum reached exceeds the tolerable level regardless of the increment rise above ambient (i.e. the increment added to the normal river temperature). This is illustrated in Figure 12.4 for a hypothetical constant flow stream with a heated-water addition. There would be an additional variation about the daily mean that is not shown.

Heat addition to the Thames River, UK resulted in a 12°C rise above

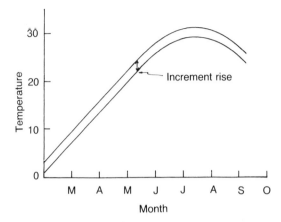

Figure 12.4. Increment rise above ambient temperature in a hypothetical constant flow river receiving a constant flow and heated effluent

the ambient daily mean with a maximum of 28°C. No change in the number of species present was observed, but the abundance of leeches, scuds (*Gammarus*) and midges decreased, whereas snails and clams increased (Mann, 1965). The same rise above ambient in the Delaware River, USA (12°C), but with a maximum of 32–35°C, resulted in an extensive reduction in the number of species as well as total abundance (Trembly, 1960). Although the area repopulated in winter, the high temperature reached in summer was nevertheless the damaging factor.

12.4.4   Community shift versus temperature level

Evidence suggests that moderate tolerance by aquatic freshwater invertebrates in general exists up to about 30°C. Figure 12.5 shows that the mode for the lethal temperatures of a large number of invertebrate species occurs at ≈35–40°C using available data. Allowing that a plotting of the preferred temperature (if available) would put the mode somewhat lower, say ≈30–35°C, means that a temperature increase to ≈30°C could result in an increase in species diversity in some environments, but that increases to levels >30°C would probably result in decreases in diversity. This can be supported by observations summarized in Table 12.3. The Delaware River case was cited previously where the decrease in species and individual abundances was greatest above 32°C (Trembly, 1960). TL_{50} values for seven species of chironomids ranged from 29 to 39°C. (Walshe, 1948). The more tolerant species were from lentic habitats indicating adaptation to warmer conditions. The optimum for normal activity, such as feeding, development and emergence, was 28°C for *Tanytarus dissimilis*, which has a TL_{50} of 33°C. This represents the

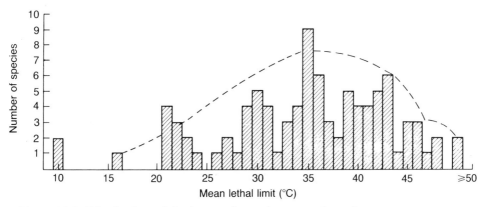

Figure 12.5. Distribution of freshwater invertebrate species tolerance to temperature. Dashed line shows approximate mode. (Bush *et al.*, 1974 with permission of the American Chemical Society)

TABLE 12.3. Observed tolerances of macroinvertebrates

	Limit for normal diversity and abundance (°C)	Upper limit for highly tolerant forms (°C)
Delaware River[a]	32	37
Chironomid midges[b] (*n* = 8)	28	34
Caddis flies[c]	28	35

[a] Trembly (1960); [b] see text; [c] Mihursky and Kennedy (1967)

median for the total of eight species tested. If the 5°C difference between optimum and the TL_{50} were typical, the upper limit for the most tolerant species of chironomids (TL_{50} = 39) may be ≈34°C.

The above limited sources of information suggest that to maintain normal diversity and abundance of most aquatic invertebrates the mean daily temperature should not exceed 30°C. However, this does not mean that the same species would necessarily be present up to 30°C.

For additional evidence to support the 30°C level, a summary of temperature tolerances of 13 invertebrates and vertebrates in Chesapeake Bay is shown in Table 12.4. In this instance damage in terms of species reduction would presumably occur at temperatures >30°C.

In the marine environment, the maximum tolerable limit of mean daily temperature is no doubt lower than in fresh water because oceanic temperature is more stable and the ambient temperature does not reach the high levels, at least in temperate oceans, that occur in fresh water. However, studies of thermal effects are scarce in the marine environment. Adams (1969) cites results from a study of a power plant on

TABLE 12.1.   Predicted species in the
suboptimal range and those lost with increases in
temperature in Chesapeake Bay

Temp (°C)	Species in suboptimal (%)	Species lost (%)
26.7	0	0
29.4	8	0
32.2	61	8
35.0	16	69
37.8	15	85

Source: Mihursky (1969)

Morrow Bay, California, USA that showed predictable shifts from cold-water to warm-water invertebrate species with a temperature increase. A rise above ambient of 5.5°C was observed at the surface as far away from the discharge as 150 m. At that point 54 species were present, 39% of which were warm-water types, which was considered normal for the area. At 90 m, of the 34 species observed, 67% were warm-water types; near the heated-water discharge only 21 species were observed, 95% of which were warm-water types.

Compared with the freshwater and estuarine examples of heated-water effects cited earlier, a 5.5°C increase seems rather small. Nevertheless, the above examples suggest that marine communities may be very sensitive to temperature change.

The importance of such shifts in species as indicated for freshwater and marine environments depends upon the nuisance aspect of the invading species and the economic importance of the species affected. A shift to warm-water species may be undesirable because generally warm-water forms grow fast and mature early but live short lives, whereas cold-water forms grow relatively slowly and mature late but live long lives. Nebeker (1971a) has shown for several insect species that even though feeding activity was high over a wide range (15 to 30°C) emergence was increasingly less successful and the longevity of adults was shortened at temperatures much above 15°C. Thus animals adapted to lower temperatures may be expected to reach larger size because they live longer. Although productivity would tend to be greater for warm-water species, the efficiency for energy transfer in the system would possibly be reduced and its overall stability impaired.

12.5   EFFECT OF FOOD SUPPLY ON MACROINVERTEBRATES

Benthic community composition depends upon the type and availability of food as well as on the chemical/thermal conditions. If the type and/or

availability of food changes, the community will respond even if pronounced changes in chemical characteristics or temperature do not occur.
Although difficult to generalize, the following indicates how changes in
food supply could cause community shifts:

1. Detrital feeders can include the net-spinning caddis flies, the aquatic
   sow bug, chironomids, clams, snails, most mayflies, some stone flies,
   and blackflies. These are largely collectors of fine particulate matter.
   Most of the mayflies, caddis flies and stone flies eat both detritus
   and algae, but usually detritus feeding is more typical (Grafius and
   Anderson, 1972).
2. Grazers of periphyton include some mayflies, case-building caddis
   flies, some beetles and snails.
3. Predators are represented by the dragonflies, some leeches, a few
   stone flies, some beetles, some midges, alder flies and a few caddis
   flies.

For example, a change in the food supply could affect stream invertebrate
composition in the following way. In a stream of low productivity the
amount of detritus resulting from primary producers is low, thus grazers
utilizing periphyton on substratum surfaces are expected to be most
abundant, with fewer collectors designed to utilize detritus accumulating
on surfaces from autochthonous production. Now, if the amount of
detritus is increased through inorganic nutrient or organic waste input,
detrital feeders would be expected to flourish downstream of those inputs
at the expense of grazers. However, grazers would be expected to
respond in the immediate area of enrichment in response to increased
autotrophic production. Such an increase in detritivores would be particularly true in the case of fine particulate inputs such as sewage. If poor
water quality resulted from the increased production, such as low DO,
and thus limited predators, which ordinarily kept prey populations low,
the tolerant detritivore prey species could reach a large biomass in the
predator's absence. Such an interaction has been noted with leeches and
tubificid worms.

A community dominated by shredders and collectors (large and small
particulate detritivores, respectively) in low-order streams with forest
cover (with $P/R < 1$) will normally shift to one dominated by grazers and
collectors in higher-order streams with less shade and more autotrophic
production (with $P/R > 1$). This has been described as a probable out-
come for streams not receiving cultural sources of enrichment (Cummins,
1974; Vannote et al., 1980).

12.6 EFFECT OF ORGANIC MATTER

12.6.1 Natural inputs

Natural inputs of allochthonous organic matter can change community structure and actually result in species increases if the physical–chemical changes are not severe and diversity was previously limited by a scarcity of food. Terrestrial sources of organic matter are leaves, needles and other forms of detritus which furnish the food supply to low-order streams as indicated above. However, moderate influxes of other types of organic matter will also change communities by increasing species richness. This was shown in results from a Californian mountain stream that received particulate organic matter input (algal detritus) from a lake. At the point where particulate matter (seston) peaked, due to enrichment from the lake, the number of species increased. Whereas 5 species were detected upstream from the lake, the number increased to 11 and 14 at two points downstream. Here the detrital-feeding blackflies (Simulidae) were very abundant and midges (chironomids) increased in abundance. These are the 'collectors' of detritus. Simulids removed an estimated 60% of the suspended plankton algae in 0.4 km of stream distance from the lake (Maciolek and Maciolek, 1968). The species number of oligochaetes (worms) can also be low if organic enrichment is low and increase with further enrichment (Milbrink, 1980).

Similar circumstances can occur in the tail waters from reservoirs without degrading stream quality. Productive stream fisheries have resulted from the increased plankton detritus produced in the reservoir that subsequently stimulates the collector insect community downstream. On the other hand, dense mats of periphytic algae may occur downstream from dams responding to the release of deep, hypolimnetic water with high dissolved nutrient content.

12.6.2 Cultural inputs

The effect of cultural wastewaters is usually more severe than the natural sources of allochthonous organic matter just described. The most obvious effect of organic wastewater is a demand on DO. In addition, the associated input of suspended solids alters the bottom substratum to one that is more depositing than eroding. Rather than an increase in species as well as biomass, the increase in biomass and productivity is usually accompanied by a reduction in species diversity.

The principal reason for the reduction in species abundance is that the increasing severity of the physical and chemical factors eliminates the intolerant species. The remaining tolerant species flourish because of increased survival, a result of the reduction of predators and a more

Figure 12.6. Distribution of species abundance in Lytle Creek, Ohio, upstream and downstream from a sewage outfall (arrow) at various seasons (Gaufin and Tarzwell, 1956)

favourable food supply, and, consequently, increased biomass results from increased food supply. But if food supply is too great for flow and DO resources, even biomass of tolerant species could decrease in septic zones. As in the example with natural organic matter, if the physical and chemical changes are not severe and the system were highly oligotrophic initially, increases in biomass as well as diversity could result from mild inputs of cultural organic matter. Thus, the degree of community shift is a function of the type of waste, the waste load, the dilution volume or rate of river flow, and turbulence (reaeration potential). Sewage effluent has produced these predictable effects in numerous cases. Results from one of the early classic, but still pertinent studies are shown in Figure 12.6. The associated BOD and DO results are shown in Figure 12.7.

Figure 12.6 shows the seasonal abundance of taxa upstream and downstream from a sewage outfall in Lytle Creek, Ohio, USA. Note the relative difference in the effect of the effluent depending on the season considered. The effect throughout the stream was much more severe in late summer and winter when flow was low. The higher BOD in August and December that resulted from the low flow are shown in Figure 12.7. Consequently maximum and minimum DO were lower during summer low flow. Also note the extension of the zone of minimum taxa abundance during the winter (Figure 12.6). This probably resulted from a reduced rate of waste decomposition during the period of low temperature and low dilution, although flow data were not present for those months.

To illustrate the effect of flow, distance and time to the maximum DO

Figure 12.7. BOD and DO in Lytle Creek, Ohio, upstream and downstream from a sewage outfall (arrow) at various seasons (Gaufin and Tarzwell, 1956)

deficit (DO sag) can be calculated for a hypothetical river using the Streeter–Phelps equation:

$$\frac{dD}{dt} = K_d L_t - K_a D_t \tag{12.1}$$

where K_d and L_t are the decomposition rate and BOD at time t (see waste characterization), respectively, and they represent the deoxygenation reaction, while K_a and D_t are the reaeration rate and DO deficit at time t, respectively, and represent the reaeration reaction. For an influent BOD of 150 mg l^{-1} and river temperature of 25°C, the time and distance to the maximum DO deficit (minimum DO) would be 1 day and 1.5 km if river flow is 5000 m^3 day^{-1}. The time to maximum deficit is still 1 day and if river flow is 10 000 m^3 day^{-1} but distance is 4.1 km and actual deficit is more than halved from 6.5 to 3.0 mg l^{-1} DO.

Hawkes (in Klein, 1962) has generalized the community response to organic waste in the form of sewage in terms of the genera that are typical for stony, rapid streams as well as for streams with primarily depositing substrata.

Decreases occur in intolerant species of stony rapids with the addition of organic waste according to the following transition:

Rithrogena → *Ephemerella* (mayfly) → *Gammarus* (scud)

Decreased competition and increased food supply results in the following sequence of elimination:

Baetis (mayfly) → *Simulium* (blackfly) → *Hydropsyche* (caddis fly) →
Limnaea (snail) → *Herpobdella* (leech)

Further increase in waste and substratum change favours dominance according to the following sequence:

Nais (worm) → *Asellus* (sow bug) → *Sialis* (alderfly) →
Chironomus (midge) → *Tubifex* (worm)

Even in depositing substrata the end-point is the same with *Sialis*, *Chironomus* and *Tubifex*, but sediment-tolerant organisms such as *Ephemera* and *Caenis* (mayflies) are characteristic of the initial depositing substrata community. The general pattern of abundance for species and individuals in response to organic waste is shown in Figure 12.8. In Figure 12.8(a) the typical effect of a single organic waste is seen. Such a pronounced effect would necessarily result from a heavy loading, particularly when a sludge deposit occurs. Abundance increases, whereas variety decreases.

Two oligochaete worms that dominate environments grossly polluted with organic waste are *Tubifex tubifex* and *Limnodrilus hoffmeisteri*. Heavy loading of organic detritus and the accompanying low DO (they can survive to 0.5 mg l^{-1}) contribute to the success of these species over all others in grossly polluted streams. Brinkhurst (1965) suggested that a principal reason for their dominance in large numbers can be largely ascribed to reproductive habits. In direct contrast to many of the less tolerant inhabitants of running water, these two species were found to breed at all times of the year. Brinkhurst suggested that the worms too are affected adversely by low DO, but are able to maintain a large biomass in response to the large food supply in spite of low DO

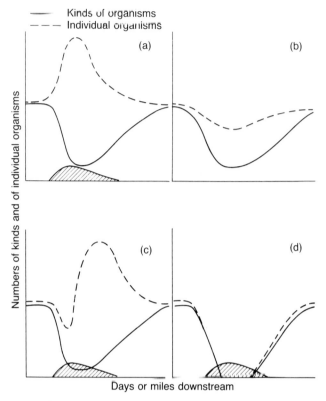

Figure 12.8. Generalized response of macroinvertebrates to (a) organic waste, (b) low levels of toxic waste, (c) organic and toxic waste, (d) high levels of organic and toxic waste. Sludge deposit is indicated by cross-hatched area (Keup, 1966)

because of their frequent breeding. The same concept of life history trait was used to explain the dominance of the polychaete worm *Capitella capitata* in organically polluted marine environments (Grassle and Grassle, 1974).

The success of tubificids in utilizing and at the same time stabilizing organic waste loads can be illustrated by the potential sludge recycling rate of a high density population. *Tubifex* has been reported to produce 170 cm of faecal pellets per day (Bartsch and Ingram, 1959). They orient with the head down in the sediment and the posterior end in the overlying water producing a mound of faecal pellets, which are often lighter in colour than the surrounding sediment. If a density of $100\,000\,\mathrm{m}^{-2}$ is assumed, and *Tubifex* has been observed at two- to three-times that amount (Gaufin and Tarzwell, 1956; Hynes, 1960), then the animals would recycle the top 3 cm of sediment per day. *Tubifex* has been observed to move to a depth of 15 cm in lakes (Milbrink, 1980). The

TABLE 12.5.   Relative composition of profundal benthic biomass in various lakes

Lake	Oligochaetes (%)	Chironomids (%)	Sphaerids (%)
Lake Washington	51	43	3
Lake Erie (1929–30)[a]	1	10	2
Lake Erie (1958)[b]	60	27	5
Cultus Lake (B.C.)	34	65	
Convict Lake (CA)	31	65	
Lake Constance (CA)	20	57	20
Lake Dorothy (CA)	23	69	3

[a] *Hexagenia* abundant;   [b] *Hexagenia* absent.   *Source:* after Thut (1969) with permission of the Ecological Society of America

computation considers that a *Tubifex* is about 4-cm long with a posterior diameter of 0.06 cm.

Cairns and Dickson (1971a) have generalized the tolerance to organic waste into three principal groups. The intolerant group includes mayflies, stone flies, caddis flies, riffle beetles and hellgrammites; the tolerant group includes sludge worms, certain midges, leeches and certain snails; and the moderately tolerant group includes most snails, sow bugs, scuds, blackflies, crane flies, fingernail clams, dragonflies and some midges.

Although this grouping may appear rather crude, particularly when the overlap in tolerance among genera within families and species within genera is relatively great, conditions of gross pollution are nonetheless often separable from no, or only moderate, pollution with these groupings. In order to detect subtle changes in community response, a more detailed approach is necessary.

12.6.3   Eutrophication in lakes

Oligochaetes increase in lakes relative to other organisms, particularly Chironomidae, as a result of cultural eutrophication. The worms also show greater abundance with depth, inversely to the trend of midges. This can be associated with oxygen depletion and sedimentation with depth, but the degree to which species competition causes the change is unknown (Thut, 1969). Note in Table 12.5 that Lake Washington compared closely with western Lake Erie and also the degree to which Lake Erie was degraded from early surveys. The density of oligochaetes can reach tens of thousands per square metre. Carr and Hiltunen (1965) found the greatest densities in Lake Erie, which averaged $6000\,m^{-2}$ in 1961, near the mouths of the Maumee, Raisin, and Detroit rivers. Densities in 1930 were $<2000\,m^{-2}$.

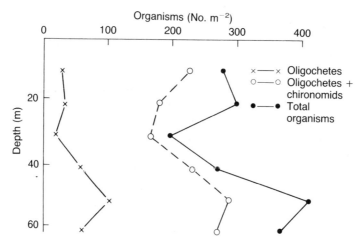

Figure 12.9. The distribution of oligochaets and chironomids with depth in Lake Washington (Thut, 1969)

Figure 12.10. Distribution of chironomids and oligochaets in Cree Lake (data from Rawson, 1959)

The distribution with depth in Lake Washington is shown in Figure 12.9. Results from oligotropic Cree Lake in Saskatchewan, Canada (Figure 12.10), did not show the proportional increase in oligochaetes with depth as occurred in Lake Washington. Thus, the proportional biomass change is an important indicator in lake eutrophication.

Species of chironomids have been associated with the trophic state of lakes. In the large lakes of Central Sweden, Wiederholm (1974) has suggested that each of four species were found to represent oligotrophy, mesotrophy, eutrophy and possibly hypereutrophy, respectively, which were in turn related to the levels of P and chlorophyll *a* (Table 12.6). Warwick (1980) has further separated lakes into trophic indices according 1: chironomid assemblages recognized by Saether (1979). Wiederholm

TABLE 12.6.   Relationship between four chironomid species and trophic
state indicators

	Heterotrissocladius subpilosus	*Micropsectra spp.*	*Chironomus anthracinus*	*Chironomus plumosus*
Chl*a* (μg l^{-1})	<3	3–10	10–20	>20
Total P (μg l^{-1})	<15	15–30	30–60	>60
P load (g m^{-2} yr^{-1})	<0.5	0.5–1.0	1.0–2.0	>2.0

Source: after Wiederholm (1974)

(1980) has further related their chironomid trophic indices to P and
chlorophyll *a* per unit mean depth. Mean depth was argued as an important factor in the response of chironomids to increased eutrophication
because more P and algal biomass per epilimnetic volume is required in a
deep lake for a given change in chironomids than in a shallow lake. The
profundal invertebrates depend on organic matter reaching the bottom
and less reaches the bottom undecomposed as depth increases. Thus
shallow lakes have a more eutrophic assemblage of chironomids than
deep lakes under similar enrichment levels.

12.7   EFFECTS OF TOXIC WASTES

The response of macroinvertebrates to toxicity is very different from that
to organic waste. The number of species decreases, as with organic waste,
but biomass remains the same or increases slightly following the mortality
of other species that could be predators or represent a significant food
source. The initial biomass can be maintained in the presence of only a
very low level of toxicity. However, in most cases biomass will decrease
as well as the number of species, and the effect of increased food supply
would be a short-term effect and probably too minimal to notice.

The degree of change in species number and biomass depends on the
resulting concentration of toxicant in relation to the range in tolerance of
the species present. Considering the foreign nature of most toxicants,
particularly the synthetic organics, an increase in species number as
predicted from temperature and organic matter increase, in relatively
cold and oligotrophic environments, respectively, would not be expected
to occur with toxicant input.

A combination of both toxicants and organic wastes result in the
predicted general community response shown in Figure 12.8. In Figure
12.8(b) the decrease in variety is similar to that which occurs with organic
waste input, but the primarily toxic waste usually has little food value
and, consequently, the tolerant forms have no extra energy with which

Figure 12.11. Mean number of invertebrate taxa related to stream concentrations of total Zn and Cu and EPA acute criteria for those metals in Prickly Pear Creek, Montana in 1980 (data from Miller *et al.*, 1983). ■—■, Zn concentration; ■---■, Zn criteria; ●—●, Cu concentration; ●---●, Cu criteria

to increase their numbers. In Figure 12.8(c) an initial toxicity exists, which reduces variety but then toxicity diminishes (complexed physically or biologically or diluted) and the full effect of the organic matter is observed. In Figure 12.8(d) the level of the toxic substance(s) does not 'diminish' and the combined effect is worse than with either waste alone.

A typical response to toxicity is shown by results from a stream receiving drainage from mine tailings ponds and spoil piles that remain from mining activities in the late 1800s and early 1900s. Six stations were established on Prickly Pear Creek, Montana, USA extending from an upstream control to the farthest (11 km) downstream sample point from the waste source (Miller *et al.*, 1983; LaPoint *et al.*, 1984). Zn and Cu were the principal toxicants being leached from the mine waste and increased to mean concentrations exceeding 1000 and $100 \mu g l^{-1}$, respectively, immediately downstream of the source (Figure 12.11). These levels were well above USEPA criteria, which varied due to hardness (Chapter 13). Cu was also above the criteria level in the impact zone. Total abundance of individuals and numbers of species decreased substantially in the maximum impact zone within 2 km downstream of the source and in direct proportion to the increase in metals. The invertebrates had still not recovered to that of the upstream control at the most downstream station.

As with DO, a wide range of tolerance among species in the different orders can be expected. However, there is an indication that midges may be more tolerant to heavy metals than are caddis flies, and mayflies may be least tolerant (Savage and Rabe, 1973; Winner et al., 1975, 1980). However, relative tolerance, as judged from species presence in waste-affected streams, may be related to the number of species in the respective families. Given the same range of tolerance within a family, there is a greater chance of encountering more tolerant species in the family with the greatest number of species.

The numbers of species of macroinvertebrates, representing a wide range in tolerance, along with a paucity of information of individual species tolerance, make the development of specific toxicant criteria largely impossible. However, it may be useful to contrast the tolerance of some invertebrate representatives with that of fish, for which there are specific criteria. With regard to metals, there is an indication that, in general, aquatic insects may be more tolerant to acute levels than fish (Warnick and Bell, 1969; Clubb et al., 1975). However, when exposure time was longer, some insects were as sensitive as, or more sensitive than, fish (Spehar et al., 1978). There are also sublethal effects on aquatic insect life stages that are analogous to those in fish. For example, moulting and emergence were reduced in several species as a result of Cd concentrations considerably below lethal levels and those of smaller size were more sensitive to acute levels than larger ones (Clubb et al., 1975). Greater sensitivity of younger, smaller specimens than older, larger ones has also been shown with pesticides (Jensen and Gaufin, 1964).

While it is important to know the concentrations of toxicants that cause damage to critical life stages of aquatic invertebrates, it is probably even more important to determine the pertinent chemical form of the toxicant and the role of interacting factors. That is particularly true for metals. The relative toxicity of a metal usually varies with pH (due to solubility), DO and temperature (due to rate of metabolism), hardness (due to competitive inhibition of Ca and Mg with the toxic metal(s)) and the amount of organic matter present. Due to less organic matter available for complexation, the effectiveness of metals may be greater in oligotrophic than eutrophic waters. The specific nature of these inter-actions will be discussed with regard to fish (Chapter 13).

12.8 SUSPENDED SEDIMENT

The sediment carried by running water can have an adverse effect on bottom-dwelling organisms. As pointed out previously, the distribution of organism types and abundance is determined to a large extent by sub-strata type. Excessive erosion from abuse of the watershed often results from overgrazing, deforestation and urban development. By removing vegetative cover, the rate of runoff increases, producing more force for dislodging soil particles and transporting them to water courses. The peak stream flow increases as the fraction of watershed impervious surface increases. Such flow increases, beyond extremes previously experienced by a stream, will result in greater within-channel erosion and sediment deposition creating depositing substrata. The greatest effect (change) on stream macroinvertebrates can be expected where increased deposition occurs on a previously eroding substratum.

Suspended sediment is an insidious and widespread pollutant. The deterioration of a stream is gradual with no dramatic kill of fish or invertebrates, as is often the case with a toxicant. Nevertheless, the community can be altered just as significantly. The principal effect of sediment is to decrease habitat of the eroding substratum community, especially by filling the interstices in rubble/gravel bottom streams. As a result, the number of species as well as the abundance of organisms should decrease because there would usually be little organic matter in the eroded material that could add to the food supply of deposit feeders, which will dominate the sedimented environment. However, a moderate increase in suspended sediment to an eroding substrata could actually result in no appreciable change or in an increase in abundance, whereas the number of species could increase if the substratum were stable. Chutter (1969) found the latter effect in an African stream where in summer the number of species in a stony stream increased from 30 to 42 and then decreased to 22 as the substratum progressed from eroding to stable depositing to unstable depositing. Abundance for the three environmental types showed the same trend (Table 12.1). Effects from logging usually result in reduced diversity and decreased abundance of some taxa (e.g. stoneflies), but increased abundance of others, e.g. worms, midges and mayflies, which are more tolerant of sediment (Graynoth, 1979; Newbold et al., 1980).

An adverse effect on bottom fauna can result from rather modest increases in sediment concentration. Such an effect was observed in Bluewater Creek, Montana, USA, an eroding substratum stream, as a result of irrigation return flows. The effect was noticed at an average level of $\approx 100\,\mathrm{mg\,l^{-1}}$. Although such a concentration of suspended sediment is not readily visible when viewed through a beaker of the stream water, a reduction in transparency would be quite visible in a natural water course where the light path is greater. There would have been no toxicity result from such a concentration in the stream, yet the abundance of organisms was depleted tenfold (Figure 12.12).

Developing criteria for sediment concentration is difficult because the effect results from the amount deposited and retained in the substratum, not on the concentration left in the water. The fraction of load deposited in turn depends upon river flow. Therefore, low-velocity streams will usually show a greater effect from a given average sediment concentration than high velocity streams.

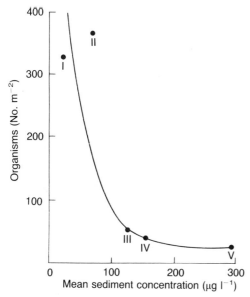

Figure 12.12. The relationship between annual mean suspended sediment concentration and the abundance of benthic organisms at five stations in Bluewater Creek, Montana. Downstream (I–V) increase in sediment was caused by irrigation return flow (data from Peters, 1960)

Effects from urban runoff are often related more to physical scouring, shifting and replacement of the eroding substratum in streams than to sediment deposition. This is due to peak flows well above those experienced during predevelopment. As a result, populations may be dislodged and transported downstream with decreased diversity and only the opportunistic species surviving. Effects are difficult to predict so evaluation of stream receiving water quality following corrective action should be an integral part of every project as is now the case for lake restoration in the USA.

12.9 RECOVERY

The recovery of stream systems is relatively rapid because in a running-water system the contaminated water is moved downstream relatively quickly following curtailment of waste input. After one hydrologic cycle, which includes high flow that would transport much of the contaminated sediment downstream (or deposit new uncontaminated sediment), the physical environment could be largely recovered. All that is necessary is repopulation, which requires recolonization of the organism-depleted area and time for regrowth to pretreatment abundance. Because most

TABLE 12.7. Recovery of macroinvertebrate and fish communities affected by different types of wastes following waste controls

Stream ecosystem	Waste	State of recovery	Waste removed	Elapsed time
Clark Fork River, Montana[a]	Acid mine	Full recovery of trout and bottom fauna	100% of acid mine water after 6 months	2 yr
Maine, several streams[b]	DDT	Full recovery of bottom fauna	100% (discontinuous applications)	2–3 yr
Thames River estuary, England[c]	Sewage	Partial recovery; from 0 to 61 fish species, from 0 to 10% minimum O_2 saturation	60% sewage through secondary treatment	7 yr
Clinch River, Virginia[d]	Acid waste spill	Complete recovery of bottom invertebrate diversity, but some mollusks still absent	100%	2 yr
Clinch River, Virginia[d]	Alkaline fly-ash spill	Complete recovery of bottom invertebrate diversity	100%	6 mo
Blue Water Creek, Montana[e]	Silt	Marked improvement in trout/rough fish from 0.64 to 3.5 and 0.14 to 1.0 at two stations, respectively	52–44% of sediment load at two stations, respectively	2–3 yr
Plum Creek (Clink River)[f]	Fuel-oil spill	Complete recovery of bottom fauna abundance	100%	5 mo

[a] Averett (1961); [b] Dimond (1967); [c] Gameson et al. (1973); [d] Cairns et al. (1971b); [e] Marcuson (1970); [f] Hoehn et al. (1974)

taxa have 1-year life cycles, it is not surprising that 1- to 2-year recoveries are common. Recovery could require a longer time in the lower reaches of rivers where, due to low velocity, sediment is not displaced readily, or where most of the tributaries are depleted of fauna, and successful recolonization would thus be delayed.

Table 12.7 lists some examples of recovery in streams from various types of wastes. Most of the recovery times are of the order of 2 years. With most of these examples it is difficult to be anything but qualitative about changes in water quality and/or the bottom fauna composition. They nevertheless convey the impression that recovery is relatively 'fast' if sources for recolonization exist and the waste input is greatly reduced. The Clinch River fish kill in 1967, following a fly-ash pond leakage, resulted in a spill of water of pH 12 moving down river killing nearly everything in its path. Opinions, at the time of the kill, of a 10-year recovery were obviously quite incorrect.

12.10   ASSESSMENT OF WATER QUALITY

The state of water quality can be effectively assessed by use of the benthic macrofauna. The advantages of this community were stressed earlier. An important question, however, is what methodology or technique should be used. Although the presence or absence of certain species has qualitative meaning to someone with experience in the relative tolerance of benthic organisms, such descriptive information is difficult to communicate or treat statistically. Thus, a knowledge of the relative tolerance of organisms present (and absent) is usually combined with a numerical expression of community structure or indicator populations to assess water quality. Numerical expressions may or may not treat the relative tolerances of one species versus another.

12.10.1   Indicator species and taxonomy

Indicator species are often employed to judge the relative quality of an environment, i.e. when a certain species is present a particular water quality condition is indicated. The absence of a particular species is also important and although this is not implied in the approach, it is nonetheless often considered.

The principal drawbacks in this approach are taxonomic and the lack of definitive information on comparative tolerance to given types of waste. On the taxonomic problem, Resh and Unzicker (1975) have stressed the importance of identification to species. They found that of 89 genera of macroinvertebrates that had known tolerances to organic waste, i.e. intolerant (I), facultative (F) or tolerant (T), 65 have species in more than one general tolerance category (I/F, I/F/T, or F/T). Although the genera of invertebrates are relatively easy to determine in most cases, there apparently exists such an overlap in tolerance that a definitive analysis is not really possible unless identification is to the species level. With many groups this is difficult and time consuming because the state of taxonomy is relatively poor with North American fauna, for example, and opinions of experts are often required.

Equally important is the known tolerance of a particular species to the various types of wastes that can be discharged into streams. Only with organic wastes, such as sewage, is there any general understanding of tolerance among species. An excellent example of the use of indicator species is that of profundal chironomids and lake trophic state described earlier.

This is not to say that nothing can be gained from a description of the organisms present, identified to some level short of species, along with their relative abundances. As has already been shown, Hawkes (1962) described a typical succession of genera within English streams of differing substrata in response to organic waste. Certainly the presence of *Tubifex* in large numbers is quickly identified as gross organic pollution. However, the problem of detecting the effect of small incremental changes in waste input is probably difficult to solve by the qualitative indicator organism approach alone.

12.10.2 Numerical indices

Formulae to express the 'state of stream quality' or 'degree of pollution' using the macroinvertebrates are numerous. They are generally of three types: biotic, diversity and similarity (Washington, 1984). The biotic ('pollution') indices require a judgment as to the relative tolerance (or saprobic valency) of the various taxa identified, usually tolerant, intolerant or facultative, and the distribution among those groups is weighted. Thus, they require considerable experience by the investigator. A major problem with these formulae is that they are largely restricted to the assessment of organic waste effects and thus lack a basis to categorize the tolerance of the taxa to the myriad of toxic wastes or physical factors (e.g. sediment, flow) that could be present (Sládeček 1973; Washington, 1984). However they are considered by many to be more biologically relevant than the other indices because they consider the physiological requirements of the taxa.

Washington (1984) has reviewed 19 biotic indices. He suggested that Beck's was the first true biotic index (Beck, 1955), which compares the abundance of species that are tolerant and intolerant to organic pollution. It includes only species abundance and not individual abundance or any other attributes besides tolerance to organic pollution. Beaks' river index (Beak, 1965), on the other hand, includes judgments regarding trophic character (feeding habits), sensitivity to organic pollution and relative abundance of individual species organized into a table. There are other indices that include one or more of these factors, such as the Trent Biotic Index, Graham's index, and Chandler's Biotic Score (see Washington, 1984). These indices generally depend on a gradient in tolerance of certain taxa to organic pollution and require identification to species in some cases, but only to order or family in others.

Hilsenhoff (1977) adapted Chutter's (1972) biotic index, developed for South African streams, to North American streams. The index uses quality (Q) values and abundance of each taxon for which a Q value is chosen. An average Q value, weighted to taxa abundance is calculated according to:

$$\text{Index} = \frac{\sum_{i=1}^{K}(n_iQ_i)}{n} \tag{12.2}$$

where n_i is the number of individuals in each of K taxa and n is the total number of individuals. Hilsenhoff's Q values range from 0–5, which is the range of final index values with <1.75 representing a clean undisturbed stream and 3.75 representing gross enrichment or disturbance.

Diversity indices do not consider the tolerances of individual species, only the number of species and number of individuals and, thus, relate more to community structure (Washington, 1984). There are many formulations for diversity (Washington lists 19), but only four will be reviewed here. A frequently used index that considers only species richness is as follows (after Margalef, 1958):

$$d = \frac{S-1}{\log_2 N} \tag{12.3}$$

where S is the number of species and N is the number of individuals. The natural log of N is also used. Practically, this formula illustrates that if one searches an environment and collects individuals, the number of species encountered increases linearly as the logarithm of the number of individuals. At some point the finite number of species is approached and the curve flattens out. This formula does not, however, account for the relative abundance of each species or evenness in distribution.

One of the most popular formulae used to express diversity, which includes richness (number of species) and evenness, is the Shannon approximate index from information theory (Shannon and Weaver, 1948):

$$H'' = -\sum_{i=1}^{s}\frac{N_i}{N}\log_2\frac{N_i}{N} \tag{12.4}$$

TABLE 12.8.   Diversity formula comparison for three hypothetical communities

Communities	N_1	N_2	N_3	N_4	N_5	ΣN	S	$\dfrac{S-1}{\ln N}$	$\bar{d}$
A	20	20	20	20	20	100	5	0.87	2.32
B	40	30	15	10	5	100	5	0.87	1.67
C	96	1	1	1	1	100	5	0.87	0.12

N = number of individuals in species 1–5; S = number of species.   *Source:* after Wilhm and Dorris (1968)

where s is the total number of species, N_i is the number of individuals in the ith species, and N is the total number of individuals. The unit for H'' is bits per individual if $\log_2$ is used, which is customary in information theory because binary systems are of interest (Zand, 1976). Approximate refers to the proportion N_i/N in the sample and not the infinite population as p_i represents in the formula for H' as is often given for the Shannon index (Kaesler *et al.*, 1978). The symbol $\bar{d}$ is also used (Wilhm, 1972) and is equivalent to H''. The maximum diversity $H''_{\max}$ from this formula is defined as $\log_2 S$. For example, $H''_{\max} = 5$ if 32 species are present.

Another formula, that describes diversity per individual is that of Brillouin (1962):

$$H = \frac{1}{N}\log\frac{N}{N_1!\,N_2!\ldots N_s!} \tag{12.5}$$

According to Zand (1976) and Kaesler *et al.* (1978), the Brillouin index describes diversity when the number of individuals in each species is low more accurately than does the Shannon index. Kaesler *et al.* (1978) gives a theoretical comparison of the two indices.

Both of the above formulae treat species richness or abundance and the evenness or equability of those species, which are the two ingredients of diversity. Thus, the environment with a more even distribution of its members or individuals, for the numbers of species present, could be thought of as being more varied and in a sense more 'balanced'. As indicated in Chapter 1, this does not necessarily mean the community is more stable and resistant to disturbance. If an unpolluted environment receives a diversity of food materials and has a diverse bottom substratum, it should also have a more equitable distribution of the abundance among species with differing requirements for food type and space.

Theoretically, one would expect the index to have more sensitivity to physical–chemical changes with evenness considered as well as richness.

A hypothetical example is shown in Table 12.8. Nevertheless, some diversity formulations have been widely criticized because they essentially define the improbability of an event and are considered to have little biological relevance (Hulbert, 1971; Washington, 1984). For example, Hulbert suggested a formula (Hulbert's PIE) that describes the probability of encounters and the relative importance of interspecific competition. Washington (1984) suggests that Hulbert's PIE has promise as a biologically relevant diversity index.

A convenient diversity index that requires little or no taxonomic knowledge is the Sequential Comparison Index (SCI) that is based on the 'theory of runs' (Cairns and Dickson, 1971). With this index, runs are enumerated on a sample and a diversity index, DI, determined:

$$\text{DI} = \frac{\Sigma \text{ no. runs/no. specimens}}{\text{no. trials}} \tag{12.6}$$

The number of taxa is then used to calculate the SCI or DI_{total}:

$$\text{SCI} = \text{DI} \times \text{no. taxa} \tag{12.7}$$

Cairns and Dixon suggested that values >12 indicate a healthy condition and <8 would indicate pollution.

As mentioned previously, diversity indices do not consider the specific taxa present among the sites being compared. Therefore, the numerical diversity index may be the same but the specific taxa representing the two communities could be completely different (and with different tolerances to the waste). Similarity, or comparative, indices are measures of community structure but compare the abundance of particular species (Washington, 1984).

Gaufin (1958) illustrated the need to keep track of the abundance of particular species in the Mad River, Ohio, USA where the pollution-tolerant snail *Physa*, the leech *Macrobdella* and the worm *Limnodrilus* were all present at the clean-water sites, but they became more abundant downstream from the waste sources. Hence, their abundance, rather than their presence only, indicated the waste effect. Comparative indices would take this aspect into account. An example of a commonly used index is that of Bray and Curtis (1957), which measures dissimilarity between two sites a and b:

$$D = \frac{\sum_{i=1}^{s} X_{ia} - X_{ib}}{\sum_{i=1}^{s} (X_{ia} + X_{ib})} \tag{12.8}$$

where $X_i a$ and X_{ib} are the number of individuals in the i_{th} species at the respective stations. Other similarity/comparative indices are given by Hellawell (1977) and Washington (1984).

TABLE 12.9. Comparison of $\bar{d}$ in response to various types of wastes

Waste	$\bar{d}$		
	Above outfall	Outfall	Downstream
Domestic and oil		0.84	1.59	3.44
Domestic and oil	3.75	0.94	2.43	3.80
Oil brines	3.36	1.58		3.84
Oil refinery		0.98	2.79	3.17
Total dissolved solids		0.55		3.01
Oil brines		1.49	2.50	
Oil brines		1.44	2.70	
Storm sewer		1.45	2.81	

Source: Wilhm and Dorris (1968)

The Shannon approximate index (H'' or $\bar{d}$) has been used most often to evaluate the effect of wastewater on diversity. Although the index has been criticized, Wilhm and Dorris (1968) demonstrated that $\bar{d}$ can be reliable in determining the effects of wastes using benthic invertebrates (Table 12.9). Near waste outfalls, $\bar{d}$ was typically ≈ 1. Intermediate values were found downstream and values farther downstream suggested recovery.

Further study of a variety of environments and waste types showed $\bar{d}$ to provide consistent results separating polluted from unpolluted conditions as follows (Wilhm and Dorris, 1968): Clean water was represented by similar values for 22 widely varying environments, mean $\bar{d} = 3.30$ (range 2.63–4.00); waste affected areas in 21 environments and widely varying waste types showed consistently low values, mean $\bar{d} = 0.95$ (range 0.42–1.60).

As a result of these observations Wilhm and Dorris suggested the following general guidelines for distinguishing the state of a stream: heavy pollution, $\bar{d} < 1.0$; Moderate pollution, $\bar{d} = 1.0$–3.0; Clean water, $\bar{d} > 3.0$.

Having absolute values for judging the conditions of streams would be highly desirable especially in situations where control sites are unavailable or pre-pollution data are lacking. However, the index has been used more often for site-to-site comparison, rather than relying too heavily on the absolute values, and in combination with other indices such as a simple abundance of species or taxa.

Diversity can be used effectively whether identification is to species or some other hierarchical level. This has been demonstrated by several workers. The effect of organic waste was easily recognized in Plum Creek, Ohio, USA as $\bar{d}$ decreased downstream from the wastewater input regardless of whether the taxonomic level used was genus, order or class

Figure 12.13. Diversity indexes for three taxonomic levels of macroinvertebrates in response to an organic waste (Egloff and Brakel, 1973)

(Egloff and Brakel, 1973) (Figure 12.13). The waste loading was rather mild as BOD increased on the average from 3.5 to 9.0 mg l^{-1} and DO was usually >6 mg l^{-1} downstream. The number of genera decreased from 16–22 upstream to 7–9 downstream. Hughes (1978) and Hellawell (1977) have also shown the close agreement among diversity values using either species, genera or family. However, sensitivity declined greatly if class or order were used similar to that shown in Figure 12.13. The family level was useful in describing conditions in two tributary streams to Lake Washington – one with a heavily developed watershed (7 invertebrate families) and the other relatively undeveloped (14 families). The range in $\bar{d}$ at three sites ($n = 9$) in the former stream was 1.75–0.29 and 2.97–1.60 ($n = 9$) in the latter (Knutzen, 1975).

The evenness factor in diversity indices may lead to difficulties in some situations. Species richness can decrease, while evenness actually increases due to other interacting factors in the environment (Hulbert, 1971). An interesting example of this is from Spring Creek, near State College, Pennsylvania, USA (Cole, 1973). The stream received a secondary treated effluent that contributed 5–20% of the stream's flow. BOD increased only slightly from 0.9–1.0 mg l^{-1} upstream to 1.0–2.2 mg l^{-1} downstream, although SRP increased from 50 to 300 μg l^{-1}. The treatment plant replaced individual septic systems and created a point source of enrichment, which apparently resulted in an increased abundance of macrophytes (*Elodea*, *Chara* and *Potamogeton*) downstream from the source producing large diurnal swings in DO. The number of species in riffles decreased from 46 to 27 (41%) downstream from the wastewater source; however, $\bar{d}$ decreased only 15% from 2.74 to 2.26 in spite of DO concentrations <3 mg l^{-1}. The author's explanation for failure of $\bar{d}$ to show an appropriate magnitude of effect was the increased habitat diversity created by the macrophytes allowing the

evenness factor to compensate for decreased species richness. Such interference with evenness by other factors has been shown by others
(Godfrey, 1978).

With regard to toxicants, Winner et al. (1975) found that the non-
formulated indices of number of species and individuals each showed a
predictable response to a gradient in Cu from 120 to $23 \mu g l^{-1}$ downstream from an experimental addition of $CuSO_4$. However, the Shannon
index (Equation 12.4) and Margelef's index (Equation 12.3) showed a
much poorer correlation with the Cu gradient. Moreover, four species
were eliminated at sites with 120 and $66 \mu g l^{-1}$ and such observations
must be considered in addition to the use of numerical indices.

The change in specific taxa is included in the Bray–Curtis dissimilarity
index (Equation 12.8) and was found to represent the response to Cu in
laboratory channels better than the Shannon index (Perkins, 1983). Other
comparative indices also gave a favourable response. The number of
species decreased linearly over a log series of Cu concentrations, while
the Bray–Curtis index increased with near linearity after 14 and 28 days'
exposure of an acclimated macroinvertebrate community. The Shannon
index, on the other hand, first increased and then decreased as Cu
increased. This behaviour was apparently related to interference by the
evenness factor. The Bray–Curtis index also provided the most meaningful description of the response to toxicity in other laboratory microcosms
(Pontasch et al., 1989).

Diversity and comparative indices were also evaluated following a
gasoline spill in Wolf Lodge Creek, Idaho, USA (Pontasch and Brusven,
1988). The Bray–Curtis index (Equation 12.8), as well as another comparative index (x^2) gave the most meaningful evaluation of the extent of
effects and recovery from the gas spill. However, as stressed by Pontasch
and Brusven, all indices have biases and none should be used to the
exclusion of sound ecological judgement.

With the myriad formulations in the literature, the investigator may
understandably be confused as to which one(s) to use to evaluate macroinvertebrate response in a stream receiving a waste that may be either
toxic, enriching, alter the habitat or some combination thereof. Much of
the criticism levelled at the diversity indices, especially Shannon's, is from
the standpoint of biological relevance. Yet, diversity indices continue
to be used, and used effectively in many cases, to describe effects of
pollution. The principal reason for their popularity is probably that they
require less knowledge about the organisms present than do biotic
indices. However, diversity (as well as similarity) indices are also more
general than biotic indices, which tend to be specific to certain geo-
graphical areas and usually to only one type of waste (organic). The
purpose here is to represent the types of indices available, emphasizing
those that have been used more frequently with some success, while

indicating their shortcomings. The investigator should not rely on any one index, but try several in an attempt to describe the population changes due to wastewater inputs to the stream in question. As Washington (1984) points out, some that have promise, such as Hulbert's PIE, should be used more frequently.

12.10.3 Sampling problems

Although the abundance of benthic invertebrates is highly variable, diversity is usually less so and should require fewer samples. Therefore, to use diversity to study waste effects in streams, stations should be selected that are of similar substrate type, velocity, etc. and random samples should be collected from either a grid or a transect in either a transverse or longitudinal direction. The sample should also be biased to high variety, or species richness, by selecting riffle areas where habitat is more diverse, thereby providing more sensitivity.

Relatively few samples are usually necessary to define diversity adequately (Wilhm, 1972). The values for $\bar{d}$ were found to approach a maximum in the first four samples out of ten in 13 different habitats. Thus, the mean diversity from three samples provided a reasonable estimate of the maximum diversity in each section of stream, assuming that the number of individuals (N) in each species is sufficiently great. Using the Brillouin index (Brillouin, 1962) (Equation 12.5), an N of only 10 was found to be adequate (95% confidence) to detect the effect of an acid spill, acid mine waste and domestic sewage, when 25 random samples were collected from a larger sample (Kaesler et al., 1978). However, H did not approach the maximum until N was >100. This was similarly shown by Hughes (1978) who successively pooled $50 \times 0.1\text{-m}^2$ samples. In this case, $\bar{d}$ approached the maximum in the whole $5\,\text{m}^2$ when $10 \times 0.3\text{-m}^2$ samples were analysed.

13
FISH

Fish are most frequently used as the so-called target organisms in developing criteria and setting standards for water quality. That is, the criteria for acceptable water quality, except for eutrophication, are usually based on fish, especially if they have economic or recreational value. Specific standards exist for dissolved oxygen (DO) and temperature that have been experimentally determined with fish and will be discussed in detail. The standards for toxicants are usually undefined, however, and the limits are determined by employing bioassays, either *in situ* or *in vitro* with the relevant water and species. That is due to variability in the effect of a given toxicant from water-to-water and species-to-species Nevertheless, published criteria for concentrations of particular toxicants are often used in the absence of on-site bioassay results to evaluate toxicity.

Fish populations are also used to assess water quality with indices similar to those used with macroinvertebrates (section 12.10). For example, Karr *et al.* (1986) developed an index of biological integrity using fish that includes trophic as well as diversity and tolerance attributes.

13.1 DISSOLVED OXYGEN CRITERIA

Recommended limits for DO were given by the National Committee on Water Quality in the USA in their 'green book' and 'blue book' (WQC, 1968, 1973) to protect communities and populations of fish and aquatic life against mortalities as well as to prevent adverse effects on eggs, larvae and population growth. A distinction has been made between cold- and warm-water environments. According to these documents, DO should be $>5 \, \text{mg} \, l^{-1}$ nearly all the time in warm water, assuming periods of much higher concentrations. For short periods, concentrations of $4-5 \, \text{mg} \, l^{-1}$ should be tolerated, but it should never become $<4 \, \text{mg} \, l^{-1}$. In cold-water environments, concentrations should be $>7 \, \text{mg} \, l^{-1}$ for successful spawning and egg and larval development. DO $>6 \, \text{mg} \, l^{-1}$ should exist for growth and $>5-6 \, \text{mg} \, l^{-1}$ for survival over short periods of time. These recommended limits have been changed recently and are published in the 'gold book' (WQC, 1986): 7-day mean minimum DOs of $4\,mg\,l^{-1}$ for warm water and $5\,mg\,l^{-1}$ for cold water with 7-day means somewhat higher at $6\,mg\,l^{-1}$ and $9.5\,mg\,l^{-1}$, respectively. As will be shown, these values may not be sufficiently restrictive.

Minimum levels for adult survival may be surprisingly low even for fish of low tolerance to DO depletion. For example, $1-3\,mg\,l^{-1}$ is sufficient for survival for a short time even with cold-water species such as salmon. These same fish could feed and reproduce at $3\,mg\,l^{-1}$ (Doudoroff and Shumway, 1967). However, the level required to sustain long-term survival and normal production of highly desired species, such as salmon and trout, is much higher.

Early field evidence by Ellis (1937) showed that DO concentrations $>5\,mg\,l^{-1}$ were associated with populations of fish of good abundance and species diversity. This was largely the basis for the early setting of the $5\,mg\,l^{-1}$ standard, which has often been misinterpreted as a 'requirement' for fish. The limit of $5-6\,mg\,l^{-1}$ that has come to be accepted in standards is probably a compromise for either warm- or cold-water fish. Requirements for some stages of a fish's life history are much higher than $5\,mg\,l^{-1}$. To ensure maximum production of many desired species of fish, a DO content near air saturation (e.g. $9.2\,mg\,l^{-1}$ at 20°C) should be maintained for the critical stages of life (Doudoroff and Shumway, 1967; WQC, 1973). However, many non-game fish species, such as goldfish (*Carassius*), have their normal activity limited at much lower concentrations of DO than do the salmonids (Macon, 1974).

Before proceeding to a discussion of effects of low DO on the various life stages and activities of fish, it would be well to make a point about the units with which to express DO. Although fish obtain oxygen from the water according to the partial pressure difference between the surrounding water and blood circulating through the gills, the desired unit is concentration rather than partial pressure or percent of saturation (Doudoroff and Shumway, 1967). This is because the metabolic demand for oxygen by fish increases as the temperature increases. At the same time the actual concentration at, e.g. 100% saturation, decreases. That is, the amount of DO necessary to maintain a constant partial pressure (say 100%) decreases as temperature increases.

That fish actually need more DO (concentration) as temperature increases to carry out normal activity, and as a result much more than a constant partial pressure, is shown in Figure 13.1, which was interpolated from data by Graham (1949). It can be seen that the standard metabolic rate of brook trout, in terms of O_2 consumed, increases dramatically as temperature increases to near the lethal limit. At the same time, the minimum DO concentration to sustain that metabolic rate increases from about 2 to $4\,mg\,l^{-1}$. The minimum saturation level to sustain metabolism,

Figure 13.1. Standard metabolism of brook trout (*Salvelinus fontinalis*) in relation to minimum DO needed in concentration and percentage saturation (modified from Graham, 1949)

of course, increases to an even greater extent. Thus, the overriding effect of temperature on metabolic demand for DO is apparent and maintenance of a constant percentage saturation as temperature increases would be detrimental to sensitive fish species.

13.1.1   Embryonic and larval development

The growth of salmonid (coho) embryos and the size of fry that emerge from the gravels in streams are limited by the supply of DO as shown by Shumway *et al.* (1964) in Figure 13.2. The dashed lines indicate embryo growth at 80 and 67% of the maximum.

Note that there is an interaction between DO and water velocity on the size of the emerging fry. Growth to a given level can be sustained by either a velocity increase with declining DO or a DO increase with declining velocity. At a constant high or low velocity, growth shows a progressive, although gradual, decrease below air-saturation values to a level of $\approx 5 \, \text{mg} \, \text{l}^{-1}$ after which the decline in growth is more rapid.

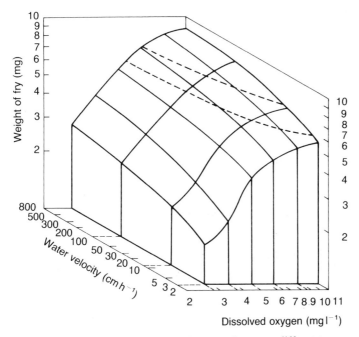

Figure 13.2. Development of coho salmon embryos at different concentrations of DO and at different velocity levels. Dashed lines indicate DO and velocity combinations producing 20 and 33% reduction in growth (Shumway *et al.*, 1964)

Subsequent work has shown a similar effect of DO on embryonic development to the larval stage (Carlson and Siefert, 1974; Carlson and Herman, 1974; Siefert and Spoor, 1973). For most species studied, which included a range from cold-water to warm-water fishes, the time to hatching increased, growth decreased and survival decreased as DO was reduced. The greatest effect occurred below $\approx$50% saturation, which was usually 5–6 mg l^{-1} at the test temperatures used. With species such as the white sucker, detrimental effects were not observed unless the DO level was <50% saturation. A greater tolerance was shown by larval survival for all species than by other indicators (Figure 13.3). Although it is clear that survival dropped off most precipitously at concentrations below $\approx$3–5 mg l^{-1} for all species, three species (salmon, trout and catfish) showed poorer survival at high concentrations. Two of those three species also showed a reduction in survival as DO was reduced below the highest level. Furthermore, delayed hatching, delayed feeding and/or slower growth were also shown with these species at levels between 100 and 50% saturation. These results, therefore, agree with those of Shumway *et al.* (1964); that is, detrimental effects on larval development will probably

Figure 13.3. Effect of DO concentration on survival of fish larvae from the embryo stages of four species. Time periods and temperature were as follows: channel catfish, 19 days at 25°C; lake trout, 131 days at 7°C; coho salmon, 119 days at 7–10°C; largemouth bass, 20 days at 20°C; and white sucker, 22 days at 18°C. (Data from Siefert and Spoor, 1973; Carlson and Siefert, 1974; Carlson *et al.*, 1974)

occur with both cold- and warm-water species when any reduction occurs in DO to levels <100% saturation even though the greatest damage occurs at concentrations <5 mg l^{-1}.

13.1.2 Swimming performance

The swimming performance of coho salmon also begins to decline progressively below 100% saturation (about 9–10 mg l^{-1}) (Figure 13.4). Again there is no clear threshold of effect, although the decrease in performance, as with larval development, is greatest at DO concentrations <5 mg l^{-1}. The points represent velocities at which the salmon failed to maintain themselves in a current (testing details are given in Davis *et al.*, 1963). Therefore, at DOs <9–10 mg l^{-1} the ability of migratory fish to traverse velocity obstacles or for them to avoid predators under any conditions would be reduced (Davis *et al.*, 1963).

13.1.3 Food consumption and growth

The amount of food consumed, as well as growth, is also progressively impaired below DO saturation (9–10 mg l^{-1} at test temperatures). This was observed for juvenile largemouth bass (a warm-water fish) as well

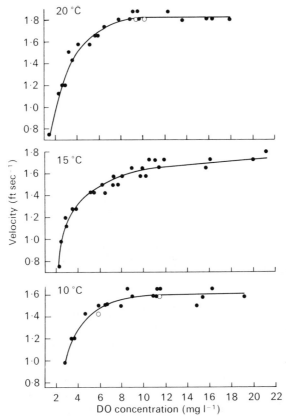

Figure 13.4. Relationships between DO concentration and swimming velocity in coho salmon at 20, 15, and 10°C. Each point represents velocity at which first underling failed to maintain orientation (Davis *et al.*, 1963)

Figure 13.5. Food consumption in largemouth bass related to DO concentration. Each curve represents a separate experiment (modification from Stewart *et al.*, 1967)

Figure 13.6. The growth rate of juvenile coho salmon fed restricted and unrestricted rations at constant and diurnally fluctuated oxygen concentrations (Doudoroff and Shumway, 1967)

as salmon (Warren, 1971). Figure 13.5 shows the relationship of food-consumption rate with DO concentration for bass (Stewart *et al.*, 1967). Again, the advantage of DO at saturation appears to hold for both warm-water and cold-water species.

Fish probably respond to the minimum daily DO rather than the daily average as shown in Figure 13.6. For juvenile salmon held at DO concentrations that fluctuated diurnally, the resulting growth was not very different from the growth of salmon held at a constant DO near the minimum for those exposed to a diurnally fluctuating DO. At unrestricted food ration, salmon clearly required a relatively high DO concentration for maximum growth. This is largely because the more food consumed, the more oxygen necessary for full utilization of that food.

The requirements for DO do vary among different species and seasons. However, the previously cited information indicates that near-saturated concentrations of DO are required during reproductive periods and periods of high food availability, consumption and growth if maximum production of cold-water as well as warm-water fish is to be realized. Levels much lower than saturation can be tolerated during periods of reduced feeding and growth is nearly independent of DO if not feeding. The minimum limits of DO at $5-6\,\mathrm{mg\,l^{-1}}$, often cited in standards, carry an assumption that normal variations result in concentrations being much higher than $5\,\mathrm{mg\,l^{-1}}$ most of the time. Doudoroff and Shumway (1967) have suggested that biologists should refrain from giving a value that is supposed to 'sustain maximum production of natural populations' unless it is air saturation ($9-10\,\mathrm{mg\,l^{-1}}$ for cold-water species). Of course, it is recognized that such high values are not needed all the time nor does DO

TABLE 13.1.   Example of recommended minimum concentrations of DO for selected levels of fish protection

Estimated natural seasonal minimum DO	Recommended minimum DO			
	Nearly maximum	High	Moderate	Low
5	5	4.7	4.2	4.0
6	6	5.6	4.8	4.0
7	7	6.4	5.3	4.0
8	8	7.1	5.8	4.3
9	9	7.7	6.2	4.5
10	10	8.2	6.5	4.6
12	12	8.9	6.8	4.8
14	14	9.3	6.8	4.9

Source: WQC (1973)

remain at or above those levels in all unpolluted water. In order to develop more specific standards, detailed seasonal requirements for various species would need to be determined.

Table 13.1 shows a working guideline based on the hypothesis of progressive detrimental effect below saturation (WQC, 1973). Minimum levels are suggested to provide high, moderate and low levels of protection based on the existing natural minimum. Note that $4 \, \text{mg} \, \text{l}^{-1}$ is suggested as a floor.

These suggested levels are based on the concept that some damage will result from any reduction in the average daily minimum DO below air saturation. From the appearance of many of the relationships (Figures 13.2–13.6), damage may not be as linearly related to DO reduction as this model (Table 13.1) would suggest. Nevertheless, this is an approach to a more reasonable guideline than the often stated minimum requirement of $5 \, \text{mg} \, \text{l}^{-1}$.

13.1.4   DO and sediment

Sediment may interact with low DO to adversely affect reproduction in salmonid fishes. Sediment enters streams from watershed erosion resulting from several types of disturbance. Of prime importance are irrigation systems, urban runoff and deforestation and the accompanying road-building activities. With the filling in of the interstices within the bottom substrata with fine sediment, the intragravel water velocities are reduced. As those velocities decrease, the supply rate of DO to developing embryos is also reduced. The expected effect is as suggested in Figure 13.2. At the same time the DO of the intragravel water is apt to decrease

TABLE 13.2.   Sediment load reduction and fish population improvement in Bluewater Creek, Montana

Station	Sediment reduction (ton day^{-1})	Reduction (%)	Pretreatment trout/rough fish	Post-treatment trout/rough fish
2	1.9	32		
3	14.0	52	39/61	78/22
4	10.5	44	12/88	51/49

Stations numbered in a downstream direction   *Source:* after Marcuson (1970)

because of the increased retention time of intragravel water (or slower replenishment of oxygen-saturated surface water) and the possible increased oxygen demand of organic matter within the sediments. The effects of reduced DO supply on developing trout eggs through sedimentation from stream-bank erosion caused by increased peak flows and sediment input from irrigation return flows were evaluated in Bluewater Creek, Montana, USA (Peters, 1967). More than 80% mortality of trout eggs buried within the stream gravels was observed at stations where water carried a mean suspended sediment content of only $200\,\mathrm{mg\,l}^{-1}$. The physical effect was similar to that caused by runoff patterns from urban watersheds with a high proportion of impervious surface (paved), although the substratum instability caused by increased flows may also result in physical damage to eggs.

Clear-cutting watersheds of small streams has resulted in DO decreases of 40% in the intragravel water with the effect persisting for at least three years after logging (Hall and Lanz, 1969; Ringler and Hall, 1975). Pre- and post-logging (three years after) intragravel DO averaged 10.5 and $6.2\,\mathrm{mg\,l}^{-1}$ respectively, during the winter-spring period. That resulted in a 73% reduction in the resident cut-throat trout population. Immediately after logging, intragravel DO dropped to $1.3\,\mathrm{mg\,l}^{-1}$ and stream temperature reached 24°C in summer. In a partially clear-cut watershed (25%), where buffer strips were left along the stream in affected areas and trees were not felled in the stream, DO decreased only $\approx 10\%$ and no damage to fish was observed. Emergence of young coho salmon was apparently unaffected by the physical changes.

In Bluewater Creek the banks were stabilized with vegetation and irrigation return-flow ditches were lined to cut down on sediment input. After 2 years of stream recovery following the improvements, the ratio of trout to rough fish increased sixfold (Table 13.2).

This illustrates the significance of stream-bank stabilization, vegetation protection and flow control, all of which should be employed to preserve the quality of urban streams. Even if inadvertently damaged, most

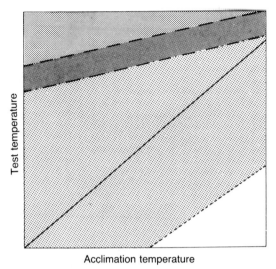

Acclimation temperature

Figure 13.7. Zones of tolerance and resistance of fish as affected by changing acclimation temperature. — —, Upper lethal limit; — · —, upper incipient lethal level; ---, lower incipient lethal level (modified from Brett, 1960)

streams can show some degree of recovery from sedimentation if proper corrections are installed. Probably the most pronounced recovery would come in those streams affected by a high discharge velocity and with only a moderate sediment load naturally. Those with a natural depositing substrata would probably show less improvement.

13.2 TEMPERATURE CRITERIA

The thermal tolerance of fish can be illustrated by Figure 13.7. A very large zone of tolerance can be defined within which all life-history stages of a species can proceed normally. The lower and upper tolerance limits, often referred to as the upper and lower incipient lethal levels, increase with increases in the acclimation temperature within the range of tolerance. At high temperatures there is a zone of resistance that is rather narrow between the incipient lethal level (long term) and the short-term lethal limit. The increase in the short-term lethal level for roach was found to be 1°C for each 3°C increase in acclimation temperature over the range of tolerance (Cocking, 1959).

13.2.1 Lethal temperature

Although the lethal limit is not very useful as a maximum limit to protect all life stages of a fish from the effects of heated water, the values are

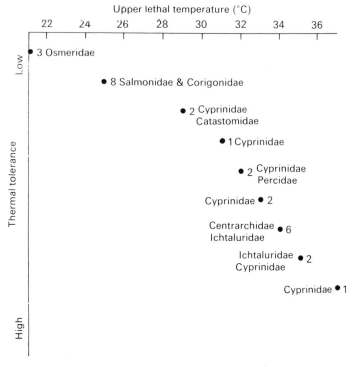

Figure 13.8 Families of fishes including number of species and lethal temperatures (compiled by Welch and Wojtalik, 1968)

determined with a high level of precision and do indicate the variation in species tolerance over a wide temperature range. Figure 13.8 shows the distribution of fish families to lethal temperatures over the range of tolerance. Several points are apparent. Smelt (Osmeridae), salmonids and white-fish (Corigonidae) are the most intolerant of high temperature, whereas catfish (Ichtaluridae) and some minnows (Cyprinidae) are the most tolerant. The minnows are, however, quite variable in their tolerance. That the members of that family are found in (and tolerate) a wide variety of thermal regimes is not surprising because there are so many species of minnows.

The lethal limit is usually expressed as an LT_{50} and is determined experimentally. Test fish are exposed at a constant temperature over time and the interpolated temperature at which 50% of the test fish survive for 96 h is the LT_{50}^{96}. If the temperature is fluctuating daily, slightly higher maximums can be tolerated for short periods if temperature is lowered below that level for longer periods. This has been demonstrated with juvenile salmon exposed to raised and lowered temperature (Figure 13.9).

Figure 13.9. Effect of exposure to three kinds of temperature cycles on mortality in juvenile salmon (Redrawn from Templeton *et al.*, 1969)

In this case, example C produced the greatest mortality, probably because the mean temperature over the exposure period was greatest even though in each case the fish were exposed to greater than the lethal temperatures for a short period.

Although fish can resist temperatures above the 96-h lethal limit for short periods, such thermal shocks have been shown to be detrimental to salmon even if not thermally lethal (Coutant, 1969). At temperatures from 28 to 29°C, salmon showed measurably shorter times to 'equilibrium loss' than to death and this equilibrium loss resulted in increased predation on the juvenile salmon.

13.2.2 Limits for normal activity and growth

Obviously, thermal standards or limits should not be set solely on lethality or 'equilibrium loss' in environments where fish are expected to grow and reproduce. The 'scope for activity' is the temperature range at which fish can optimally feed, swim and avoid predators. The scope for activity (as determined by respiration rate) occurs over a temperature range that is much less than the lethal limit (25°C) for brook trout, whereas for bullhead catfish the maximum scope occurs near the lethal limit, which is 37°C. This example illustrates a difference between cold- and warm-water fishes (Figure 13.10). The position of the maximum scope for activity for brown trout is similar to that for the bullhead, but that trout's lethal limit is near 25°C, similar to brook trout. Brown trout are generally considered more tolerant of higher sublethal temperatures than brook trout (Brett, 1956). This may partly explain why brown trout often tend to dominate in apparently marginal waters.

Warren (1971) has discussed how the 'scope for growth' for salmon should be maximum at temperatures much less than the lethal limit, but the scope for growth is also a function of food availability. Growth will decrease irrespective of temperature when food is restricted; however, the temperature range for the optimum scope will become narrower when food is restricted than if unrestricted. Averett (1969) has supported this hypothesis with experiments using juvenile coho salmon, showing that the optimum temperature or maximum scope for growth is in the range

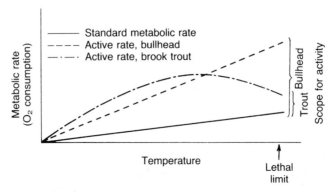

Figure 13.10. Relation of temperature to metabolic rate and scope for activity in brook trout and bullhead catfish. For brook trout, lethal temperature is 25°C and peak scope for activity is at 19°C. For bulhead, lethal and peak scope temperatures are at 37°C. For these species, scope for activity is difference between active and standard metabolic rate. Note that scope for activity is much greater for bullhead than trout near the lethal limit (from Brett, 1956)

Figure 13.11. Partitioning of food energy consumed in coho salmon at varying temperatures. Crosshatched area shows energy as growth, diagonal line indicates total energy consumed (modified from Averett, 1969)

14–17°C (Figure 13.11). At unrestricted ration, the amount for growth, after the necessary metabolic demands were met, decreased above 17°C, which is 7°C below the lethal limit for coho. The normal metabolic demands or losses were cost of food handling, standard metabolism and excretion.

Although the temperatures required for maximum scope for growth and activity would be desirable as criteria for all important and sensitive species, such extensive data may be difficult to obtain. A more easily obtained index experimentally is 'preferred temperature', or the temperature willingly selected by a species exposed to an experimental temperature gradient. That temperature (or range) tends to approximate the temperature range for the maximum scope for growth and activity. Table 13.3 shows lethal and preferred temperatures for cold- and warm-water species. Note that the preferred temperature for brook trout is

TABLE 13.3.   Final temperature preferred and
some upper lethal temperatures for a range in
species tolerance

Species	Final preference (°C)	Upper lethal (°C)
Lake trout	12	
Lake whitefish	12.7	
Rainbow trout	13.6	
Brook trout	14–16	25
Brown trout	12.4–17.6	
Yellow perch	21	
Burbot	21.2	
Muskellunge	24	
Yellow perch	24.2	32
Grass pickerel	26.6	
Small-mouth bass	29	
Goldfish	28.1	34
Pumpkinseed	31.5	
Carp	32	
Large-mouth bass	30–32	34
Blue-gill	32.3	34

Source: Welch and Wojtalik (1968)

14–16°C, very close to the temperature for maximum scope for growth of coho salmon (14–17°C). These data are plotted in Figure 13.12 and show an interesting phenomenon; the preferred temperature becomes closer to the lethal limit as thermal tolerance increases from cold- to warm-water species.

That fish select a preferred temperature in nature was demonstrated in a study of fish distribution in the Wabash River, Indiana, USA in response to a heated-water discharge (Gammon, 1969). The effluent was on average 8°C higher than the river temperature and the river-effluent mixed temperature within 1.2 km downstream was 2–3°C above the ambient (upstream) temperature. Good agreement was observed between temperature preference determined in the laboratory and the distribution of species in thermal zones in the river (Table 13.4). Although a clear segregation of species was observable during the warm season in relation to mean temperature, during winter all species were attracted to the effluent area because the resulting temperature there was closer to their preferred temperature than that of the normal river.

The abundance of all species in the river in relation to the mean daily maximum temperature is shown in Figure 13.13. The number of species was substantially reduced in the zone of maximum temperature increase. Although the temperature considerably downstream of the maximum

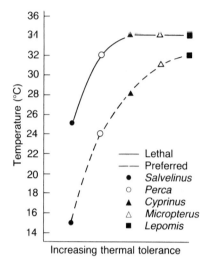

Figure 13.12. Preferred and lethal temperatures of five genera of fishes (data compiled by Welch and Wojtalik, 1968)

TABLE 13.4.   Distribution of representative species of fish in the Wabash River among several definable thermal zones

| Avoided all increases | Thermal gradient in the river | | |
	Normal to 27°C	32°C	35°C
Golden redhorse	Sauger	Gizzard shad	All fish left the
Shorthead redhorse	White bass	Carp	area, no mortality
Smallmouth bass	Northern river	Buffalo fish	
Spotted bass	carpsucker	Longnose gar	
		Shortnose gar	
		Channel cat	
		Flathead cat	

Source: data from Gammon (1969)

heated zone was higher than the normal river, more species were actually found there. Fish do not necessarily exist only in areas of their preferred temperature nor are they excluded from environments where temperatures are less than preferred. Although all species sampled were present in the river before heat addition, most apparently preferred a slightly elevated temperature than the normal. An important point for management in this case, however, is that the most desirable species was one of the most sensitive. Smallmouth bass, a highly prized game species, avoided all temperature increases.

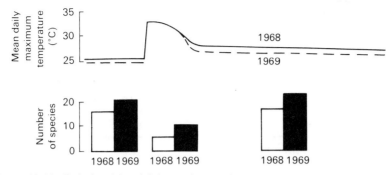

Figure 13.13. Relationship of fish species number to temperature increase in the Wabash River, Indiana, caused by heated water from a power plant (after Gammon, 1969)

Figure 13.14. Predicted species loss rate from a cold-water system (Columbia River, maximum mean daily temperature 21°C) and a warm-water system (Tennessee River, maximum mean daily temperature 30°C) with increasing temperature (Bush *et al.*, 1974)

Based largely on preferred and lethal temperatures, the prediction of fish community changes in the Columbia River (cold water) and the Tennessee River (warm water) were hypothesized with increasing (mean daily) maximum temperature (Bush *et al.*, 1974). If the percentage of species lost is plotted versus temperature for both rivers and compared with the normal maximum temperatures (Figure 13.14), it is again indicated that fish in warm-water environments are living closer to their lethal limit than most species in cold water (Figure 13.12). That is, a 2°C increase in the Tennessee River would result in more species lost than

TABLF 13.5. Thermal limits for salmon reproduction

Species	Limits for successful hatching and survival of young (°C)
Sock-eye	4.4–5.8 to 12.8–14.2
Chinook	5.6–9.4 to 14.4
All salmon	5.8 to 12.8

Source: Brett (1956)

the same increase in the Columbia River. In reality then, a greater 'assimilative capacity' for heat with respect to fish tolerance does not occur in warm water just because those species are more tolerant. On the other hand, there is not that much assimilative capacity available in cold water either because the maximum scope for growth and activity tends to be much less than the lethal limit in contrast to warm-water species. Moreover, salmon would be the first to disappear with a 2°C increase in the Columbia River. Nevertheless, more species would be lost by an equivalent increase in warm water, although the margin available for temperature increase in cold water may be only a few degrees, because of a few important species like salmon and trout.

This type of analysis may have value in assessing the potential effects of global warming on aquatic organisms. As previously indicated, fish tend to represent a narrower range of thermal tolerance than invertebrates. Increases in the average temperature of 3–4°C may cause extinction of several fish species in the Great Plains, for example, because migration north could be blocked and insufficent time would likely exist for genetic adaptation (Matthews and Zimmerman, 1990).

Thermal limits for reproduction have been determined rather precisely for salmon but less data are available for warm-water species. The results given in Table 13.5 for salmon were summarized by Brett (1956). The following are upper limits (maximum weekly means) for spawning suggested for the representative warm-water species (WQC, 1986): large-mouth bass, 21°C; small-mouth bass, 17°C; yellow perch, 12°C; northern pike, 11°C.

The critical nature of the reproductive life-stage is evident in the seasonal thermal requirements for pacific salmon (Figure 13.15) summarized by Brett (1960). Note that the most restricted range is during egg development and that a gradual increase is recommended in spring for normal growth to result. The temperature for optimum growth is suggested to remain rather constant after the larval stage, but the lethal tempera-ture for post-larval fish increases in spring and summer, probably as a result of acclimation to rising temperature and a longer photoperiod.

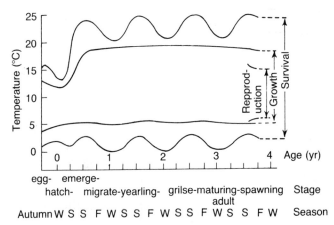

Figure 13.15. Schematic representation of thermal requirements for different life processes that characterize Pacific salmon (Brett, 1960)

13.3 TEMPERATURE STANDARDS

The standards for temperature vary from state to state in the USA, but in general are not greatly different from those recommended in WQC (1968) and will be discussed here. Standards for cold-waters are usually stated separately from those for warm-water environments. Warm water was defined by a >25°C maximum daily mean temperature, which is actually the upper lethal limit for most salmonids. The recommendations were that the rise above ambient (normal, before heat addition) should not exceed 2.8°C in streams and 1.7°C in lakes. The maximum temperature at any time should not exceed 32–34°C, depending on area. The rate of rise should not exceed $3°C\,h^{-1}$. Cold water was conversely defined by a <25°C maximum daily mean temperature. The rise above ambient should not exceed 2.8°C in streams and 3°C in lakes. The maximum temperature at any time should not exceed 20–21°C.

More recently, species-specific criteria were recommended for survival, growth and reproduction (WQC, 1986). For growth the maximum weekly mean temperatures were calculated as the optimum temperature plus one-third of the ultimate incipient lethal level minus the optimum). These values would fall roughly one-third of the distance between the lower and upper curves in Figure 13.12.

13.4 COMMENTS ON STANDARDS

The recommended maximum temperatures of 34°C and 21°C actually fall between the lethal and preferred temperatures for warm- and cold-water species as indicated above. If the limit for warm water (34°C) persisted

TABLE 13.6. Effect of flow and climate change on the flow-weighted temperature in a heated portion of a river

	Flow ($m^3 s^{-1}$)	°C	Flow-weighted	Mixed °C
Normal conditions during spring reproduction period				
Effluent	2	38	$\dfrac{76 + 84}{6}$	= 26.7
River	4	21		
Sudden increase in flow only				
Effluent	2	38	$\dfrac{76 + 252}{14}$	= 23.4
River	12	21		
Sudden flow increase and normal river temperature drop				
Effluent	2	38	$\dfrac{76 + 216}{14}$	= 21.9
River	12	18		

constantly for as long as a day or two, mortality could result or, more likely, many species would leave the area. However, fish would not be exposed to 34°C constantly. If the maximum daily temperature were kept below that level, temperature would be lower at night.

The maximum rise above ambient, or ΔT, is the second most critical factor and must be set from consideration of the following.

1. If the maximum temperature at any time is not allowed to exceed the recommended limit (e.g. 34°C), the length of exposure to that summer maximum will increase and may become critical as ΔT increases. Reduced growth may result if they are exposed too long to above the maximum scope for growth.
2. Reproduction may be offset from food availability if the ΔT is too high so that the optimum temperature for reproduction would occur earlier than the onset of food production which, if dependent on phytoplankton production, may be controlled by light intensity.
3. The amount of day-to-day fluctuation that would occur as a result of power-plant problems, flow change or climate change could result in day-to-day fluctuations that progressively increase.

Considering point 3, a hypothetical effect of flow and climate change is illustrated in Table 13.6. Such rapid temperature changes (drops) are possible in heated portions of rivers and would create stress in developing fry. Such changes have been shown to cause warm-water fish, such as smallmouth bass, to desert their nests (Rawson, 1945).

A logical approach to developing standards for temperature, as well as for DO, would be to make them less general and more specific for a purpose. The specific purpose would probably be protection of the most

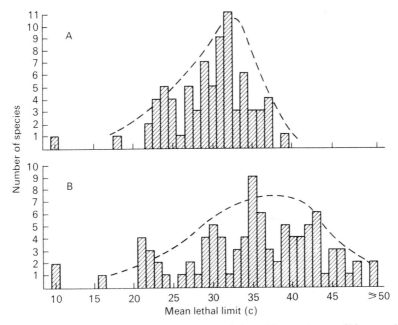

Figure 13.16. Distribution of freshwater fish (A) and invertebrates (B) according to tolerance to temperature. Dashed line shows approximate mode (Bush *et al.*, 1974 with permission of the American Chemical Society)

recreationally/economically important fish species, which is effectively what is usually done. The aquatic community in general would probably be protected by protecting those most sensitive fish, which has been suggested (Mount, 1969; Becker, 1972; Bush *et al.*, 1974). This seems reasonable if one considers the position of invertebrates compared to fish in the species frequency distribution with lethal temperatures. The lethal temperature mode for freshwater invertebrates is to the right of that for fish in Figure 13.16. Assuming that standards to protect fish would also protect the aquatic community in general, specific guidelines, such as those suggested in Table 13.7, could be utilized in setting limits for maximum temperature and ΔT in different environments.

With respect to points one and two earlier, one could then use these maxima with an ambient temperature cycle, which normally increases gradually from the winter low to the summer high, to judge a safe increment rise above ambient (ΔT) for individual species environments. Using this approach for the Tennessee River, it can be shown that 2.8°C may be a safe limit for ΔT while 5.6°C is not (Figure 13.17).

A ΔT rise above ambient of 5.6°C appears excessive for smallmouth bass, a highly prized sports species, in the Tennessee River during high-

TABLE 13.7. Maximum temperatures for various life stages of important fish species

Maximum temperature (°C)	Life stages
34	Growth: catfish, gar, white and yellow bass, buffalo, carpsucker, shad
32	Growth: largemouth bass, drum, bluegill, and crappie
29	Growth: pike, perch, walleye, smallmouth bass, and sauger
26.7	Spawning and egg development of largemouth bass, white and yellow bass, and spotted bass
20	Growth or migration routes of salmonids and for egg development of perch and smallmouth bass
12.8	Spawning and egg development of salmon and trout
9	Spawning and egg development of lake trout, walleye, northern pike, sauger, and Atlantic salmon

Source: WQC (1968)

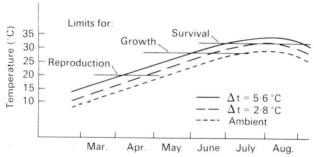

Figure 13.17. Effect of 2.8°C and 5.6°C temperature increments on the normal temperature cycle of the Tennessee River (Welch, 1969b). Periods for life-cycle activities that could be affected are shown according to the standards for smallmouth bass and yellow perch (*WQC*, 1968)

temperature, low-flow years. The periods of reproduction and growth, according to temperature maxima for those activities in Table 13.7 (20°C and 29°C maxima, respectively), would occur from one to two months earlier and either mortality, substantial weight loss or a complete evacuation of the species from the area could be expected in summer even if the maximum were restricted to 34°C. In fact, Horning and Pearson (1973) have noted that whereas the best growth for smallmouth bass occurred at 26–29°C, a temperature even lower was preferred, so accordingly 29 or 30°C would be the maximum tolerable temperature for the species. This technique of an optimized thermal regime based on a consideration of the various life-stage requirements has also been illustrated similarly for largemouth bass in WQC (1973).

Interestingly enough, a ΔT of 5.6°C (10°F) was accepted by most states in the Tennessee Valley in the late 1960s, but under pressure from the then federal Water Pollution Control Administration, the standard was reduced to 2.8°C (5°F) which seems appropriate from the above analysis. Subsequently, a series of whole-ecosystem, 112-m-long artificial streams were constructed at the Browns Ferry Power Station to develop temperature criteria for populations of smallmouth bass (Wrenn, 1980). Survival and growth of bass through a full year (1977–1978) was not different among four treatment temperatures: ambient (maximum = 30°C), ambient +3°C, ambient +6°C and ambient +9°C. As a result, a maximum weekly mean temperature of 32–33°C for growth and a short-term maximum of 35°C were recommended for smallmouth bass (Wrenn, 1980). These limits contrast significantly with the previously reported range for optimum (25–29°C) and the reported upper lethal temperature (35°C). Apparently, a greater tolerance to raised temperature was shown for this species when exposed throughout a full year in a whole-stream environment than had been the case when exposed under laboratory conditions.

13.5 TOXICANTS AND TOXICITY

The problems caused by toxicants in wastewater and their effects on fish and aquatic invertebrates have plagued fisheries managers and pollution control agencies in the USA for over 50 years. Recently, the threat of toxicants and other hazardous substances to humans and the clean-up of abandoned waste dumps have received much attention. The pathway to the human population from these toxic waste dumps is usually through the groundwater, so aquatic organisms in surface waters are often not an issue at hazardous waste sites. Nevertheless, the >1000 abandoned sites deemed sufficiently hazardous to require clean-up (Miller, 1988) represent a more recent addition to society's legacy of less than effective control of toxicants.

While many problems of toxicity in surface waters have been controlled (e.g. treatment of point-source industrial effluents and banning the use of some pesticides), there are continually new ones caused by the increased production of synthetic organic chemicals. Although detailed analysis of the effects of each of these compounds is required before marketing is approved, aspects of their long-term effects and fate/recycling in the environment are often not revealed for many years. Food-chain bioaccumulation effects of chlorinated hydrocarbon insecticides (DDT, Dieldrin, etc.), polychlorinated biphenyls (PCBs) and methylated mercury (CH_3Hg^+) are examples of unforeseen problems.

Because of the variation in tolerance among species to be protected and variation in effects caused by interacting environmental factors, toxic effects are usually not predictable in any quantitative sense. That is, the response of fish populations cannot be predicted, especially in the long term, from given inputs of toxicants to a specific stream. Moreover, acceptable levels for toxicants, as is the case for temperature, DO and pH, are usually not given in enforcement standards for water quality. What is usually offered in place of specific standards is a statement to the effect that 'no contaminant (pollutant) can be added to waters that is detrimental to fish and aquatic life or other water uses'. The onus is left on the wastewater discharger and/or enforcement agency to assess each problem on its own merits. Such an assessment most appropriately employs an experimental approach, but may only entail a simple comparison of existing levels of toxicants in the water in question with published criteria (e.g. WQC, 1986). Application of test results from one water to another, with differing chemical characteristics, may not be appropriate without correction or qualification as will be discussed later. Criteria for some toxicants (e.g. Cu, Zn, Cd, ammonia) are interpreted as standards in some states.

13.5.1 Determination of toxicity

Controlling waste loading to waters to prevent detrimental effects without specific standards usually necessitates some type of investigation into the toxicity existing in the water or the potential toxicity from wastes either added already or planned for addition. The investigation usually entails a standardized toxicity bioassay, employing the waste, the receiving water and an experimental animal, usually a fish, but may include a crustacean or alga. The results from such a bioassay (or test) provide information about the acute, or short-term, toxicity, from which a lethal concentration can be determined. Based on the lethal concentration and some knowledge of chronic effects, a 'safe level' can be estimated.

Procedures for acute toxicity bioassays (or tests) are given in Standard Methods (APHA, 1985). Typically, the exposure period is 2 to 4 days to a series of concentrations with the number of surviving animals recorded daily or for shorter intervals. Response is commonly plotted as either percentage survival (arithmetic scale or transformed, e.g. as probits) with time being held constant (4 days), or as time to 50% survival, on the ordinant and concentration of toxicant on the abscissa (Figure 13.18). With time constant, the concentration producing 50% mortality (survival) is interpolated and is termed the lethal concentration producing 50% survival after 96 h or the 96-hour LC_{50}. If log concentration is plotted, the relationship is more linear.

With time variable, the 'lethal threshold concentration' can be estimated by approximating the toxicant concentration at which the resistance time becomes long (asymptotic) at an additional, relatively small decrease in toxicant concentration (Figure 13.18).

Figure 13.18. Two graphical methods of determining the lethal concentration of a toxicant. LC_{50}^{96} (medium lethal concentration after 96-h exposure) and LTC (lethal threshold concentration) indicated by broken-line arrows

LC values at other percentage survivals (e.g. LC_1, LC_{10}) can also be estimated from the data by plotting results on probit paper on which survival versus concentration is a straight line (APHA, 1985). However, the 50%, or median, survival level is most commonly used because that estimate is least variable.

Bioassays can be either static, with no exchange of test water during the 96-h period, or continuous flow, entailing either several solution changes per day or simple renewal of the bulk solution daily. The continuous-flow procedure is considered more appropriate, because in nature organisms are continuously exposed to renewed solutions. In static laboratory solutions, animals can effectively alter the toxicant concentrations and either detoxify the solutions or cause self-contamination with waste products. The lethal concentrations from continuous-flow experiments are apt to be lower than from static ones. For example, Brungs (1969) reported a 96-hour LC_{50} for Zn and fathead minnows (*Pimaphales promelas*) in static tests of $12–13\,mg\,l^{-1}$, but it was only $8.4\,mg\,l^{-1}$ in continuous-flow tests.

Knowing the median lethal concentration, or the concentration that kills 50% of the fish, is not an acceptable goal to control toxicity in a receiving water. Conventionally, an application factor (AF) is used to proportionately reduce the lethal concentration to one that would allow complete survival of all stages of the animal's life history. The 'safe concentration' is the product of the 96-hour LC_{50} and the AF.

AFs vary from 0.5 to 0.01, depending on the toxicant. Although in many instances the AF value is arbitrary, AFs have been developed for many toxicants that are based on long-term (1–2 months), chronic bioassays using such life-history functions as growth, fecundity (egg production), hatching success, food conversion and behaviour as the response. With the chronic effect level determined for the most sensitive life history stage, the AF can be estimated as:

$$AF = chronic\ LC_{50}/LC_{50}^{96}$$

Figure 13.19. Effect of Zn^{2+} on fecundity in fathead minnows (Brungs, 1969)

Results of AF determinations for such toxicants as pentachlorophenol, cyanide, dieldrin, 2,4-D, malathion, copper and pulp-mill effluent, using growth and reproductive success most often, have ranged from 0.02 to 0.5 (Warren, 1971). Additional values are given in WQC (1973) for LAS (detergent), chromium, chlorine, sulphide, nickel, lead and zinc. As an example of the procedure, Brungs (1969) showed that fecundity in fathead minnows was more sensitive to chronic levels of Zn^{2+} than growth or egg hatching success (Figure 13.19). Using the calculated 50% reduction in fecundity and the associated concentration of Zn^{2+}, the estimated AF could be calculated:

$$AF = 0.088/9.2 \, mg \, l^{-1} \, LC_{50}^{96} = 0.009$$

In the case of Zn, the more conventional AF of 0.1 would be tenfold too high.

This procedure of estimating a safe level is very effective in accounting for the variability in water quality characteristics within a receiving water, as well as the variability among waters. Those variations are accounted for in the short-term acute bioassay using the receiving water in question. The more difficult to obtain AF is determined from long-term chronic bioassays conducted earlier.

Faster procedures for determining chronic toxicity have been developed for testing complex effluents. These are a 7-day subchronic test using larval and embryo-larval stages of fathead minnows (Norberg and Mount, 1985), and a three-brood, life-cycle test using the cladoceran *Ceriodaphnia* (Mount and Norberg, 1984). Chronic concentrations for the larval fathead minnow tests are estimated as either the LC_1 or the mean of the concentrations giving no observable effect and the lowest observ-

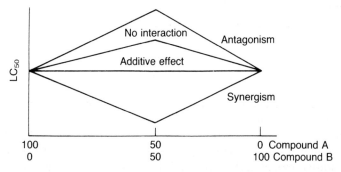

Figure 13.20. Relative effect of toxicant mixtures on the LC_{50}

able effect using growth and survival over 7 days. Precision was greater and threshold concentrations lower for LC_1 estimates than for means of no effect and lowest effect using two toxicants (Pickering, 1988). Short-term chronic tests such as these were developed for use in the permission system for wastewater discharge.

There is still the problem of extrapolation from one test species to the myriad species in the receiving water. Because it is not possible to conduct long-term (or even short-term) chronic tests with every species in a receiving water, there are three alternatives (Woelke, 1968): select the most sensitive species with which to determine the AF; assume proportionality in AFs between an easily studied test species (e.g. fathead minnow) and the most sensitive species in the receiving water; or select the most economically important species even if it may not be the most sensitive. Woelke (1968) used the reproductive stage of the economically important Pacific oyster and subsequently showed that it is as sensitive as many other shellfish species to pulp-mill waste as well as oil-spill dispersants and is much more sensitive than adult fish. In that case, the oyster is an ideal test species being both sensitive and economically important.

Another serious problem if the $LC_{50} \times AF$ procedure is used is the complexity of most wastes which often contain numerous toxicants that can interact. Mixtures of toxicants can display either additive, antagonistic or synergistic effects (Figure 13.20). If strictly additive, there is no difference in the LC_{50} whether the mixture is mostly compound A or B. If antagonistic in their effect, the more equal the mixture of A and B, the higher the LC_{50} (or lower the toxicity). Antagonism exists between Zn^{2+} and S^-, where toxicity of the mixture is least when the stochiometric ratio is one (Hendricks, 1978). Synergism results in a lower LC_{50} (more toxicity) as the mixture of A and B is equalized. Cu^{2+} and Zn^{2+} together is a classic example of synergism. There is a no interaction area where the effect is due to either A or B.

Assuming an additive or slightly synergistic effect is one approach to estimating the effect of a mixture of toxicants. An additive model has been used to predict lethal effects of mixtures of toxicants (Brown, 1968; Brown et al., 1970):

$$C_1/LC_{50(1)} + C_2/LC_{50(2)} + C_n/LC_{50(n)} = 1.0$$

where C_1, $C_2 \ldots C_n$ are the concentrations of various toxicants in the mixture. Thus, in a mixture where the fractions of all toxicant concentrations, relative to their respective LC_{50} values, sum to 1.0, 50% mortality should result. In actual applications, Brown et al. (1970) found that the additive model underestimated the acute toxicity in two polluted rivers and an estuary. That is, the respective fractions produced 50% mortality when the sum was >1.0 – of the order of two-thirds. Therefore, one might wish to stipulate that the sum of the fractions of the respective toxicants should not exceed some value <1.0, say 0.5–0.8. A model to approximate long-term safe concentration of a wastewater with a mixture of toxicants in a receiving water would be (WQC, 1973):

$$C_1/LC_{50(1)}AF_1 + C_2/LC_{50(2)}AF_2 \ldots C_n/LC_{50(n)}AF_n < 1.0 \text{ (e.g. 0.8)}$$

A simple hypothetical example of estimating an acceptable volume addition of a wastewater using the above model is as follows. River X flows at $5\,m^3\,s^{-1}$ during summer low flow and effluent Y flows at $0.1\,m^3\,s^{-1}$ with a Cu content of $1.0\,mg\,l^{-1}$ and a Zn content of $1.5\,mg\,l^{-1}$. Actual short-term bioassays are usually conducted with a sensitive species and the receiving water in question.

The mixed, effluent-river concentrations ($0.1/5 \times 1.0$ and 1.5) are $0.02\,mg\,l^{-1}$ Cu and $0.03\,mg\,l^{-1}$ Zn. Assuming slight synergism and determined LC_{50} values of 0.08 and $0.5\,mg\,l^{-1}$ for Cu and Zn, respectively, suggests that short-term mortality should not result:

$$0.02/0.08 + 0.03/0.5 = 0.31$$

But there could be a chronic effect assuming AFs for Cu and Zn of 0.1 and 0.01, respectively:

$$0.02/(0.08 \times 0.1) + 0.03/(0.5 \times 0.01) = 8.5$$

Therefore, a recommendation would be to reduce the effluent flow during low river flow by 90% $(1 - 0.8/8.5)$. Or the discharger could opt to remove the toxic metals through treatment or to pond the wastewater and wait for higher river flows such that the mixed river concentrations do not exceed $1.9\,\mu g\,l^{-1}$ Cu and $2.8\,\mu g\,l^{-1}$ Zn ($0.8/8.5 \times 20\,\mu g\,l^{-1}$ and $30\,\mu g\,l^{-1}$).

13.5.2　Heavy metals

Elements such as Zn, Cd, Cu, Pb, Cr, Hg, Ag and Ni are well-known heavy metals that can occur in a variety of wastes and cause either acute or chronic effects on organisms in receiving waters. Although one or more of these metals can originate from a variety of industrial activities, they can also occur in significantly high concentrations in municipal effluents, urban runoff and with Pb and Hg through atmospheric deposition. Recommended criteria for maximum, 4-day mean concentrations for each of these metals (as total acid digestable) are given in Table 13.8. These values are based on rather extensive chronic bioassays and adherence to these criteria should provide minimal risk to even the most sensitive species.

The toxic effect of a given concentration of a metal can vary greatly depending on the form of the metal, which depends on pH, temperature and the presence of dissolved organic substances. The mode of action of metals that produces the lethal effect is an irritation to the fish's gills, causing a secretion of mucus and internal deterioration of the gill lamellae. Death is due to suffocation as the capacity of the gill system to effectively exchange gases between the ambient environment and the fish's blood is impaired. Anionic imbalance can also result as excretion and ion exchange across the gill membrane are adversely affected. Lloyd (1960) showed with rainbow trout killed by exposure to $ZnSO_4$ that most of the Zn was contained in gills and mucus. The mucus, however, was produced from the body surface and not the gills. Cytological breakdown of the gill tissue was observed within a few hours at high concentrations. Tissue hypoxia resulting from cytological breakdown of the gills has been further demonstrated as the cause for acute toxicity from Zn (Burton et al., 1972).

Usually the effect of heavy metals is considered to be due more to the dissolved ionic form of a metal (e.g. Cu^{2+} or Zn^{2+}) than to other hydroxide complexes (e.g. $Cu(OH)^+$, $Cu(OH)_2$, $Zn(OH)^+$) or carbonate complexes. The relative amount in the ionic form increases with a decrease in pH. For example, Cairns (1971) showed for a range of Zn concentrations from $10-32\,mg\,l^{-1}$ that mortality of only 0–10% occurred at pH values from 7.3 to 8.8, but mortality was complete at pH values from 5.7 to 7.0. However, Mount (1966) found that toxicity of Zn to fathead minnows increased with increasing pH. The LC_{50} values increased by a factor of nearly 3 when pH was decreased from 8.6 to 5.0. This was thought to be due to an effect of continuous flow bioassay systems maintaining precipitated Zn in suspension. A direct effect of pH on the toxicity to fish from Cd, Zn and Cu was also shown by Cusimano et al. (1986). Toxicity was reduced at pH 4.7 compared to pH 7 and was explained as H^+ acting as a competitor for binding sites on the gill surface (see discussion below).

Thus, there is considerable uncertainty in estimating the toxic effect of a metal (or metals) based on determinations of total acid-digestible content of either filtered or unfiltered water. The latter (i.e. unfiltered) would tend to overestimate availability of the ionic form while the former

TABLE 13.8.   Recommended maximum
concentrations for representative toxicants
computed as 4-day means

Toxicant	Maximum $(\mu g\, l^{-1})$	AF
Metals		
Al	100	–
Cd[a]	1.1	–
Cr (VI)	11	–
Cu[a]	12	0.1
Pb[a]	32	0.03
Hg	0.012	–
Ni[a]	160	0.02
Zn[a]	110	0.01
Other inorganics		
NH$_3$	20	0.05
Cl$_2$	11	–
H$_2$S	2	–
HCN	5.2	0.05
Organics		
Phenol	100	0.05
LAS (detergent)	200	0.05
Organo Cl pesticides	0.001–0.004	–

[a] At $100\,\mathrm{mg\,l^{-1}}$ hardness.   Source: WQC (1973; and 1986)

may underestimate availability from equilibrium considerations or over-estimate availability if an organic complex is involved. The criteria listed in Table 13.8 are based largely on experiments employing the ionic soluble fraction of the metal, but for lack of an appropriate fast analytical technique are interpreted based on total metal concentrations (WQC, 1986).

As an alternative, McKnight (1981) used a specific ion electrode to determine the effect of the copper ion, following a CuSO$_4$ treatment of a water supply reservoir in which there was a complexing interference to toxicity from humic substances. Laegrid et al. (1983) also used the specific ion electrode to show that the free metal ion accounted for the toxicity of Cd in waters with high humic substances, but also found that com-plexation with low-molecular weight organic compounds, apparently secreted by phytoplankton, increased toxicity beyond that due to Cd^{2+} alone. A similar approach with Cd was used by Sherman et al. (1987), who found that Cd additions to a pond environment did not produce the expected toxicity based on Cd^{2+} concentrations in laboratory experi-ments, because high pond pH caused precipitation as CdCO$_3$.

Another important factor that affects the toxicity of metals is hardness (Ca + Mg), which reduces the toxic effect of a metal through competitive inhibition at the gill surface. The non-toxic Ca^{2+} and Mg^{2+} ions compete with the toxic metals for binding sites. If Ca or Mg occupy the sites, the gill lamellae are protected from deterioration. Pagenkopf (1983) has shown that the competitive interaction (inhibition) factor (CIF) can be estimated according to:

$$CIF = S^{n-}/S_T = 1/(1 + K_M[M^{2+}])$$

which represents the available fraction (S^{n-}) of the total (S_T) gill sites. K_M and $[M^{2+}]$ are, respectively, the equilibrium solubility coefficient and molar metal concentration. The effective toxicant concentration (for Cu in this case, ETC_{Cu}) includes the CIF and the available toxic metal concentration (the ionic form):

$$ETC_{Cu} = [Cu_T]\alpha Cu_i/(1 + K_M[M^{2+}])$$

where Cu_T and αCu_i are total and fractional coefficient for available $[Cu^{2+} + Cu(OH)^+ + Cu(OH)_2]$ Cu concentrations. Pagenkopf considered competition to be effective if hardness was >1 mM (100 mg l^{-1} $CaCO_3$). By correcting LC_{50} values for hardness by computing ETC values from several data sets using Cu, Zn and Cd, the variability among experiments was reduced by an average of 35%.

Criteria for maximum, 4-day mean concentrations for metals (Table 13.8) can be corrected for hardness using empirical equations (WQC, 1986):

$$Cd = e^{(0.7852[\ln \text{ hard}]-3.49)}$$

$$Cu = e^{(0.9422[\ln \text{ hard}]-1.464)}$$

$$Zn = e^{(0.8473[\ln \text{ hard}]+0.7614)}$$

Using the above equations, recommended maximum concentrations for Cd, Cu and Zn for hardness levels of 50, 100 and 200 mg l^{-1} are, respectively: 0.66, 1.1 and 2.0; 6.5, 12 and 21; and 59, 110 and 190, all in μg l^{-1} (WQC, 1986). The values listed in Table 13.8 are calculated for 100 mg l^{-1} hardness.

Temperature can affect the toxicity of metals and other toxicants as well (DeSylva, 1969). LC_{50} values can be expected to decrease with increased temperature even within the animals optimum range. This may be due the increase in the fish's metabolic rate, enhancing the rate of toxicant uptake or the stress from increased temperature may lower the fish's resistance.

Toxicity can also be increased by low DO. Pickering (1968), for example, showed a lowering in the 96-hour LC_{50} for bluegill sunfish

TABLE 13.9.   Comparison of acute toxic levels (designated) for different organisms and tests.

Species/test	Acute toxic level ($\mu g\,l^{-1}$)				Index
	Cr	Cd	Cu	Zn	
WQC (1986)	16	4	18	120	Acute max. for $100\,mg\,l^{-1}$ hardness
Rainbow trout	17	10	80	500	LC_{50}
Daphnia	50	5	60	1000	LC_{50} (approx.)
Selenastrum	238	57	54	51	EC_{50}
Microtox[a]	651	236	8	107	$EC_{50} \times 10^{-3}$
Techan[b]	124	140	54	30	$EC_{50} \times 10^{-3}$

[a] Bacterial luminescent activity after 5 min. [b] Marine alga O_2 uptake and luminescent activity after 1 h. From WQC (1973, 1986), Turbak et al. (1986) and McFeters et al. (1983)

exposed to Zn; a LC_{50} of $10-12\,mg\,l^{-1}$ at a DO content of $5.6\,mg\,l^{-1}$ compared to a LC_{50} near $7\,mg\,l^{-1}$ at a DO of $1.8\,mg\,l^{-1}$.

At this point it may be interesting to compare the response of different species/tests to representative metals (Table 13.9). The acute criteria (WQC, 1986) are recommended maximum, 1-h mean concentrations and are not much lower than the LC_{50} values for rainbow trout. LC_{50} values for Daphnia are quite similar to those for trout. Several authors have shown the similar response of fish and Daphnia (Atwater et al., 1983; Doherty, 1983). The sensitivity of Selenastrum to metals, comparable to that for fish and Daphnia, is not surprising considering that Cu (as $CuSO_4$ and organic complexes) and Zn (organic complexes) are used as algicides. Some of the relatively fast tests used for toxicity screening for potential toxicity, such as the Microtox and Techan procedures, are relatively insensitive compared to the responses by other organisms.

Over the long term, however, metals can be taken up at sublethal concentrations and accumulated within body tissues. In general, uptake and accumulation of a metal at low concentration can occur more readily if it is complexed in organic form and relatively non-ionic. An example of the relative effectiveness of an ionic and non-ionic form was shown with Hg. Ethyl mercury phosphate (EMP) showed a marked difference in acute toxicity to fish whether added with or without Cl^- as described by Amend et al. (1969):

Test chemicals	Mercurial produced	Percent mortality
EMP + H_2O	$CH_3CH_2Hg^+$	0–5
EMP + NaCl	CH_3CH_2HgCl	25–60

EMP + CaCl$_2$	CH$_3$CH$_2$HgCl	15-65
EMP + Ca(NO$_3$)$_2$	—	0
NaCl, CaCl$_2$ + Ca(NO$_3$)$_2$	—	0

As the above indicates, and as suggested by Laegrid *et al.* (1983), such an effect is most pronounced by low-molecular weight compounds. With Cl$^-$ added, an unionized organic mercury complex was produced, which can be taken up more readily and therefore produce a greater mortality.

Compounds of intermediate water solubility seem to be most toxic, i.e. produce an effect at a lower concentration. If a compound is very water-soluble (polar), it is readily available because of its solubility in water and, if in high enough concentration, it can produce toxicity through an irritation effect on gills. However, the ionized form does not move across lipid membrane surfaces as readily as non-polar low-water-soluble compounds and they would not be readily absorbed. The non-polar low-water-soluble substance, on the other hand, will not be as available for uptake. Because of the ease of uptake of organic complexes, high body concentrations have been shown to result in fish exposed to very low concentrations in water.

One of the most toxic compounds is methyl mercury (CH$_3$Hg$^+$) which is an organic Hg complex. This compound can be produced in nature through microorganism-mediated reactions in the sediment regardless of the form in which Hg is added. Once an environment is contaminated with Hg, the sediments, through the continued microbial production of CH$_3$Hg$^+$, can contaminate fish for many years. It would not be unusual for a receiving water to require from 20 to 30 years to recover from the detrimental effects of continued recycling of CH$_3$Hg$^+$ from sediments to organisms even after the Hg input has ceased (Håkanson, 1975; Shin and Krenkel, 1976). Although organisms will lose Hg from tissues once the absorption rate has decreased, with a half-life body decay of about 2 years, CH$_3$Hg$^+$ nevertheless can be concentrated in tissue by a factor of thousands of times from water concentrations as low or lower than 1 μg l^{-1} (Peakall and Lovett, 1972). The sediment effect of CH$_3$Hg$^+$ production could last from 10 to 100 years, but normally the sedimentation process should cover the Hg reserve sufficiently to result in recovery to acceptable levels in organisms in from 15 to 30 years (Håkanson, 1975).

13.5.3 Other inorganic toxicants

Examples of common toxic non-metal, inorganic compounds include ammonia (NH$_3$), hydrogen cyanide (HCN), hydrogen sulphide (H$_2$S) and

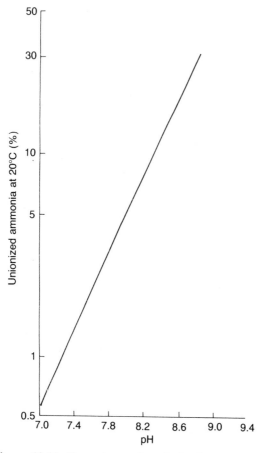

Figure 13.21. Percentage of un-ionized ammonia (NH_3) as a function of pH (*WQC*, 1973)

chlorine (Cl_2). All of these toxicants are most toxic in their undissociated, molecular state.

The quantity of the toxic undissociated (unionized) ammonia (NH_3) form is a function of pH, as shown in Figure 13.21. Ammonium (NH_4^+) is the most abundant form and accordingly analyses should be reported as NH_4^+-N. The pH controlled equilibrium is:

$$K = [NH_4^+][OH^-]/[NH_3]$$

Therefore, as pH increases so does the unionized ammonia fraction.

The lethal concentration for unionized NH_3 is $\approx 1\,mg\,l^{-1}$ for most species of fish. Concentrations of even total NH_4^+ seldom reach that level naturally. However, ammonium may reach or exceed that quantity in anoxic eutrophic environments where nitrification is impeded. Also, several milligrams per litre are common in domestic sewage and some industrial wastes. In fact, ammonium-N in effluents from secondary sewage treatment plants can approach $20 \, \text{mg} \, l^{-1}$. However, one can readily see the overriding importance of pH from Figure 13.21. About $200 \, \text{mg} \, l^{-1}$ ammonium would be required for lethality ($1 \, \text{mg} \, l^{-1} \, NH_3$) at pH 7, while $<3 \, \text{mg} \, l^{-1}$ would be sufficient at pH 9. Hence, the importance of algal blooms (eutrophication) in waters receiving high ammonium wastewater, due to the potential for photosynthetically caused high pH (Chapter 4.4).

The recommended maximum limit for NH_3 as a 4-day mean is $0.02 \, \text{mg} \, l^{-1}$. This value may be unnecessarily low, if calculated from the ambient water pH, because fish tend to lower the pH of water surrounding their gills by respiring CO_2 (Szumski et al., 1982). However, Solbe and Shurben (1989) have shown that rainbow trout eggs, if exposed from shortly after fertilization to 2.5 months, sustained high mortality at NH_3 concentrations as low as $0.027 \, \text{mg} \, l^{-1}$.

HCN and H_2S are also undissociated forms that are controlled by pH and are also the toxic fraction (Doudoroff et al., 1966). In contrast to NH_3, the undissociated HCN and H_2S fractions increase with decreasing pH. For example, at pH 7 one half is the toxic H_2S and one half is HS^-, but at pH 5, 99% is H_2S (WQC, 1986). H_2S is unstable in the presence of O_2 and readily converts to SO_4^{2-} (Chapter 4.3). Thus, H_2S would usually not present a problem where DO is otherwise adequate for fish. However, a critical condition could occur for developing eggs and larvae near sediments where H_2S would originate. Recommended maximum, 4-day mean concentrations for HCN and H_2S are 5.2 and $2 \, \mu g \, l^{-1}$, respectively (WQC, 1986).

Chlorine became increasingly important during the early and mid-1970s as investigators discovered that effluents from sewage treatment plants were toxic to fish due to maintained residuals of chlorine for disinfection. The toxicity of free and/or combined (chloramines) chlorine, referred to as total residual chlorine ($TRCl_2$), is also controlled by pH. The most toxic form, hypochlorous acid (HOCl), increases as pH decreases.

The problem was reviewed by Brungs (1973), who cited work in Michigan, USA with caged fish held downstream from treatment-plant effluents. When the effluent was not chlorinated, no mortality resulted, but when chlorinated the LC_{50} values for 96 h ranged from 14 to $29 \, \mu g \, l^{-1}$ as $TRCl_2$. In some areas, such as the lower James River, massive fish kills were reported from effluent chlorine residuals as high as $2.2 \, \text{mg} \, l^{-1}$ (Bellanca and Bailey, 1977). Similarly, Tsai (1973) investigated streams receiving treated sewage effluent from 149 plants and found fishless waters where total residual chlorine was $0.37 \, \text{mg} \, l^{-1}$ or greater and observed levels as high as $1.5 \, \text{mg} \, l^{-1}$.

The recommended maximum, 4-day mean concentration of TRCl$_2$ is 11 μg l^{-1} (WQC, 1986). Hematological work has shown that 3 μg l^{-1} is the maximum TRCl$_2$ concentration that would be safe for coho salmon (Buckley *et al.*, 1976; Buckley, 1977).

13.5.4   Organic toxicants

The myriad synthetic organic chemicals produced annually may pose the major threat, as well as challenge, to society's ability to prevent the degradation of environmental quality resulting from toxicity. The threat of organic toxicants may be greatest because of their more neutrally charged state and therefore greater mobility to transfer across cellular membranes (i.e. more soluble in lipids), as well as a high resistance of many compounds to degradation. Such characteristics create the problem of long-lasting residues and food-chain magnification that has rendered populations of fish unsafe for human consumption.

The most famous compound in this regard is the organochlorine insecticide DDT and its metabolites DDE and DDD. Because DDT is fat-soluble, but rather insoluble in water ($\approx$1 μg l^{-1}), it can be readily absorbed by organisms from water or in food resulting in body residues that are 10^6-times the water concentration (Reinart, 1970). While DDT can be absorbed from water or ingested with food, the two pathways are not additive (Chadwick and Brocksen, 1969). While other organochlorine insecticides are famous for their toxicity and are even more toxic to fish than DDT (e.g. endrin and dieldrin), DDT has received relatively more attention and has had a greater toxic impact due to its greater use. Over one-half million metric tons of DDT have been used in the USA and it continues to be used in other countries (Hileman, 1988).

The recommended maximum, 4-day mean concentrations for organo-chlorine insecticides range from 0.001 to 0.004 μg l^{-1} (Table 13.8). Although such concentrations are one or two orders of magnitude below the lethal concentrations, they should not be considered over restrictive considering the biomagnification potential of these substances. For example, fish in Lake Michigan apparently concentrate DDT and dieldrin from the part per trillion level in water to the part per million level (Reinart, 1970).

The most detrimental subacute effect of organochlorine compounds to fish is on the reproduction stage. That was first demonstrated by Johnson (1967) for the Japanese medaka (*Oryzias latipes*). He found that whereas adults exposed to endrin concentrations of 0.3 μg l^{-1} or less survived in apparent good health, the hatching fry from those fish showed severe behavioural changes and poor survival which were directly related to the concentration and time of adult exposure. That was true even though the eggs were incubated in endrin-free water, indicating that residues in the adult fish were transferred to the eggs.

Pesticides have created major problems in the environment because most of what is applied never reaches the target organism (Younos and Weigmann, 1988), but are available to non-target species. Many years of education and persistence on the part of fisheries and pollution control agencies (and dead fish) were required before the documented damage caused by application of organochlorine insecticides had an impact on their use. As a result, relatively shorter-lived and less toxic insecticides have since replaced many of the organochlorine and organophosphate compounds, whose use has been greatly reduced or curtailed altogether since the 1970s. Nevertheless, much of their residues still remain in the environment due to their relative resistance to degradation. Residue levels of organochlorine pesticides (e.g. DDT and dieldrin) in fish and birds have decreased significantly in the USA following curtailment of use – demonstrating that, even for highly resistant and biomagnifying substances, availability in the environment and body residues will decrease if the input to the environment decreases. Nevertheless, highly toxic organochlorine insecticides such as mirex and chlordane (MCs of 0.0043 and $0.001\,\mu g\,l^{-1}$, WQC, 1986) can still be used and a thousand metric tons of pesticides in general are used in agriculture daily in the USA (Younos and Weigmann, 1988).

PCBs, which were used extensively in industry and appear to have similar effects and persistence as the organochlorine insecticides (WQC, 1973), have become increasingly important in the 1970s and 1980s. Although their use was banned in the late 1970s, their residues in landfills, river and lake sediments and transformers continue to present problems of dispersal in the environment and biomagnification in the food chain, e.g. the Hudson River (Brown et al., 1985) and the Great Lakes (Hileman, 1988).

In contrast to the organochlorine insecticides, most herbicides, such as 2,4-D, 2,4,5-T and endothal, are less acutely toxic and degrade more rapidly. The herbicide 2,4-D persists in some trophic levels and sediment for one to three months but apparently has produced little adverse effect on aquatic animals (Wojtalik et al., 1971). On the other hand, 2,4,5-T may contain dioxin and presents a residue problem (Galston, 1979).

Wastewater from the bleaching of pulp and paper contains organochlorine compounds that are highly toxic, including dioxin. Leach and Thakore (1975) found that chlorinated compounds derived from resin acids in Kraft pulp-mill wastewater were the cause of most of the effluent toxicity. Five identified compounds had 96-hour LC_{50} values ranging from 0.32 to $1.5\,mg\,l^{-1}$. Organochlorine compounds from Kraft mill bleaching effluents have been distributed over broad areas in the northern Baltic

Sea (Sodergren *et al.*, 1988) (Figure 3.13). Detrimental effects on perch (*Perca fluviatilis*) reproduction and physiology were demonstrated in laboratory experiments to collaborate the increased tissue levels of organochlorine compounds and reduced biomass levels of perch observed up to 10 km from effluent sources.

13.5.5 Acidification

Acidification of lakes and streams due to acid precipitation is a global pollution problem affecting large areas of the eastern USA, eastern Canada, Scandinavia and northern Europe. Acidification is the loss of acid neutralizing capacity (ANC) or alkalinity (Chapter 4.4). The loss is caused by the replacement of HCO_3^-, which is essentially equivalent to ANC (Alk) in waters with pH < 8.3, with either SO_4^{2-} or NO_3^-. The latter are the strong acid anions in precipitation resulting from the combustion of fossil fuels and the resulting atmospheric emission of SO_2 and NO_x, which are ultimately oxidized to the strong acids (H_2SO_4 and HNO_3) and can be deposited far downwind from sources. When acid in precipitation is deposited on to a watershed, it becomes neutralized (H^+ is consumed by bases) by weathering processes to a greater or lesser degree. In watersheds dominated by sedimentary bedrock rich in $CaCO_3$, acidity is readily neutralized. However, in those dominated by igneous bedrock such as granite and basalt, which is overlaid by a thin soil mantle, the supply of neutralizing bases through weathering is low (Wright and Snekvik, 1976; Henriksen, 1980). The effectiveness of neutralization through weathering decreases as the ratio of lake area to watershed area increases, because the relative supply of bases decreases. Due to its more conservative nature, SO_4^{2-} represents the major cause for lake and stream acidification (Henriksen, 1980; Brakke *et al.*, 1988).

The chemical process of acidification with strong acid can be described by considering the cation–anion charge balance in freshwater:

$$2[Ca^{2+}] + 2[Mg^{2+}] + [Na^+] + [K^+] + [H^+]$$
$$= 2[CO_3^{2-}] + [HCO_3^-] + [Cl^-] + 2[SO_4^{2-}] + [OH^-] + [NO_3^-] \quad (13.1)$$

To obey neutrality, the cation equivalents must equal the anion equivalents. If H_2SO_4 is added to a water body, H^+ and SO_4^{2-} will increase at the expense of HCO_3^- according to:

$$H_2SO_4 + Ca(HCO_3^-) \rightarrow CaSO_4 + 2CO_2 + 2H_2O \quad (13.2)$$

Equation 13.1 can be simplified to the following in most dilute freshwaters with near neutral pH:

$$[HCO_3^-] = 2[Ca^{2+} + Mg^{2+}] - 2[SO_4^{2-}] \quad (13.3)$$

Na^+ is relatively scarce without marine salt contamination and K^+ and NO_3^- are relatively low. Thus in unacidified waters, Ca^{2+} and Mg^{2+} are the principal cations and HCO_3^- is the principal anion so that significant departures from a near $1:1$ relation between the non-marine equivalents of $Ca^{2+} + Mg^{2+}$ and HCO_3^- has indicated the extent of acidification (Henriksen, 1980). As SO_4^{2-} increases, HCO_3^- decreases (Equation 13.3). This led to the following equation (in equivalents) to evaluate how much acidification has occurred (Wright, 1983):

$$\Delta_{ALK} = 0.91(Ca^* + Mg^*) - Alk + H^+ + Al \qquad (13.4)$$

where Alk on the right-hand side is the existing level and above pH 5, H^+ and Al are insignificant and an asterisk denotes non-marine.

The following was used to estimate the amount of non-marine SO_4 added:

$$\text{net } SO_4^* = SO_4^* - [0.09(Ca^* + Mg^*) + Na^* + K^*] \qquad (13.5)$$

where SO_4^* on the right-hand side is existing SO_4^*, and the bracketed term is background SO_4^* before acidification. Thus, net SO_4^* should approximate Δ_{ALK} and Wright (1983) showed that in 206 Adirondack lakes net SO_4^* averaged $119\,\mu\text{equiv.}\,l^{-1}$ and Δ_{ALK} averaged $103\,\mu\text{equiv.}\,l^{-1}$. That is, for the average lake, $\approx 1\,\mu\text{equiv.}\,l^{-1}$ alkalinity (HCO_3^-) was removed by the addition of $\approx 1\,\mu\text{equiv.}\,l^{-1}$ SO_4^* through acid deposition.

While this is an easily explained equilibrium model that has been effective at understanding and predicting the general pattern of acidification, there are problems when individual lakes are considered. For example, $Ca^* + Mg^*$ may increase with acidification through weathering and thereby compensate for some of the SO_4^* added. However, the maximum effect of such compensation is considered to be $\approx 40\%$ (Wright and Henriksen, 1983).

The effects of acidification are due to decreased pH and increased Al. As ANC is depleted, pH decreases; very slowly at first because of the buffering effect of HCO_3^- (H^+ is consumed), but rapidly as ANC nears depletion. Usually, at a pH of 5.2, all the ANC is gone and a further decrease in pH occurs with a continued increase in strong acid. Also as pH approaches 5, Al is dissolved from watershed soils and rocks and the concentration begins to increase in runoff water while providing some buffering against further decreases in pH. Al may occur in various forms, but the free Al^{3+} and hydroxides, $Al(OH)^{2+}$ and $Al(OH)_2^+$, are principal forms of the labile monomeric Al fraction that is considered most toxic (Driscoll et al., 1980; Henriksen et al., 1984). Organic-Al complexes are not considered toxic.

Damage to fish populations may occur if lake pH drops below ≈ 6 although effects are more pronounced at pH 5.5 and below. While a lake or stream with pH 5 would be considered completely acidified (all ANC gone), damage to fish populations occurs at pH values >5. Much of the adverse effect at pH values between 6 and 5 observed in natural waters is probably due to Al toxicity as well as to H^+ (Baker and Schofield, 1982). For example, Muniz and Leivestad (1980) showed that Cl^- loss from the blood of brown trout did not increase until pH was decreased below 4.6, while Cl^- loss was much greater at pH 5.1 and 5.5 than at 4.0 in the presence of $900 \mu g \, l^{-1}$ Al. Others have found toxic effects due to Al to be greatest around pH 5.2–5.4, possibly due to Al hydroxides precipitating on gill surfaces (Schofield and Trojnar, 1980; Baker and Schofield, 1982). Henriksen et al. (1984) showed mortalities to juvenile salmon when labile monomeric Al concentrations increased to between 50 and $70 \mu g \, l^{-1}$ following stream-flow-increase events even though pH did not drop below 5. Jensen and Snekvik (1972) considered the limit for damage due to pH effects alone to be between 4.5 and 5.0, although Schindler et al. (1986) reported long-term effects on lake trout production resulting from food web modifications at pH values <6 without Al. Also, Rahel and Magnuson found some fish species to be missing from Wisconsin lakes with pHs below 6.2.

The problems caused by acid precipitation are considered global in proportions; effects on fish populations have been most pronounced in southern Norway (Figure 13.22) and south-western Sweden, eastern Canada and north-eastern USA, especially New York state. Although acid precipitation affects a much larger area, acidification is most damaging in waters of low buffering (ANC) capacity (see above). Western Sweden and Norway have many lakes in watersheds composed of resistant rock and thin soils and are thus highly sensitive to acid precipitation (i.e. have

Figure 13.22. Salmon-catch records for 79 Norwegian rivers unaffected by acid rain and for a southern Norway acidic river during 1880–1970 (Wright et al., 1976)

low ANC). For example, Wright and Snekvik (1978) found that 80% of 700 lakes surveyed in southern Norway during 1974–75 were acidified to the point that they were fishless or had sparse fish populations. Figure 13.22 shows the dramatic decline and elimination of salmon from the Tovdal River. Leivestad and Muniz (1976) found that brown trout mortality in the Tovdal River was inversely related to Cl^- content of the blood. A resurvey of the southern Norwegian lakes in 1986 showed that pH had changed little in the intervening years, but that sulphate had decreased and nitrate had increased significantly (Henriksen et al., 1988).

An extensive, stratified lake survey in acidification-sensitive regions of the USA was reported by Landers et al. (1988). Results showed that 4.6% (326) of the 7096 lakes of 4–200 ha in the north-east had ANC <0, while 138 (11%) of Adirondack lakes had <0 ANC (results based on statistical projections). Fewer (148, 1.7%) of the 8501 lakes in the upper midwest USA had ANC <0. The Adirondack lakes had lower median ANC and pH and higher Al than lakes in other areas. Although SO_4^{2-} in the Adirondacks was not higher than other areas, the median level was still of the order of ANC (see results above by Wright, 1983). There is general agreement that acidification of north-eastern and upper mid-western lakes in the USA has been caused largely by SO_4^{2-} deposition and replacement of ANC. That is indicated by trends of decreasing ANC/(Ca + Mg) associated with increasing H^+ deposition (Brakke et al., 1988; Eilers et al., 1988).

Based on the above discussion of effects of pH and Al, the 474 north-eastern and midwestern USA lakes estimated to have ANC <0 would also be expected to have damaged fish populations. Damage could be expected in even more lakes because, as indicated above, adverse effects of pH and Al have been observed at pH levels between 5 and 6 before ANC is exhausted, especially if Al is present. For example, while only 3.4% (240) of the north-eastern lakes had pH < 5, 12.9% (916) had pH < 6. In the upper midwest, only 1.5% (130) had pH < 5, while 9.6% (818) of the lakes had pH < 6. Also, small lakes, which are known to be more susceptible to acidification, were under-represented in the eastern survey.

Appendices

Appendix A
DESCRIPTION OF BENTHIC MACROINVERTEBRATES AND COMMENTS ON THEIR BIOLOGY

The following is a brief account of the taxonomy and biology of the life stages of macroinvertebrate fauna that inhabit fresh and, to some extent, brackish water. For more details see Usinger (1956).

A. Body with three pairs of legs (Insecta); abdomen with two or three long tail-like cerci; thorax and abdomen usually with plate-, feather-, tassel-, or finger-like tracheal gills.

1. Mayflies (Ephemeroptera);
   (a) Taxonomy: tarsi with one claw; gills plate-, feather- or tassel-like and present on one or more of first abdominal segments;
   (b) Food habits: herbivorous grazers and detrital feeders;
   (c) Habitat: depositing substrata, Ephemeridae (burrowers); eroding substrata, Heptageniidae (clingers); eroding and depositing substrata, Baetidae (sprawlers on bottom, agile, free-ranging and trash- and moss-inhabiting);
   (d) Life history: three-stage cycle (egg $\rightarrow$ naiad $\rightarrow$ adult); usually <1–1 year.
2. Stone flies (Plecoptera);
   (a) Taxonomy: tarsi with two claws; gills usually present and finger-like and may occur on abdomen, thorax or labium
   (b) Food habits: herbivorous mostly, some carnivorous (Perlidae, Perlodidae);
   (c) Habitat: mostly swift currents ($\geq 1\,\mathrm{ft\,s^{-1}}$), eroding substrata;
   (d) Life history: three-stage cycle; 1–3 years.

B. Body worm-like, elongate and cylindrical to short and obese, or flattened and oval with appendages and sclerotized head capsule.

1. Caddis flies (Trichoptera);
    (a) Taxonomy. thorax with three pairs of jointed legs and segment with a pair of hook-bearing appendages; antennae inconspicuous, one-segmented;
    (b) Food habits: herbivorous (including detrital) and carnivorous (Rhyacophilidae);
    (c) Habitat: usually eroding substrata (free living, net spinners and case builders); depositing substrata (some case builders);
    (d) Life history: four-stage cycle (egg → larvae → pupae → adult); 1–2 years.
2. True flies (Diptera);
    (a) Taxonomy: thorax without jointed legs but often with fleshy prolegs; head capsule distinct, at least anteriorly; abdomen with gills or a breathing tube at posterior end;
    (b) Food habits: detrital herbivorous feeders;
    (c) Habitat: eroding substrata (cling by permanent attachment); vegetation (cling with hooks); depositing substrata (tube builders);
    (d) Life history: four-stage cycle; <1–1 year.

C. Body not worm-like, wing rudiments as external flap-like appendages, mouthparts consisting of long and scoop-like labium, which covers other mouthparts when folded.

1. Dragonflies and damselflies (Odonata);
    (a) Food habits: all predaceous;
    (b) Habitat: depositing substrata and vegetation (sprawlers, climbers and burrowers);
    (c) Life history: three-stage cycle; 1–5 years.

D. Animals living within hard carbonate shell: Mollusca (snails and clams).

1. Snails (Gastropoda): shells entire, usually spiral, but may be 'coolie-hat'-shaped;
    (a) Food habits: mostly detrital feeders and grazers;
    (b) Habitat: mainly in depositing substrata, but also in eroding substrata;
    (c) Life history: three-stage cycle, <1–1 year.
2. Clams, bivalves (Pelecypoda): shells consisting of two hinged halves;
    (a) Food habits: detrital and herbivorous filter feeders;
    (b) Habitat: mainly depositing substrata;
    (c) Life history: three-stage cycle; <1–1 year.

E. Body worm-like and divided into many small segments much wider than long; bristles frequently present.

1. Worms and leeches (Annelida);
   (a) Food habits: mostly detrital feeders, some predators and parasites;
   (b) Habitat: usually depositing substrata;
   (c) Life history: two- or three-stage cycle; <1–1 year.

F. Body not worm-like, with at least five pairs of legs (Crustacea).

1. Body flattened laterally (Amphipoda, scuds);
   (a) Food habits: detrital;
   (b) Habitat: eroding and depositing substrata with vegetation;
   (c) Life history: three-stage cycle; <1 year.
2. Body flattened horizontally: Isopoda (aquatic sow bugs), eyes not on stalks; Decapoda (crayfish), eyes on stalks; biology similar to Amphipoda.

SHORT-ANSWER QUESTIONS

1. Explain how *two phenomena* that occur as a result of increasing primary productivity of diatoms and/or green algae due to eutrophication would tend to favour blue-green algae.
2. Explain verbally, with a steady-state equation, why the inflow concentration of limiting nutrient to a continuous culture of algae is so closely related to the concentration of biomass in the outflow. Also, with a steady-state equation, explain why growth rate increases with dilution rate.
3. Eutrophic lakes A and B have similar areas and wind speed and fetch, but have mean depths of 10 m and 3 m, respectively. Internal loading of phosphorus was shown by mass balance to be of a similar magnitude in both lakes.
   (a) Describe two sediment release mechanisms for each lake that may explain their internal loading.
   (b) Why might you expect concentrations of algae to be greater in lake B than A?
4. Calculate the expected net internal phosphorus loading in a lake with the following characteristics:
   flushing rate = 2.25/year; mean depth = 15 m; external loading = 2.0 g/m^2 yr and mean annual TP = 60 µg/l (mg/m^3).
   Would you expect this lake's summer epilimnetic TP to be greater or less than 60 µg/l? Why?
5. Give two explanations why eutrophic lakes tend to have more small than large bodied zooplankton. What is the consequence of that condition in terms of algal abundance and transparency?
6. Two streams have similar incident light, temperature and rubble substratum, but have different average velocities; A = 35 cm/s and B = 10 cm/s. In which stream would you expect the concentration of

SRP (soluble reactive phosphorus) limiting periphytic algal growth to be lowest? Why?

7. Is Lake Tahoe, as a whole, more or less sensitive to a given inflow concentration of phosphorus (assuming P is limiting) than Lake Washington? Explain with an equation. Water detention times: L. Tahoe = 700 years; L. Washington = 3 years.

8. A eutrophic lake has a mean summer chlorophyll a concentration of 25 μg/l and a transparency of 1.0 m. The lake has a maximum depth of 10 m, a mean depth of 6 m, and it stratifies thermally with the thermocline at about 3 m. Is it possible to reduce algal biomass so that the average is less than 10 μg/l by completely mixing the lake with compressed air during the summer? Assume nutrients will not limit.

9. The following data are available from a stream receiving wastewater from a point source.

	upstream	waste ↓	1 km	5 km
AFDW/chl a	200		400	1000
No. species (macroinvertebrates)	25		5	10
No./m^2 (macroinvertebrates)	500		200	5000

Describe the type(s) of wastewater that could produce such an effect and give your reasons with regard to each variable.

10. Thick mats of *Cladophora*, a filamentous alga, were observed immediately downstream from a wastewater effluent input. Given the alternatives of (a) secondary treated sewage, (b) untreated sewage, (c) pulp mill waste liquor, and (d) an organic insecticide, which type(s) of waste would most likely cause this problem and why? Which would not likely be the cause and why?

11. (a) Determine, using the steady state version of Vollenweider's model (or a modified version) how effective (in %) an in-lake treatment should be at reducing internal loading to a loading level that would recover the lake to a mesotrophic state (i.e. ≤25 μg/l TP). The lake has the following characteristics:

$$L_{ext} = 300 \text{ mg/m}^2 \cdot \text{y}, \quad \rho = 0.4/\text{y}, \quad \overline{TP} = 35 \text{ μg/l (mg/m}^3), \quad \bar{z} = 15 \text{ m}.$$

(b) What is the major assumption you must make (other than completely mixed, constant volume, accuracy of external load, etc.) in order to use the model in this way?

12. Give two physical processes that reduce the biomass and diversity

(no, of species) of invertebrate animals in streams with urbanized watersheds and briefly describe their effect.

13. Calculate the steady state biomass in a nutrient limited continuous culture system, which is characterized by the following parameters: maximum growth rate = 1.3/day; half saturation constant = 0.3; inflow nutrient concentration = 4.0 mg/l; dilution rate = 1.0/day; and the biomass to nutrient yield ratio = 64. Would a higher or lower maximum biomass result if these parameters were applied to a batch culture and why?

14. Based on the turbidity of inflow water, a proposed reservoir is projected to have a Secchi disc transparency averaging 2 m during the spring and summer. Assuming complete mixing, how deep should the reservoir be on average to prevent the maximum algal biomass (chl *a*) from exceeding 10 μg/l? Also assume that the Secchi disc disappears at 15% surface intensity (I_0), the respiration: P_{max} ratio is 0.1 and the light extinction coefficient for chl *a* is $0.025 \, m^2/mg$.

15. Explain why buoyancy, possessed by some blue green algae such as *Aphanizomenon*, should provide a competitive advantage in a relatively shallow, polymictic eutrophic lake.

16. Compare the expected effects of wind on the internal loading of phosphorus (release of P from sediments) in two 300 ha (750 acre) eutrophic lakes; A is stratified and B is polymictic.

17. Is the expected relationship between the LC_{50}s (with fish) determined for heavy metals and water hardness inverse or direct? Give an explanation for the phenomenon.

18. For a soft water stream receiving wastewater containing the heavy metals Cu and Zn that raised stream levels to double the maximum acceptable concentrations (MACs = 7 μg Cu/l; 59 μg Zn for 50 mg/l hardness), speculate on the direction of effect (increase, decrease, no change) and a reason for your choice for each of the following variables.
    (a) AFDW/chl *a*;
    (b) no. sp. macroinvertebrates;
    (c) $\bar{d}$ for macroinvertebrates;
    (d) *D* for macroinvertebrates.

19. The lower Deschutes River in Oregon has a summer flow that is maintained, partly due to releases from a dam, at around $100 \, m^3/s$ (3500 cfs), and the mid-summer water temperature reaches a maximum of about 17°C. Explain the probable effects of increases of 3°C and 7°C above ambient on growth and survival of the highly prized rainbow trout population.

20. Explain 'top-down' control of plankton algal biomass in lakes and give two methods by which it may be enhanced to improve the quality of eutrophic lakes.

MULTIPLE-CHOICE QUESTIONS

1. Net productivity at steady state is
   (a) dx/dt
   (b) μx
   (c) $\mu x - Dx$
   (d) $\dfrac{dx/dt}{x}$

2. In the nitrogen budget of an eutrophic lake, the important losses are
   (a) denitrification and sedimentation
   (b) denitrification and fixation
   (c) nitrification and fixation
   (d) nitrification and sedimentation.

3. In the P budget of an eutrophic lake, in which external loading = 30 kg, gain in the lake = 50 kg, and outflow = 20 kg, internal loading is
   (a) -40 kg
   (b) -60 kg
   (c) $+60$ kg
   (d) $+40$ kg

4. Lake acidification can be most simply defined as
   (a) a decrease in pH
   (b) an increase in H^+
   (c) a replacement of HCO_3^- with strong acid
   (d) a replacement of $Ca^{2+} + Mg^{2+}$ with strong acid

5. Iron redox reactions at the sediment surface in lakes result in
   (a) ferrous iron and soluble P under oxic conditions
   (b) ferrous iron and soluble P under anoxic conditions
   (c) ferric iron and particulate P under anoxic conditions
   (d) ferric iron and soluble P under oxic conditions.

6. The Q_{10} rule predicts, for a 10°C rise in temperature, that biological activity (e.g. growth rate) will increase by
   (a) 200%
   (b) 100%
   (c) 50%
   (d) 25%

7. Photosynthesis in the water column of lakes tends to be directly related to light intensity at
   (a) intensities $<I_K$
   (b) intensities $>I_K$
   (c) rates $<P_{max}$
   (d) rates $>P_{max}$

8. Compensation depth is the depth at which
   (a) light intensity is enough for $P > R$.

(b) light intensity $= 1\% I_0$

(c) $P - R$ with mixing

(d) light intensity is sufficient for $P = R$

9. Lake ageing (or filling) may accelerate as rooted macrophtes establish because they

(a) outcompete plankton algae for nutrients

(b) increase recycling of nutrients from sediments

(c) prevent resuspension of particulate matter

(d) encourage bottom covering projects by lake-shore residents.

10. $R = \dfrac{1}{1 + \sqrt{\rho}}$ represents an estimate of

(a) P sedimentation rate

(b) relative thermal resistance to mixing

(c) P sedimentation rate coefficient

(d) P retention coefficient

11. Match the process names with the appropriate reactions by placing the correct letters in the blanks.

_____ nitrification

_____ sulfate reduction

_____ denitrification

_____ nitrogen fixation

_____ bacterial photosynthesis

(a) $CO_2 + 2H_2S \rightarrow CH_2O + H_2O + 2S$

(b) $NH_4^+ + 2O_2 \rightarrow NO_3^- + 2H^+ + H_2O$

(c) $SO_4^{2-} + 2CH_2O + 2H^+ \rightarrow H_2S + 2CO_2 + 2H_2O$

(d) $5S + 6NO_3^- + 2H_2O \rightarrow 5SO_4^{2-} + 3N_2 + 4H^+$

(e) $2N + 3H_2 \rightarrow 2NH_3$

12. Acidification of lakes, defined as alkalinity reduction, frequently occurs in nearly direct proportion to

(a) pH in precipitation minus uptake in the watershed

(b) pH in precipitation

(c) SO_4^{2-} in precipitation

(d) $NO_3^- + SO_4^{2-}$ in precipitation

13. Algae may self limit their productivity in lakes because

(a) CO_2 accumulates faster than it can escape to the atmosphere

(b) O_2 becomes supersaturated

(c) CO_2 decreases and pH increases

(d) CO_2 increases and pH increases

14. Phosphorus is usually limiting in lakes because

(a) P occurs in smaller concentration than N or C

(b) N/P ratios are usually greater than 10/1 in oligotrophic–mesotrophic lakes

(c) N/P ratios are usually less than 10/1 except in hypereutrophic lakes

(d) N/P ratios are usually greater than 20/1 in hypereutrophic lakes

15. Algal growth (yield) of non-N fixers will likely be limited by nitrogen if the N:P ratio in water is
    (a) more than that in algal cells
    (b) less than that in algal cells
    (c) more than 10:1
    (d) more than 20:1

16. A mass balance of phosphorus in lakes will indicate a net internal loading if
    (a) output exceeds input
    (b) output is less than input
    (c) output plus lake P increase exceeds input
    (d) output plus lake P increase is less than input

17. dx/dt represents net productivity of plankton algae when there are
    (a) no losses due to dilution or grazing
    (b) losses due to dilution or grazing
    (c) no respiration losses
    (d) nutrient limitations

18. Plankton algal growth is said to be light limited if the critical depth
    (a) exceeds the compensation depth
    (b) exceeds the mixing depth
    (c) is less than the compensation depth
    (d) is less than the mixing depth

19. Photosynthesis in the water column is considered to be light limited at intensities less than
    (a) $50\% \, I_0$
    (b) $100\% \, I_0$
    (c) I_k
    (d) μ_{max}

20. P_{max} (mg C/mg chl-hr) in the water column is subject to limitation by
    (a) temperature and nutrients
    (b) temperature but not nutrients
    (c) nutrients but not temperature
    (d) grazing

21. The growth of individual species of plankton algae in response to temperature is best described by
    (a) the RGT rule
    (b) the Q_{10} rule
    (c) a nearly linear model
    (d) an exponential model

22. Carbon
    (a) never limits plankton algal growth
    (b) may limit when photosynthesis is high and alkalinity is low
    (c) may limit when nighttime respiration is low

    (d) may limit when photosynthesis is high and pH is low
23. According to the Vollenweider model (completely mixed, constant volume, etc.), lakes with
    (a) short detention times are tolerant of high inflow phosphorus concentrations
    (b) short detention times are tolerant to any inflow phosphorus concentration
    (c) long detention times are intolerant to high inflow phosphorus concentrations
    (d) long detention times are tolerant to high inflow phosphorus concentrations
24. Lakes may theoretically age (fill in) and 'die' due to cultural eutrophication at rates of
    (a) 1–5 m per 1000 years
    (b) 1–5 m per 100 years
    (c) 1–5 m per 10 years
    (d) 1–5 m per 10 000 years
25. The capacity of zooplankton to control phytoplankton biomass in lakes is indicated by a ratio of
    (a) zooplankton production : phytoplankton production = 1.0
    (b) zooplankton production : phytoplankton production > 0.1
    (c) zooplankton consumption : phytoplankton production = 1.0
    (d) zooplankton consumption : phytoplankton production > 0.1
26. *Sphaerotilus* growth in streams is largely dependent on
    (a) the limiting nutrient content, whether N or P
    (b) DOC content above 15 mg/l
    (c) light availability
    (d) DOC content from 1–30 mg/l
27. Barring large, nonalgal sources of particulate matter, transparency can usually be improved more per unit chl *a* reduction in
    (a) oligotrophic than eutrophic lakes
    (b) eutrophic than oligotrophic lakes
    (c) mesotrophic than oligotrophic
    (d) eutrophic than mesotrophic
28. Compared to the P concentration producing a nuisance biomass level in an eutrophic lake ($\sim$10 mg chl a/m^3), a nuisance biomass level in streams ($\sim$150 mg chl a/m^2) could probably occur at a P concentration
    (a) greater than that in the eutrophic lake
    (b) less than that in the eutrophic lake
    (c) about the same as that in the eutrophic lake
    (d) that is uncomparable to that in the eutrophic lake
29. If the hypolimnetic DO is 10 mg/l on 31 May and 2 mg/l on 31 August, and the hypolimnetic area and volume are, respectively, 1.25×10^6 m^2 and 5×10^6 m^3, the ODR in mg/m$^2 \cdot$ day is about

(a) 250
(b) 200
(c) 350
(d) 400

30. N/P ratios can be expected to decrease as lakes become eutrophic because
    (a) wastewater has an N/P $\simeq$ 3 and nitrification becomes more important
    (b) wastewater has an N/P $\simeq$ 13 and denitrification becomes more important
    (c) wastewater has an N/P $\simeq$ 13 and nitrification becomes more important
    (d) wastewater has an N/P $\simeq$ 3 and denitrification becomes more important

31. The expected flux rate of TP to the sediment in a lake with a TP loading of 512 mg/m$^2 \cdot$ y, a detention time of two years, a $\overline{TP}$ concentration of 42 mg/m^3, and a mean depth of 10 m, is about
    (a) 300 mg/m$^2 \cdot$ y
    (b) 0.71/y
    (c) 212 mg/m$^2 \cdot$ y
    (d) 604 mg/m^2

32. Organisms most characteristic of a septic zone in a stream receiving organic waste are
    (a) *Sphaerotilus* and protozoans
    (b) *Sphaerotilus* and facultative bacteria
    (c) facultative bacteria and protozoans
    (d) facultative bacteria and fecal bacteria

33. Organisms most characteristic of the recovery zone of a stream receiving organic waste are
    (a) filamentous algae
    (b) *Sphaerotilus*
    (c) facultative bacteria
    (d) protozoans

34. Rooted, submersed macrophytes have been shown to contribute internal loading of P in lakes primarily by
    (a) root uptake and leaf excretion
    (b) root uptake and plant senescence
    (c) leaf uptake and leaf excretion
    (d) leaf uptake and plant senescence

35. Macroinvertebrate communities normally found in streams of high order tend to be composed of
    (a) burrowers and small particulate selectors
    (b) clingers and large particulate shredders
    (c) burrowers and large particulate shredders

    (d) clingers and small particulate collectors

36. Phosphorus internal loading in the shallow unstratified lakes may occur as a result of
    (a) iron *reduction* under *oxic* conditions at the sediment–water interface
    (b) high pH in the *photic zone*
    (c) excretion of SRP from *zooplankton*
    (d) microbial decomposition of organic matter at the sediment–water interface

37. The concept of cascading trophic interactions suggests that algal biomass may be controlled by
    (a) increased piscivory resulting in decreased planktivory
    (b) increased planktivory resulting in decreased piscivory
    (c) decreased piscivory resulting in increased planktivory
    (d) decreased planktivory resulting in increased piscivory

38. Blue green algal buoyancy occurs with
    (a) increased CO_2, decreased pH and decreased light
    (b) decreased CO_2, increased pH and decreased light
    (c) decreased CO_2, decreased pH and decreased light
    (d) decreased CO_2, increased pH and increased light

39. N/P ratios in lakes decrease with increasing eutrophication because
    (a) nitrification and sediment phosphorus release increase under anoxic conditions
    (b) denitrification and sediment phosphorus release increase under oxic conditions
    (c) denitrification and sediment phosphorus release increase under anoxic conditions
    (d) nitrification and sediment phosphorus release increase under oxic conditions

40. Phosphorus is considered the macronutrient that usually limits primary production in most lakes because
    (a) there is little recycling of phosphorus
    (b) lakes are a net sink for phosphorus
    (c) lakes have more N and C than phosphorus
    (d) N and C have atmospheric sources

41. Oxygen in stratified lakes decreases as hypolimnetic depth
    (a) increases and as eutrophication increases
    (b) decreases and as eutrophication increases
    (c) decreases and as eutrophication decreases
    (d) increases and as eutrophication decreases

42. Blue green algae that fix nitrogen are species of
    (a) Oscillatoria and Anabaena
    (b) Anabaena and Aphanizomenon
    (c) Oscillatoria and Microcystis

(d) *Microcystis* and *Aphanizomenon*

43. The greatest increase in periphytic algae in streams would most likely occur from a $10\,\mu g/l$ increase in SRP if current velocity were
    (a) $\leqslant 15\,cm/s$ and initial SRP were $20\,\mu g/l$
    (b) $>15\,cm/s$ and initial SRP were $2\,\mu g/l$
    (c) $>15\,cm/s$ and initial SRP were $20\,\mu g/l$
    (d) $\leqslant 15\,cm/s$ and initial SRP were $2\,\mu g/l$

44. The threshold concentration of BOD_5 that produces nuisance biomass levels of filamentous bacteria in streams is around
    (a) $1\,mg/l$
    (b) $5\,mg/l$
    (c) $10\,mg/l$
    (d) $50\,mg/l$

45. By suggesting that the number of invertebrate species will increase as the maximum daily mean temperature increases from $15°$ to $30°C$ assumes that
    (a) there is ample recruitment of new species that are tolerant to the new temperature
    (b) the more tolerant species are already present
    (c) the present species acclimated to a $15°$ maximum will readapt to a $30°$ maximum
    (d) invertebrates cannot acclimate

46. Submersed, rooted macrophytes colonize lakes to a maximum depth where I_z (radiation at depth) is
    (a) $\geqslant 1\%\,I_O$
    (b) $\geqslant 50\%\,I_O$
    (c) $\geqslant 10\%\,I_O$
    (d) $\geqslant 75\%\,I_O$

47. Chl a in periphytic algal mats in streams may not correlate well with TP in the water, as algal chl a does in lakes, because
    (a) N usually limits in streams
    (b) the chl a/area is correlated with supply rate (TP $\times$ flow)
    (c) only SRP is utilized
    (d) the chl a/area is not restricted to the P in a given volume

48. Filamentous bacterial biomass will dominate the periphyton in streams if there is a
    (a) DOC (dissolved organic carbon) source even without N and P
    (b) N and P source even without DOC
    (c) DOC source with N and P
    (d) DOC source with N

49. ODR (oxygen deficit rate) is related to phosphorus loading in stratified lakes because
    (a) P is related to algal biomass, which demands oxygen when it sinks into the hypolimnion

(b) P is related to algal biomass, which demands oxygen in the epilimnion

(c) phosphate can be oxidized

(d) P is related to algal and macrophyte biomass, which demands oxygen when it sinks into the hypolimnion

50. Some aquatic worms, especially *Tubifex*, are thought to tolerate, and even thrive in, organically enriched streams primarily because they
    (a) can live anaerobically
    (b) resist low *DO*
    (c) resist low *DO* and reproduce rapidly
    (d) are not preyed upon

51. The depositing substrata in a stream selects for
    (a) stoneflies, caddisflies and mayflies
    (b) worms, midges, and burrowing mayflies
    (c) midges and burrowing stoneflies
    (d) worms, snails and caddisflies

52. There is evidence that organic enrichment of streams, resulting in mats of *Spahaeratilus*, has
    (a) always benefited fisheries
    (b) sometimes been a benefit and sometimes a detriment to fisheries
    (c) never benefited fisheries
    (d) benefited some fisheries but not trout

53. Regarding Vollenweider's steady state model for phosphorus, sedimentation flux rate
    (a) and the sedimentation rate coefficient (σ) decrease with an increase in flushing rate
    (b) increases, but the sedimentation rate coefficient (σ) decreases, with an increase in flushing rate
    (c) and the sedimentation rate coefficient (σ) increase with an increase in influshing rate
    (d) decreases, but the sedimentation rate coefficient (σ) increases, with an increase in flushing rate

54. Submersed macrophyte populations often decrease as highly eutrophic or hypereutrophic states are reached in lakes as a result of
    (a) shading by planktonic and periphytic algae
    (b) shading by planktonic algae only
    (c) increasing organic content of the sediments
    (d) increasing toxicants in sediments

55. The autotrophic index (AFDW/chl a)
    (a) decreases with inorganic nutrient enrichment because heterotrophs increase
    (b) increases with organic enrichment because heterotrophs increase
    (c) increases with organic enrichment because autotrophs increase
    (d) decreases with organic enrichment because autotrophs increase

56. The ΔT experienced by invertebrate organisms entrained in cooling waters is
    (a) more important to their survival than the final temperature
    (b) of equal importance with final temperature
    (c) less important than the final temperature
    (d) of little concern in the receiving stream
57. The net increase in biomass of periphytic algae in streams is equal to
    (a) net productivity
    (b) net productivity minus losses to grazing and scouring
    (c) gross productivity
    (d) gross productivity minus losses to grazing and scouring
58. Dissolved oxygen concentration (mg/l) should be used as the unit in water quality standards rather than per cent saturation because
    (a) transport into the fish's blood is a function of concentration rather than partial pressure
    (b) analytical methods are more appropriate and criteria research results are largely in concentration
    (c) % saturation gives waste dischargers too much of an advantage
    (d) fish need more DO as temperature increases
59. The toxicity of heavy metals to fish is affected physiologically by H^+ in a manner
    (a) proportional to that of hardness ions
    (b) opposite to that of hardness ions
    (c) consistent with that of hardness ions
    (d) opposite to that of dissolved oxygen
60. In stratified lakes, trout are limited by eutrophication more than bass, crappie and pike because of (directly or indirectly)
    (a) oxygen
    (b) temperature
    (c) pH
    (d) NH_3
61. Excess *Cladophora* biomass in streams and nearshore areas in lakes and even estuaries is most likely caused by increases in
    (a) DOC
    (b) N
    (c) P
    (d) BOD
62. Excess *Sphaerotilus* biomass in streams is most likely caused by increases in
    (a) treated sewage
    (b) treated pulp mill waste
    (c) untreated pulp mill waste
    (d) agricultural runoff
63. Low level enrichment of an oligotrophic stream

   (a) may benefit fish growth, but will always decrease macroinvert
       ebrate (M) diversity
   (b) may benefit fish growth and increase M diversity
   (c) will probably not affect either fish growth or M diversity
   (d) will probably not affect fish growth, but will always decrease M
       diversity

64. To use diversity indices (e.g. d, $\bar{d}$) requires
   (a) an experimental control and taxonomic knowledge to species
   (b) a taxonomic knowledge to sp., but no control
   (c) a control, but no taxonomic knowledge to sp.
   (d) neither a control or taxonomic knowledge to sp.

65. Macrophyte distribution and abundance in lakes have been associated
    with
   (a) sediment type and light availability more than nutrients
   (b) sediment type and nutrients more than light
   (c) light and nutrients more than sediment type
   (d) nutrients more than sediment type and light

66. A DO concentration decrease from 9 mg/l to 6 mg/l may result in
   (a) mortality in juvenile bluegill sunfish
   (b) mortality in juvenile coho salmon
   (c) growth impairment in juvenile coho salmon
   (d) growth impairment in common carp

67. The cause for mortality of fish exposed to lethal concentrations of
    heavy metals is asphyxiation due to
   (a) gill tissue deterioration
   (b) gill tissue deterioration and mucous accumulation
   (c) cellular respiration failure internally
   (d) a reduction in red blood cell count

68. Natural factors that may favour high diversity in macroinvertebrates
    are
   (a) eroding substratum and moderate organic food supply
   (b) eroding substratum and high organic food supply
   (c) depositing substratum and moderate organic food supply
   (d) depositing substratum and high organic food supply

ANSWERS

Short-answer questions

1. The two phenomena of choice are:
   (a) increased algal abundance increases turbidity and light attenua-
       tion, thus decreasing the photic zone depth;
   (b) increased productivity (photosynthesis) raises the pH and lowers
       the CO_2 concentration.

The lower CO_2 level favours BGs, because of their lower half saturation constant and buoyancy is also increased as a result of reducing photosynthesis at depth where CO_2 and light are low and N and P high – vacuoles expand due to reduced turgor pressure or consumed photosynthate (polysaccharide) ballast. The resulting buoyancy provides more light to BGs in an otherwise low-light environment.

Alternatively, there are the hypotheses of Si depletion with increased production of diatoms eventually limiting their capacity to use the increased N and P. For this to work one still needs the light/CO_2 aspect to restrict green algae. Declining N/P ratios also occur with eutrophication, due to increased P internal loading and denitrification, with low NO_3 favouring N fixers, but not all BGs that dominate in eutrophic lakes are N fixers (e.g. *Microcystis*). Best correlations of BGs with nutrient changes have involved transparency as well. (Nonetheless credit was given for these answers)

2.
$$X = Y\left[N_i - \frac{K_n D}{\mu_m - D}\right] = Y(N_i - N)$$

This shows that biomass X is about equal to the difference between inflow and outflow nutrient concentration with the biomass having converted the available soluble nutrient into biomass; most of N_2 is removed and converted to biomass so long as D is low enough to permit enough time for uptake.

$$N = \frac{K_n D}{\mu_m - D}$$

This shows that as D increases, N increases, which results in higher μ according to

$$\mu = \frac{\mu_m N}{K_n + N}$$

or at steady state $\mu X = DX$, so as D increases, μ increases.

3. (a) A stratifies resulting in an anoxic hypolimnion, thus P can release from sediment due to iron reduction and PO_4 solubilization and diffusion into anoxic overlying water and/or release from bacterial cells in surficial sediment under anoxia.

B does not stratify although periods of calm weather can result in anoxic water layer overlying sediment accumulating P due to iron reduction with subsequent mixing entraining the high-P layer. Higher temperature at sediment surface due to complete mixing enhances microbial decomposition of organic matter. BG migration from oxic sediments, macrophyte decomposition and P released that was translocated from roots, and high pH caused by intense photosynthesis if such high pH water reaches the sediment surface are all possibilities. High pH can act together with wind mixing solubilizing P from entrained particulate P.

(b) The thermocline may act as a barrier to P entrainment to the photic zone, which is probably restricted to the epilimnion in eutrophic lakes (due to high light attenuation), while in un-stratified lakes, internally loaded P is readily available for uptake.

4.
$$60 = \left[\frac{L_{ext}}{15 \times 2.25} \frac{1}{1 + 1/\sqrt{2.25}} \right] + \frac{L_{int}}{15 \times 2.25}$$

$$60 = 35.6 \frac{L_{int}}{15 \times 2.25}$$

$$L_{int} = 825 \, mg/m^2 \cdot y$$

Less, because external input generally declines with low flow in summer and having $15 \, m \, \bar{z}$, internally loaded P in the hypolimnion would have only a moderate impact to the epilimnion until turnover, unless storms eroded the metalimnion.

5. Eutrophication promotes dominance by blue green algae which are inefficiently grazed resulting in increased amounts of detritus and bacteria that favour small bodied zooplanktons.

Increased planktivory can also remove the large bodied zoo-planktons because they are selected over the small bodied ones. Low DO in hypolimnion may deny the large zooplankton a refuge during the daylight hours.

Removal of the large-bodied, efficient-grazing zooplankton results in higher algal biomass and poorer transparency, which in turn favours more blue greens.

6. In A, because with higher velocity, the boundary layer surrounding the periphyton mat would be smaller affording a large gradient in SRP for uptake and, thus, uptake efficiency per unit ambient SRP would be greater.

7. $TP = \dfrac{TP_i}{1 + \sqrt{\tau}}$ shows that L. Tahoe is less sensitive because long τ allows for much greater loss of inflow TP due to sedimentation.

$$TP = TP_i \frac{1}{1 + \sqrt{700}} = TP_i \, 0.036$$

$$= TP_i \frac{1}{1 + \sqrt{3}} = TP_i \, 0.366$$

Tahoe can tolerate 10X the TP_i as Lake Washington.

8. $\ln 100/15 = [(0.025 \times 25) + K_w] \, 1$
   $1.9 = 0.625 + K_w \qquad K_w = 1.27$

$$C_{max} = \frac{1}{K_c}\left[\frac{F}{r24Z_m} - K_w\right] = \frac{1}{0.025}\left[\frac{2.7 \times 12}{0.1 \times 24 \times Z_m} - 1.27\right]$$

$$= 3.2\,\mu g/l \text{ for } 10\,m \text{ and } 39 \text{ for } 6\,m$$

alternatively $\quad Z_{cr} = \dfrac{2.7 \times 12}{0.1 \times 24[(10 \times 0.025) + 1.27]} = 8.9\,m$

$$= 7.4 \text{ for } 10\,hr \text{ light}$$

Thus, Z_{mix} needed to have $<10\,\mu g/l$ chl a is about $9\,m$. Although $39\,\mu g/l$ chl a represents the maximum that can occur with the mean being somewhat lower, it is unlikely that the mean can be maintained at $10\,\mu g/l$.

9. The waste is toxic initially, because both sp. no. and biomass decline (biomass would not decline if strictly organic) – the effect is mild because populations still exist – typical effect of inhibition exceeding beneficial effects of increased food. Organic is secondary effect, because although both sp. no. and biomass increase, biomass increases much more than sp. no., which is still less than control – now increased food effect on biomass, while adverse physical–chemical environment maintains low sp. no. (or still could be lingering effect of toxicants on intolerant forms). Increased AI could be due to toxicity to autotrophs in first stage, then the large increase above the 400 level due to organic response of heterotrophs.

10. *Cladophora* mats could be caused by 'a', due to the inorganic nutrients (produces autotrophic growth) or by insecticide that would have killed grazers allowing algae to proliferate. Being an autotroph, *Cladophora* would not respond to organic substances for energy like heterotrophs. And it would not be adversely affected by an insecticide – the effect of periphytic algae proliferating if grazers are killed with insecticide was shown in artificial channels and has been observed in streams so affected.

   Cladophora mats would not result from untreated sewage immediately downstream because of competition with faster growing filamentous bacteria, nor from a pulp mill waste for the same reason, plus there is little N or P in p.m.w. Recall the course of events in the Manawatu R., NZ.

11. (a) $L_{ext} + L_{int} = TPZ(p + p^{0.5}) = 35 \times 15(0.4 + 0.4^{0.5})$
    $\qquad\qquad\quad = 542 \text{ mg/m}^2\text{·y}$
    $\quad L_{int} = L_{int+ext} - L_{ext} = 542 - 300 = 242 \text{ mg/m}^2 \cdot \text{y}$

$L_{int} + L_{ext} = 25 \times 15(0.4 + 0.4^{0.5}) = 387 \, \text{mg/m}^2 \cdot \text{y}$ for mesotrophy

$100 - [(387 - 300)/242] - 64\%$

Answer is the same if you use $TP = L_{ext}/[Z(\rho + \rho^{0.5})] + L_{int}/Z\rho$

(b) That the empirical sedimentation rate coefficient $\rho^{0.5}$ applies to the lake in question.

12. Scouring of bottom substrata by high peak flows – unstable gravels/rubbles is undesirable habitat (even depositing substrata adversely affected).

Deposition of high sediment loads transported from impervious surfaces reduces the habitat in substrata for diverse communities.

Toxicants, BOD, DO, etc are chemical effects

13.
$$x = y\left[N_i - K_N\frac{D}{\mu_{max} - D}\right] = 64\left[4 - 0.3\frac{1.0}{1.3 - 1.0}\right]$$
$$= 192 \, \text{mg/l}$$

Higher, because $1 \, \text{mg/l}$ nutrients in the continuous culture remains unused due to D being so near μ_{max}. In a batch culture, more time would be available for biomass to convert nearly all soluble nutrient into biomass.

14. $\quad I_Z = I_0 e^{-K_w Z}$ $\qquad C = \dfrac{1}{K_c}\left[\dfrac{Fi\lambda}{r24Z_{mix}} - K_w\right]$

$\ln\dfrac{I_Z}{I_O} = -K_w Z$ $\qquad CK_c + K_w = \dfrac{Fi\lambda}{r24Z_{mix}}$

$\ln\dfrac{100}{15} = -K_w 2$ $\qquad Z_{mix} = \dfrac{Fi\lambda}{r24(CK_c + K_w)}$

$K_w = 0.95 \, /\text{m}$ $\qquad = \dfrac{2.7 \cdot 12}{0.1 \cdot 24(10 \cdot 0.025 + 0.95)}$

$\qquad\qquad\qquad\qquad = 11.3 \, \text{m}$

$\qquad\qquad\qquad\qquad 9.4 \, \text{m}$ for $\lambda = 10 \, \text{hr}$

15. Buoyant algae would obtain more light than non-buoyant species in eutrophic lakes where transparency is low and optimum light would occur only near the surface. Soluble nutrients would be depleted near the surface, during weakly stratified periods, but exit at higher concentrations near the bottom and/or below the photic zone. Blue greens can easily traverse the water column in a shallow lake in less than a day optimizing their growth conditions, while non-buoyant species would sink out of the photic zone.

16. Wind will increase internal loading in B, but not in A, by entraining water overlying sediments that have become anoxic and enriched with P during stable conditions preceding the wind. By replacing high-P water overlying sediment with low-P water in that wind-off,

wind-on process, sediment P release rates are increased by increasing the sediment-water concentration gradient. This same sediment-water interface renewal process occurs if cause for internal loading is high pH or temperature controlled microbial decomposition. Thermocline erosion in a stratified lake entrains P into the epilimnion and increases the availability of internally loading P, but that does not increase internal loading *per se*.

17. The relationship is direct; that is LC_{50s} increase (solution becomes less toxic) as hardness increases. An explanation is that the hardness ions (Ca^{2+} and Mg^{2+}) compete for binding sites on the gill surface, blocking access of heavy metals to the sites, resulting in less gill deterioration and irritation (= less mucus buildup).

18. Double the MACs should result in stressful conditions producing low survival in some sp. – also the effect is compounded because of the $\geq$ additive effect of Zn and Cu.
    (a) Increase, probably, because algae affected more than heterotrophs; could see no change if both heterotrophs and algae affected equally.
    (b) Decrease, because metals are inhibitors – sensitive sp. will be eliminated – only most tolerant will remain.
    (c) Decrease – richness (no. sp.) declines, although the magnitude of change may be less than # sp. due to evenness.
    (d) Increase – the decreased no. of sp. and abundance changes of individual sp. increases the dissimilarity of the two stations.

19. Growth will be reduced w/+3°, because the max. temp. is already at the upper range of the growth scope for salmonids. Growth will be reduced much more at +7° because the fish will be stressed and lose weight for most of the period of insect emergence (although insect growth/emergence may adjust to earlier in the year) and also because 24° is right at their 4-day lethal temp. In both cases they will seek lower temp.

20. Refers to control of planktivorous fish by either reducing their populations or providing refuges for large efficiently grazing zooplankton, thereby causing a reduction in algal content. Aerate hypolimnion to provide a refuge for large zoops or introduce high density of piscivorous fish. Poisoning all fish of course will work, but is usually not compatible with recreational use.

Multiple-choice questions

1. b	35. b
2. a	36. d
3. d	37. a
4. c	38. b

5. b
6. b
7. a
8. d
9. b
10. d
11. (a) b
    (b) c
    (c) d
    (d) e
    (e) a
12. d
13. c
14. b
15. b
16. c
17. a
18. b
19. c
20. a
21. c
22. b
23. d
24. a
25. c
26. d
27. a
28. b
29. c
30. d
31. a
32. d
33. a
34. b

39. c
40. d
41. b
42. b
43. b
44. b
45. a
46. c
47. d
48. c
49. a
50. c
51. b
52. b
53. d
54. a
55. b
56. c
57. b
58. d
59. a
60. b
61. c
62. c
63. b
64. c
65. a
66. c
67. b
68. a

Appendix C
GLOSSARY*

Absorption penetration of one substance into the body of another.

Acclimation the process of adjusting to change, e.g. temperatures, in an environment.

Acute involving a stimulus severe enough to rapidly induce a response; in bioassay tests, a response observed within 96 hours typically is considered an acute one.

Adaptation a change in the structure, form, or habit of an organism resulting from a change in its environment.

Adsorption the taking up of one substance at the surface of another.

Aerobe an organism that can live and grow only in the presence of free oxygen.

Aerobic the condition associated with the presence of free oxygen in an environment.

Algae (alga) simple plants, many microscopic, containing chlorophyll. Most algae are aquatic and may produce a nuisance when conditions are suitable for prolific growth.

Algicide a specific chemical highly toxic to algae. Algicides are often applied to water to control nuisance algal blooms.

Allocthonous said of food material reaching an aquatic community from the outside in the form of organic detritus.

Alluvial transported and deposited by running water.

Amphipods, *See* Scuds.

Anadromous fish fish that typically inhabit seas or lakes but ascend streams at more or less regular intervals to spawn; e.g. salmon, steelhead, or American shad.

Anaerobe an organism for whose life processes a complete or nearly complete absence of oxygen is essential.

Anaerobic the condition associated with the lack of free oxygen in an environment.

Annelids segmented worms, as distinguished from the nonsegmented

* Modified from WQC, 1968, 1973.

roundworms and flatworms. Most are marine; however, many live in soil or fresh water. Aquatic forms may establish dense populations in the presence of rich organic deposits. Common examples of segmented worms are earthworms, sludgeworms, and leeches.

Anoxic depleted of free oxygen; anaerobic.

Antagonism the power of one toxic substance to diminish or eliminate the toxic effect of another; interactions of organisms growing in close association, to the detriment of at least one of them.

Application factor a factor applied to a short-term or acute toxicity test to estimate a concentration of waste that would be safe in a receiving water.

Assimilation the transformation and incorporation of substances (e.g. nutrients) by an organism or ecosystem.

Autochonous said of food material entering an environment from inside in the form of photosynthetic (autotrophic) production.

Autotrophic organism an organism capable of constructing organic matter from inorganic substances.

Benthic region the bottom of a body of water. This region supports the benthos.

Benthos aquatic bottom-dwelling organisms including: (1) sessile animals, such as the sponges, barnacles, mussels, oysters, some worms, and many attached algae; (2) creeping forms, such as insects, snails, and certain clams; and (3) burrowing forms which include most clams and worms.

Bioaccumulation uptake and retention of environmental substances by an organism from its environment, as opposed to uptake from its food.

Bioassay a determination of the concentration or dose of a given material necessary to affect a test organism under stated conditions.

Biomagnification *see* Trophic accumulation.

Biomass the living weight of a plant or animal population, usually expressed on a unit area basis.

Biota all living organisms of a region.

Biotic index a numerical index using various aquatic organisms to determine their degree of tolerance to differing water conditions.

Biotoxin toxin produced by a living organism; the biotoxin which causes paralytic shellfish poisoning is produced by certain species of dinoflagellate algae.

Bivalve an animal with a hinged two-valve shell; examples are the clam and oyster.

Black liquor waste liquid remaining after digestion of rags, straw, and pulp.

Bloom an unusually large number of organisms per unit of water, usually algae, made up of one or a few species.

Blue-green algae   a group of algae with a blue pigment, in addition to the green chlorophyll. A stench is often associated with the decomposition of dense blooms of blue-green algae in fertile lakes.

Body burden   the total amount of a substance present in the body tissues and fluids of an organism.

Buffer capacity   the ability of a solution to maintain its pH when stressed chemically.

Carrying capacity   the maximum biomass that a system is capable of supporting continuously.

Catadromous fishes   fishes that feed and grow in fresh water, but return to the sea to spawn. The best known example is the American eel.

Chelate   to combine with a metal ion and hold it in solution preventing it from forming an insoluble salt.

Chronic   involving a stimulus that lingers or continues for a long period of time, often one-tenth of the life span or more.

Clean Water Association   an association of organisms, usually characterized by many different kinds (species). These associations occur in natural unpolluted environments.

Coagulation   a water treatment process in which chemicals are added to combine with or trap suspended and colloidal particles to form rapidly settling aggregates.

Coarse or rough fish   those species of fish considered to be of poor fighting quality when taken on tackle and of poor food quality. These fish may be undesirable in a given situation, but at times may be classified differently, depending upon their usefulness. Examples include carp, goldfish, gar, sucker, bowfin, gizzard shad, goldeneye and mooneye.

Coelenterate   a group of aquatic animals that have gelatinous bodies, tentacles, and stinging cells. These animals occur in great variety and abundance in the sea and are represented in fresh water by a few types. Examples are hydra, corrals, sea anemones, and jellyfish.

Cold-blooded animals (poikilothermic animals)   animals that lack a temperature regulating mechanism that offsets external temperature changes. Their temperature fluctuates to a large degree with that of their environment. Examples are fish, shellfish, and aquatic insects.

Coliform bacteria   a group of bacteria inhabiting the intestines of animals including man, but also found elsewhere. It includes all the aerobic, nonspore forming, rod-shaped bacteria that produce from lactose fermentation within 48 hours at 37°C.

Conservative pollutant   a pollutant that is relatively persistent and resistant to degradation, such as PCB and most chlorinated hydrocarbon insecticides.

Consumers   organisms that consume solid particles of organic food material. Protozoa are consumers.

Crustacea   mostly aquatic animals with rigid outer coverings, jointed appendages, and gills. Examples are crayfish, crabs, barnacles, water fleas, and sow bugs.

Cumulative   brought about or increased in strength by successive additions.

Daphnia   mostly microscopic swimming crustaceans, often forming a major portion of the zooplankton population. The second antennae are very large and are used for swimming.

Demersal   living or hatching on the bottom, as fish eggs than sink to the bottom.

Dermatitis   any inflammation of the skin. One type may be caused by the penetration beneath the skin of a cercaria found in water; this form of dermatitis is commonly called 'swimmers' itch'.

Detention time   the time necessary for lake water replacement.

Detritus   unconsolidated sediments comprised of both inorganic and dead and decaying organic material.

Diurnal   occurring once a day, i.e. with a variation period of 1 day; occurring in the daytime or during a day.

Diversity   the abundance in numbers of species in a specified location.

Dystrophic   said of brownwater lakes and streams usually with a low lime content and a high organic content; often lacking in nutrients.

Ecology   the science of the interrelations between living organisms and their environment.

Emergent aquatic plants   plants that are rooted at the bottom but project above the water surface. Examples are cattails and bulrushes.

Environment   the sum of all external influences and conditions affecting the life and the development of an organism.

Epilimnion   the surface waters in a thermally stratified body of water; these waters are characteristically well mixed.

Epiphytic   living on the surface of other plants.

Estuary   commonly an arm of the sea at the lower end of a river. Estuaries are often enclosed by land except at channel entrance points.

Euphotic zone   the lighted region that extends vertically from the water surface to the level at which photosynthesis fails to occur because of ineffective light penetration.

Eurytopic organisms   organisms with a wide range of tolerance to a particular environmental factor. Examples are sludgeworms and bloodworms.

Eutrophic   abundant in nutrients and having high rates of productivity frequently resulting in oxygen depletion below the surface layer.

Eutrophication   the intentional or unintentional enrichment of water.

Eutrophic waters waters with a good supply of nutrients. These waters may support rich organic productions, such as algal blooms.

Facultative able to live under different conditions, as in facultative aerobes and facultative anaerobes.

Facultative aerobe an organism that although fundamentally an anarobe can grow in the presence of free oxygen.

Facultative anaerobe an organism that although fundamentally an aerobe can grow in the absence of free oxygen.

Fall overturn a physical phenomenon that may take place in a body of water during the early autumn. The sequence of events leading to fall overturn include: (1) Cooling of surface waters, (2) density change in surface waters producing convection currents from top to bottom, (3) circulation of the total water volume by wind action, and (4) vertical temperature equality, 4°C. The overturn results in a uniformity of the physical and chemical properties of the water.

Fauna the entire animal life of a region.

Fecal coliform bacteria bacteria of the coliform group of fecal origin (from intestines of warm-blooded animals) as opposed to coliforms from non-fecal sources.

Finfish that portion of the aquatic community make up of the true fishes as opposed to invertebrate shellfish.

Flatworms (Platyhelminthes) nonsegmented worms, flattened from top to bottom. In all but a few of the flatworms, complete male and female reproductive systems are present in each individual. Most flatworms are found in water, moist earth, or as parasites in plants and animals.

Floating aquatic plants plants that wholly or in part float on the surface of the water. Examples are water lilies, water shields, and duckweeds.

Flocculation the process by which suspended colloidal or very fine particles are assembled into larger masses or floccules which eventually settle out of suspension.

Flora the entire plant life of a region.

Flushing rate the fraction of lake water replaced per time.

Food chain the transfer of food energy from plants or organic detritus through a series of organisms, usually four or five, consuming and being consumed.

Food web the interlocking pattern formed by a series of interconnecting food chains.

Free residual chlorination chlorination that maintains the presence of hypochlorous acid ($HOCl$) or hypochlorite ion (OCl^-) in water.

Fry (sac fry) the stage in the life of a fish between the hatching of the egg and the absorption of the yolk sac. From this stage until they attain a length of 1 inch, the young fish are considered advanced fry.

Fungi (fungus) simple or complex organisms without chlorophyll. The simpler forms are one-celled; the higher forms have branched filaments and complicated life cycles. Examples of fungi are moulds, yeasts, and mushrooms.

Fungicide   substances or a mixture of substances intended to prevent, destroy, or mitigate any fungi.

Game fish   those species of fish considered to possess sporting qualities on fishing tackle. These fish may be classified as undesirable, depending upon their usefulness. Examples of fresh water game fish are salmon, trout, grayling, black bass, muskellunge, wall-eye, northern pike, and lake trout.

Green algae   algae that have pigments similar in colour to those of higher green plants. Common forms produce floating algal mats in lakes.

Gross production   total energy fixed including that used for respiration.

Half-life   the period of time in which a substance loses half of its active characteristics (used especially in radiological work); the time required to reduce the concentration of a material by half.

Hemostasis   the cessation of the flow of blood in the circulatory system.

Herbicide   substances or a mixture of substances intended to control or destroy any vegetation.

Herbivore   an organism that feeds on vegetation.

Heterotrophic organism   organisms that are dependent on organic matter for food.

Higher aquatic plants   flowering aquatic plants. (These are separately categorized herein as Emergent, Floating, and Submerged Aquatic Plants.)

Histopathologic   occurring in tissue due to a diseased condition.

Holomictic lakes   lakes that are completely circulated to the deepest parts at time of winter cooling.

Hydrophobic   unable to combine with or dissolve in water.

Hydrophytic   growing in or in close proximity to water; e.g. aquatic algae and emergent aquatic vascular plants.

Hypertrophy   nontumorous increase in the size of an organ as a result of enlargment of constituent cells without an increase in their number.

Hypolimnion   the region of a body of water that extends from below the thermocline to the bottom of the lake; it is thus removed from much of the surface influence.

Insecticide   substances or a mixture of substances intended to prevent, destroy, or repel insects.

Invertebrates   animals without backbones.

Labile   unstable and likely to change under certain influences.

Lentic or lenitic environment standing water and its various intergrades. Examples of lenitic environments are lakes, ponds, and swamps.

Lethal involving a stimulus or effect causing death directly.

Life cycle the series of stages in the form and mode of life of an organism: i.e. the stages between successive recurrences of a certain primary stage such as the spore, fertilized egg, seed, or resting cell.

Limnetic zone the open-water region of a lake. This region supports plankton and fish as the principal plants and animals.

Lipophilic having an affinity for fats or other lipids.

Littoral zone the shallow shoreward region of a body of water having light penetration to the bottom; frequently occupied by rooted plants.

Lotic environment running waters, such as streams or rivers.

Macronutrient a chemical element necessary in large amounts for the growth and development of plants.

Macro-organisms plants, animals, or fungal organisms visible to the unaided eye.

Macrophyte the larger aquatic plants, as distinct from the microscopic plants, including aquatic mosses, liverworts and larger algae as well as vascular plants; no precise toxonomic meaning; generally used synonymously with aquatic vascular plants.

Marl an earthy, unconsolidated deposit formed in freshwater lakes, consisting chiefly of calcium carbonate mixed with clay or other impurities in varying proportions.

Median lethal concentration (LC50) the concentration of a test material that causes death to 50 per cent of a population within a given time period.

Median lethal dose (LD50) the dose of a test material, ingested or injected, that kills 50 per cent of a group of test organisms.

Median tolerance limit (TL50) the concentration of a test material in a suitable diluent (experimental water) at which just 50 per cent of the test animals are able to survive for a specified period of exposure.

Meromictic lakes lakes in which dissolved substances create a gradient of density differences in depth, preventing complete mixing or circulation of the water.

Mesotrophic having a nutrient load resulting in moderate productivity.

Metabolites products of metabolic processes.

Methylation combination with the methyl radical (CH_2).

Micronutrient chemical element necessary in only small amounts for growth and development; also known as trace elements.

Microorganism any minute organism invisible or barely visible to the unaided eye.

Molluscicide substances or a mixture of substances intended to destroy or control snails.

Mollusc (mollusca)   a large animal group including those forms popularly called shellfish (but not including crustaceans). All have a soft unsegmented body protected in most instances by a calcareous shell. Examples are snails, mussels, clams, and oysters.

Moss   any bryophytic plant characterized by small, leafy, often tufted stems bearing sex organs at the tips.

Motile   exhibiting or capable of spontaneous movement.

Nekton   swimming organisms able to navigate at will.

Nematoda   unsegmented roundworms or threadworms. Some are free living in soil, fresh water, and salt water; some are found living in plant tissue; others live in animal tissue as parasites.

Net productivity   energy fixed into organic matter by green plants and available for transfer in food web. Respiration of plants excluded.

Neuston   organisms resting or swimming on the surface film of the water.

Nonconservative pollutant   a pollutant that is quickly degraded and lacks persistence, such as most organophosphate insecticides.

Nutrients   organic and inorganic chemicals necessary for the growth and reproduction of organisms.

Oligotrophic   having a small supply of nutrients and thus supporting little organic production, and seldom if ever becoming depleted of oxygen.

Organic detritus   the particulate remains of disintegrated plants and animals.

Oxygen-debt   a phenomenon that occurs in an organism when available oxygen is inadequate to supply the respiratory demand. During such a period the metabolic processes result in the accumulation of breakdown products that are not oxidized until sufficient oxygen becomes available.

Parasite   an organism that lives on or in a host organism from which it obtains nourishment at the expense of the latter during all or part of its existence.

Pelagic zone   the free-water region of a sea. (Pelagic refers to the sea, limnetic refers to bodies of fresh water.)

Periphyton   associated aquatic organisms attached or clinging to stems and leaves of rooted plants or other surfaces projecting above the bottom of a water body.

Pesticide   any substance used to kill plants, insects, algae, fungi, and other organisms; includes herbicides, insecticides, algalcides, fungicides, and other substances.

Photosynthesis   the process by which simple sugars and starches are produced from carbon dioxide and water by living plant cells, with the aid of chlorophyll and in the presence of light.

Phytoplankton plant plankton that live unattached in water.

Piscicide substances or a mixture of substances intended to destroy or control fish populations.

Plankton (plankter) organisms of relatively small size, mostly microscopic, that have either relatively small powers of locomotion or that drift in that water with waves, currents, and other water motion.

Pool zone the deep-water area of a stream, where the velocity of current is reduced. The reduced velocity provides a favourable habitat for plankton. Silt and other loose materials that settle to the bottom of this zone are favourable for burrowing forms of benthos.

Producers organisms, for example, plants, that synthesize their own organic substance from inorganic substances.

Productivity the rate of storage of organic matter in tissue by organisms including that used by the organisms in maintaining themselves.

Profundal zone the deep and bottom-water area beyond the depth of effective light penetration. All of the lake floor beneath the hypolimnion.

Protozoa organism consisting either of a single cell or of aggregates of cells, each of which performs all the essential functions in life. They are mostly microscopic in size and largely aquatic.

Pycnocline a layer of water that exhibits rapid change in density, analogous to thermocline.

Rapids zone the shallow-water area of a stream, where velocity of current is great enough to keep the bottom clear of silt and other loose materials, thus providing a firm bottom. This zone is occupied largely by specialized benthic or periphytic organisms that are firmly attached to or cling to a firm substrate.

Red tide a visible red-to-orange coloration of an area of the sea caused by the presence of a bloom of certain 'armoured' flagellates.

Reducers organisms that digest food outside the cell wall by means of enzymes secreted for this purpose. Soluble food is then absorbed into the cell and reduced to a mineral condition. Examples are fungi, bacteria, protozoa, and nonpigmented algae.

Refractory resisting ordinary treatment and difficult to degrade.

Riffle a section of a stream in which the water is usually shallower and the current of greater velocity than in the connecting pools; a riffle is smaller than a rapid and shallower than a chute.

Rotifers (rotatoria) microscopic aquatic animals, primarily free-living, fresh water forms that occur in a variety of habitats. Approximately 75 per cent of the known species occur in the littoral zone of lakes and ponds. The more dense populations are associated with a substance of submerged aquatic vegetation. Most forms ingest fine organic detritus for food, whereas others are predaceous.

Safety factor a numerical value applied to short-term data from other organisms in order to approximate the concentration of a substance that will not harm or impair the organism being considered.

Scuds (amphipods) macroscopic aquatic crustaceans that are laterally compressed. Most are marine and estuarine. Dense populations are associated with aquatic vegetation. Great numbers are consumed by fish.

Secchi disc a device used to measure visibility depths in water. The upper surface of a circular metal plate, 20 centimetres in diameter, is divided into four quadrants and so painted that two quadrants directly opposite each other are black and the intervening ones white. When suspended to various depths of water by means of a graduated line, its point of disappearance indicates the limit of visibility.

Seiche a form of perodic current system, described as a standing wave, in which some stratum of the water in a basin oscillates about one or more nodes.

Sessile organisms organisms that sit directly on a base without support, attached or merely resting unattached on a substrate.

Seston suspended particles and organisms between 0.0002 and 1 mm in diameter.

Shellfish a group of molluscs usually enclosed in a self-secreted shell; includes oysters and clams.

Sludge a solid waste fraction precipitated by a water treatment process.

Smolt a young fish, usually a salmonid, as it begins and during the time it makes its seaward migration.

Sorption a general term for the processes of absorption and adsorption.

Species (both singular and plural) a natural population or group of populations that transmit specific characteristics from parent to off-spring. They are reproductively isolated from other populations with which they might breed. Populations usually exhibit a loss of fertility when hybridizing.

Sphaerotilus a slime-producing, nonmotile, sheathed, filamentous, attached bacterium. Great masses are often broken from their 'hold-fasts' by currents and are carried floating downstream in gelatinous flocks.

Sponges (porifera) one of the sessile animals that fasten to piers, pilings, shells, rocks, etc. Most live in the sea.

Spore the reproductive cell of a protozoan, fungus, alga, or bryophyte. In bacteria, spores are specialized resting cells.

Spring overturn a physical phenomenon that may take place in a body of water during the early spring. The sequence of events leading to spring overturn include: (1) Melting of ice cover, (2) warming of surface waters, (3) density change in surface waters producing con-vection currents from top to bottom, (4) circulation of the total water volume by wind action, and (5) vertical temperature equality, 4°C. The overturn results in a uniformity of the physical and chemical properties of the water.

Standing crop   quantitative abundance of organisms, either mass, volume or numbers, present at any one time.

Stenotopic organisms   organisms with a narrow range of tolerance for a particular environmental factor. Examples are trout, stonefly nymphs, etc.

Stoichiometric   the mass relationship in a chemical reaction.

Stratification   the phenomenon occurring when a body of water becomes divided into distinguishable layers.

Subacute   involving a stimulus not severe enough to bring about a response speedily.

Sublethal   involving a stimulus below the level that causes death.

Sublittoral zone   the part of the shore from the lowest water level to the lower boundary of plant growth.

Submerged aquatic plant   a plant that is continuously submerged beneath the surface of the water. Examples are the pondweed and coontail.

Succession   the orderly process of community change in which a sequence of communities replaces one another in a given area until a climax community is reached.

Swimmers' itch   a rash produced on bathers by a parasitic flatworm in the cercarial stage of its life cycle. The organism is killed by the human body as soon as it penetrates the skin; however, the rash may persist for a period of about 2 weeks.

Symbiosis   two organisms of different species living together, one or both of which may benefit and neither is harmed.

Symbiosis   two organisms of different species living together, one or both of which may benefit and neither is harmed.

Synergistic   interactions of two or more substances or organisms producing a result that any was incapable of independently.

Thermocline   a layer in a thermally stratified body of water in which the temperature changes rapidly relative to the remainder of the body.

Tolerant association   an association of organisms capable of withstanding adverse conditions within the habitat. It is usually characterized by a reduction in species (from a clean water association) and an increase in individuals representing a particular species.

Trophic accumulation   passing of a substance through a food chain such that each organism retains all or a portion of the amount in its food and eventually acquires a higher concentration in its flesh than in the food.

Trophic level   a scheme of categorizing organisms by the way they obtain food from primary producers or organic detritus involving the same number of intermediate steps.

Trophogenic region   the superficial layer of a lake in which organic production from mineral substances takes place on the basis of light energy.

Tropholytic region   the deep layer of a lake, where organic dissimilation predominates because of light deficiency.

Turnover rate   fractional replacement of standing crop per time

Turnover time   time necessary to replace the standing crop.

Warm and cold-water fish   warm-water fish include black bass, sunfish, catfish, gar, and others; whereas cold-water fish include salmon and trout, whitefish and others. The temperature factor determining distribution is set by adaptation of the eggs and larvae to warm or cold water.

Zooglea   bacteria embedded in a jellylike matrix formed as the result of metabolic activities.

Zooplankton   protozoa and other animal microorganisms living unattached in water. These include small crustacea, such as *Daphnia* and *Cyclops*.

REFERENCES

Adams, J. R. (1969) Ecological investigations related to thermal discharges. *Pacific Gas and Electric Company Report*, Emeryville, CA.

Ahlgren, G. (1977) Growth of *Ocillatoria agardii* Gom. in chemostat culture. I. Investigation of nitrogen and phosphorus requirements. *Oikos*, **29**, 209–24.

Ahlgren, G. (1978) Response of phytoplankton and primary production to reduced nutrient loading in Lake Norrviken. *Verh. Int. Verein. Limnol.*, **20**, 840–5.

Ahlgren, G. (1987) Temperature functions in biology and their application to algal growth constants. *Oikos* **49**, 177–90.

Ahlgren, I. (1977) Role of sediments in the process of recovery of a eutrophicated lake. In *Interactions Between Sediments and Freshwater* (ed. H. L. Golterman), Dr W. Junk, The Hague, pp. 372–7.

Ahlgren, I. (1978) Response of Lake Norrviken to reduced nutrient loading. *Verh. Int. Verein. Limnol.*, **20**, 846–50.

Ahlgren, I. (1979) Lake metabolism studies and results at the Institute of Limnology in Uppsala. *Arch. Hydrobiol. Beih.*, **13**, 10–30.

Ahlgren, I. (1980) A dilution model applied, to a system of shallow eutrophic lakes after diversion of sewage effluents. *Arch. Hydrobiol.*, **89**, 17–32.

Ahlgren, I. (1988) Nutrient dynamics and trophic state response of two eutrophicated lakes after reduced nutrient loading. *Eutrophication and Lake Restoration: Water Quality and Biological Impacts* (ed. G. Balvay), Thonon-les-Bains, pp. 79–97.

Ahlgren, I., Bostrom, B. and Petersson, A.-K. (1988a) Seasonal variation of the microbial community in the sediments of a hypereutrophic lake. *Verh. Int. Verein. Limnol.*, **23**, 460–1.

Ahlgren, I., Frisk, T. and Kamp-Nielsen, L. (1988b). Empirical and theoretical models of phosphorus loading, retention and concentration vs. lake trophic state. *Hydrobiologia*, **170**, 285–303.

Allison, E. M. and Walsby, A. E. (1981) The role of potassium in the control of turgor pressure in a gas vacuolate blue-green alga. *J. Exp. Bot.* **32**, 241–9.

Ambühl, H. (1959) Die bedeutung der strömung als ökologischer factor. *Schweiz. Z. Hydrol.*, **21**, 133–264.

Amend, D., Yasutake, W. and Morgan, R. (1969) Some factors influencing susceptibility of rainbow trout to the acute toxicity of an ethyl mercury phosphate formulation (Timsan). *Trans. Am. Fish Soc.*, **98**, 419–25.

Andersen, J. M. (1975) Influence of pH on release of phosphorus from lake sediments. *Arch. Hydrobiol.*, **76**, 411–19.

Anderson, J. B. and Mason, W. T. (1968) A comparison of benthic macroinvertebrates collected by dredge and basket sampler. *J. Water Pollut. Control Fed.*, **40**, 252–9.

Andersson, G., Berggen, H., Cronberg, G. and Gelin, C. (1978) Effects of planktivorous and benthivorous fish on organisms and water chemistry in eutrophic lakes. *Hydrobiologia*, **59**, 9–16.

Antia, N. J., McAllister, C. D., Parsons, T. R., Stephens, K. and Strickland, J. D. H. (1963) Further measurements of primary production using a large volume plastic sphere. *Limnol. Oceanogr.*, **8**, 166–83.

APHA (1985) *Standard Methods for the Examination of Water and Wastewater*, 16th edn, APHA, Washington, DC, 1136 pp.

Arnold, D. E. (1971) Ingestion, assimilation, survival, and reproduction by *Daphnia pulex* fed seven species of blue-green algae. *Limnol. Oceanogr*, **16**, 906–20.

Arrhenius, S. (1989) Uber die Reakionsgeschwindigkeit bei der Inversion von Rohrzucker durch Sauren. *Z. Phys. Chem.*, **4**, 226–34.

Atwater, J. W., Jasper, S., Mavinic, D. S. and Koch, F. A. (1983) Experiments using *Daphnia* to measure landfill leachate toxicity. *Water Res.*, **17**, 1855–61.

Averett, R. C. (1961) Macroinvertebrates of the Clark Fork River, Montana – a pollution survey. *Montana Board of Health and Fish and Game Department Report.*, No. 61–1.

Averett, R. C. (1969) Influence of temperature on energy and material utilization by juvenile coho salmon. PhD thesis, Oregon State University, Corvallis.

AWWA (1987) *Current Methodology for the Control of Algae in Surface Reservoirs* Am. Water Works Assn. Res. Found., Denver.

Ayles, B. G., Lark, J. G. I., Barica, J. and Kling, H. (1976) Seasonal mortality of rainbow trout (*Salmo gairdnerii*) planted in small eutrophic lakes of Central Canada. *J. Fish Res. Board Canada.*, **33**, 647–55.

Baalsrud, K. (1967) Influence of nutrient concentration on primary production. in *Pollution and Marine Ecology* (eds T. A. Olson and F. J. Burgess), John Wiley and Sons, New York, pp. 159–69.

Bahr, T. G., Cole, R. A. and Stevens, H. K. (1972) *Recycling and Ecosystem Response to Water Manipulation*, Technical Report No. 37, Inst. Water Res., Michigan State University, East Lansing.

Baker, J. P. and Schofield, C. L. (1982) Aluminium toxicity to fish in acid waters. *Water, Air and Soil Poll.*, **18**, 289–309.

Barbiero, R. P. (1991) Sediment-water transport of phosphorus by the blue-green alga *Gloeotrichia echinulata*. PhD dissertation, Dept Civil Eng., University of Washington, Seattle.

Barko, J. W. (1983) The growth of *Myriophyllum spicatum L.* in relation to selected characteristics of sediment and solution. *Aquat. Bot.*, **15**, 91–103.

Barko, J. W. and Smart, R. M. (1986) Sediment-related mechanisms of growth limitation in submersed macrophytes. *Ecology*, **67**, 1328–40.

Bartsch, A. F. and Ingram, W. M. (1959) Stream life and the pollution environment. *Public Works*, **90**, 104–10.

Beak, T. W. (1965) A biotic index of polluted streams and the relationship of pollution to fisheries. *Adv. Water Pollut. Res., Proc. 2nd Int. Conf.*, **1**, 191–210.

Beck, W. M. (1955) Suggested method for reporting biotic data. *Sewage Ind. Wastes*, **27**, 1193–7.

Becker, D. (1972) Columbia River thermal effects study: reactor effluent problems. *J. Water Pollut. Cont. Fed.*, **45**, 850–69.

Beeton, A. M. (1965) Eutrophication of the St. Lawrence and Great Lakes. *Limnol. Oceanogr.*, **10**, 240–54.

Beeton, A. M. and Edmondson, W. T. (1972) The eutrophication problem. *J. Fish Res. Board Canada*, **29**, 673–82.

Bělehrádek, J. (1926) Influence of temperature on biological processes. *Nature (Lond.)*, **118**, 117–18.

Bělehrádek, J. (1957) Physiological aspects of heat and cold. *Ann. Rev. Physiol.*, **198**, 59–82.

Bellanca, M. A. and Bailey, D. S. (1977) Effects of chlorinated effluents on aquatic ecosystem in the lower James River. *J. Water Pollut. Control Fed.*, **49**, 639–45.

Bengtsson, L., Fleischer, S., Lindmark, G. and Ripl, W. (1975) The Lake Trummen restoration project. I. Water and sediment chemistry. *Verh. Int. Verein. Limnol.* **19**, 1080–7.

Benndorf, J. (1990) Conditions for effective biomanipulation; conclusions derived from whole-lake experiments in Europe. *Hydrobiologia*, **200/201**, 187–203.

Benndorf, J., Kneschke, H., Kossatz, K. and Penz, E. (1984a) Manipulation of the pelagic food web by stocking with predacious fishes. *Int. Rev. Ges. Hydrobiol.*, **69**, 407–28.

Benndorf, J. and Putz, K. (1987) Control of eutrophication of lakes and reservoirs by means of pre-dams. II. Validation of the phosphate removal model and size optimization. *Water Res.*, **21**, 839–42.

Bernhardt, H. (1981) Recent developments in the field of eutrophication prevention. *Z Wasser Abwasser Forsch.*, **17**, 14–26.

Best, M. D. and Mantai K. E. (1978) Growth of *Myriophyllum spicatum*: sediment or lake water as a source of nitrogen and phosphorus. *Ecology*, **59**, 1075–80.

Bierman, V. J. (1976) Mathematical model of selective enrichment of blue-green by nutrient enrichment, in *Modeling of Biochemical Processes in Aquatic Ecosystems* (ed. R. P. Canale), Ann Arbor Science Publ., Inc., Ann Arbor, MI.

Bierman, V., Verhoff, V., Poulson, T. and Tenney, M. (1973) Multinutrient dynamic model of algal growth and species competition in eutrophic lakes. In *Modeling the Eutrophication Process* (ed. J. M. Middlebrooks), Water Research Laboratory, Utah State University, pp. 89–109.

Biggs, B. J. (1985). Algae – a blooming nuisance in rivers. *Soil Water*, **21**, 27–31.

Biggs, B. J. F. (1988) Algal proliferations in New Zealand's shallow, stony foothills–fed rivers; toward a predictive model. *Verh. Int. Verein. Limnol.*, **23**, 1405–11.

Biggs, B. J. F. (1989) Biomonitoring of organic pollution using periphyton, South Branch, Canterbury, New Zealand. *N.Z.J. Marine Freshwater Res.*, **23**, 263–74.

Biggs, B. J. F. and Close, M. E. (1989) Periphyton biomass dynamics in gravel bed rivers: the relative effects of flows and nutrients. *Freshwater Biol.* **22**, 209–31.

Biggs, B. J. F. and Price, G. M. (1987) A survey of filamentous algal proliferations in New Zealand rivers. *N.Z.J. Marine Freshwater Res.*, **21**, 175–91.

Brillouin, L. (1962) *Science and Information Theory*, 2nd edn, Academic Press, New York.

Birch, P. B. (1976) The relationship of sedimentation and nutrient cycling to the trophic status of four lakes in the Lake Washington drainage basin. Ph.D. Dissertation, University of Washington, Seattle.

Björk, S. (1974) *European lake Rehabilitation Activities* Institute of Limnology Report, University of Lund, 23 pp.

Bjork, S. (1985) Lake restoration techniques, in *Lake Pollution and Recovery*,

International Congress European Water Pollution Control Association, Rome, pp. 293–301.

Björk, S. *et al.* (1972) Ecosystem studies in connection with the restoration of lakes. *Verh. Int. Verein. Limnol.*, **18**, 379–87.

Bloomfield, J. A., Park, R. A., Scavia, D. and Zahorcak, C. S. (1973) Aquatic modeling in the EDFB, US-IBP, in *Modeling the Eutrophication Process* (eds E. J. Middlebrooks, D. H. Falkenborg and T. W. Maloney) Ann Arbor Science Publ., Inc., Ann Arbor, MI, pp. 139–58.

Bole, J. B. and Allan, J. R. (1978) Uptake of phosphorus from sediment by aquatic plants, *Myriophyllum spicatum* and *Hydrilla verticillata*. *Water Res.*, **12**, 353–8.

Booker, M. J. and Walsby, A. E. (1981). Bloom formation and stratification by a planktonic blue-green alga in an experimental water column. *Br. Phycol. J.*, **16**, 411–21.

Borman, F. H. and Likens, G. E. (1967) Nutrient cycling. *Science*, **155**, 474–29.

Boström, B. (1984) Potential mobility of phosphorus in different types of lake sediments. *Int. Rev. Ges. Hydrobiol.*, **69**, 454–74.

Boström, B., Ahlgren I. and Bell, R. T. (1985) Internal nutrient loading in a eutrophic lake, reflected in seasonal variations of some sediment parameters. *Verh. Int. Verein. Limnol.*, **22**, 3335–9.

Boström, B., Janson, M. and Forsberg, C. (1982) Phosphorus release from lake sediments. *Arch. Hydrobiol. Beih. Ergebn. Limnol.*, **18**, 5–59.

Bothwell, M. L. (1985) Phosphorus limitation of lotic periphyton growth rates: an intersite comparison using continuous-flow toughs (Thompson River System, British Columbia). *Limnol. Oceanogr.*, **30**, 527–42.

Bothwell, M. L. (1988) Growth rate responses of lotic periphytic diatoms of experimental phosphorus enrichment: the influence of temperature and light. *Can. J. Fish. Aquat. Sci.*, **45**, 261–70.

Bothwell, M. L. (1989) Phosphorus-limited growth dynamics of lotic periphytic diatom communities: areal biomass and cellular growth rate responses. *Can. J. Fish Aquat. Sci.*, **46**, 1293–301.

Bray, J. R., and Curtis, J. T. (1957) An ordination of the upland forest communities of southern Wisconsin. *Ecol. Monogr.*, **27**, 325–49.

Brett, J. R. (1956) Some principles in the thermal requirements of fishes. *Q. Rev. Biol.*, **31**, 75–87.

Brett, J. R. (1960) Thermal requirements of fish – three decades of study, 1940–70, in *Biological Problems in Water Pollution*. Trans. 2nd Seminar 1959, USPHS, Cincinnati, Ohio, pp. 111–17.

Brezonik, P. L., and Lee, G. F. (1968) Denitrification as a nitrogen sink in Lake Mendota, Wisc. *Environ. Sci. Technol.*, **2**, 120–5.

Brillouin (1962) *Science and Information Theory*, 2nd edn, Academic Press, New York.

Brinkhurst, R. O. (1965) Observations on the recovery of a British River from gross organic pollution. *Hydrobiolia*, **25**, 9–51.

Bristow, J. W. and Whitcombe, M. (1971) The role of roots in the nutrition of aquatic vascular plants. *Am. J. Bot.*, **58**, 8–13.

Brock, D. B. (1970) *Biology of Microorganisms*. Prentice-Hall, Englewood Cliffs, NJ.

Brooks, J. L. and Dodson, S. (1965) Predation, body size and composition of plankton. *Science*, **150**, 28–35.

Brown, M. P., Werner, M. B., Sloan, R. J. and Simpson, K. W. (1985) Polychlorinated biphenyls in the Hudson River. *Environ. Sci. Technol.*, **19**, 656–61.

Brown, V. M. (1968) The calculation of the acute toxicity of mixtures of poisons to rainbow trout. *Water Res.*, **2**, 723–33.

Brown, V. M., Shurben, D. G. and Shaw, D. (1970) Studies on water quality and the absence of fish from some polluted English rivers. *Water Res.*, **4**, 363–82.

Brungs, W. (1969) Chronic toxicity of zinc to the Fathead Minnow, *Pimephales promelas* Rafinesque. *Trans. Am. Fish Soc.*, **98**, 272–9.

Brungs, W. (1973) Effects of residual chlorine on aquatic life. *J. Water Pollut. Control Fed.*, **45**, 2180–93.

Bryan, A., Marsden, K. and Hanna, S. (1975) A summary of observations on aquatic weed control methods. British Columbia Department of Environment, unpublished report, 73 pp.

Buckley, J. A. (1971) Effects of low nutrient dilution water and mixing on the growth of nuisance algae. MS thesis, University of Washington, Seattle, 116 pp.

Buckley, J. A. (1977) Heinz body hemolytic anemia in coho salmon (*Oncorhynchus kisutch*) exposed to chlorinated waste water. *J. Fish Res. Board Canada*, **34**, 215–24.

Buckley, J. A., Whitmore, C. M. and Matsuda, R. I. (1976) Changes in blood chemistry and blood-cell morphology in coho salmon (*Oncorhynchus kisutch*) following exposure to sublethal levels of total residual chlorine in municipal waste water. *J. Fish Res. Board Canada*, **33**, 776–82.

Burns, C. W. (1968) The relationship between body size of filter-feeding cladocera and the maximum size of particle ingested. *Limnol. Oceanogr.*, **13**, 675–8.

Burns, C. W. (1969) Relation between filtering rate, temperature, and body size in four species of *Daphnia*. *Limnol. Oceanogr.*, **14**, 693–700.

Burns, C. W. (1987) Insights into zooplankton–cyanobacteria interactions derived from enclosure studies. *N.Z.J. Marine Freshwater Res.*, **21**, 477–82.

Burton, D. J., Jones, A. H. and Cairns, J., Jr. (1972) Acute zinc toxicity to rainbow trout (*Salmo gairdnari*): confirmation of the hypothesis that death is related to tissue hypoxia. *J. Fish Res. Board Canada*, **29**, 1463–6.

Bush, R. M., Welch, E. B. and Buchanan, R. J. (1972) Plankton associations and related factors in a hypereutrophic lake. *Water, Air Soil Pollut*, **1**, 257–74.

Bush, R. M., Welch, E. B. and Mar, B. W. (1974) Potential effects of thermal discharges on aquatic systems. *Environ. Sci. Technol.*, **8**, 561–8.

Butcher, R. W. (1933) Studies on the ecology of rivers. I. On the distribution of macrophytic vegetation in the rivers of Britain. *J. Ecol.*, **21**, 58–91.

Butkus, S. R., Welch, E. B., Horner, R. R. and Spyridakis, D. E. (1988) Lake response modeling using biologically available phosphorus. *J. Water Pollut. Control Fed.*, **60**, 1663–9.

Cairns, J., Jr. (1956) Effects of increased temperatures on aquatic organisms. *Ind. Wastes.*, **1**, 150–2.

Cairns, J., Jr. (1971) The effects of pH, solubility, and temperature upon the acute toxicity of zinc to the bluegill sunfish (*Lepomis macrochirus* Raf.). *Trans. Kans. Acad. Sci.*, **74**, 81–92.

Cairns, J., Jr., Crossman, J. S., Dickson, K. L. and Herricks, E. E. (1971) The recovery of damaged streams. *Assoc. Southeastern Biol. Bull.*, **18**, 79–106.

Cairns, J., Jr. and Dickson, K. L. (1971) A simple method for the biological assessment of the effects of waste discharges on aquatic bottom-dwelling organisms. *J. Water Pollut. Control Fed.*, **43**, 755–72.

Canfield, D. E., Jr and Bachmann, R. W. (1981) Prediction of total phosphorus concentrations, chlorophyll *a*, and Secchi depths in natural and artificial lakes. *Can. J. Fish. Aquat. Sci.*, **38**, 414–23.

Canfield, D. E. Jr., Langeland, K. A., Linda S, R and Haller, W. T. (1985) Relations between water transparency and maximum depth of macrophyte colonization in lakes. *J. Aquat. Plant Management*, **23**, 25–8.

Carignan, R. and Kalff, J. (1980) Phosphorus sources for aquatic weeds, water or sediments. *Science*, **207**, 987–9.

Carlson, A. R. and Herman, L. J. (1974) Effects of lowered dissolved oxygen concentrations on channel catfish (*Ictalurus punctatus*) embryos and larvae. *Trans. Am. Fish Soc.*, **103**, 623–6.

Carlson, A. R. and Siefert, R. E. (1974) Effects of reduced oxygen on the embryos and larvae of lake trout (*Salvelinus namaycush*) and largemouth bass (*Micropterus salmoides*). *J. Fish Res. Board Canada*, **31**, 1393–6

Carlson, K. L. (1983) The effects of induced flushing on water quality in Pelican Horn, Moses Lake, Wa. MS thesis, University of Washington, Seattle.

Carlson, R. E. (1977) A trophic state index for lakes. *Limmol. Oceanogr.*, **22**, 361–8.

Carmichael, W. W. (1986) Algal toxins. *Adv. Bot. Res.*, **12**, 47–101.

Carpenter, S. R. (1980a) Enrichment of Lake Wingra, Wisconsin by submersed macrophyte decay. *Ecology*, **61**, 1145–55.

Carpenter, S. R. (1980b) The decline of *Myriphyllum spicatum* in a eutrophic Wisconsin USA lake. *Can. J. Bot.*, **58**, 527–35.

Carpenter, S. R. (1981) Submersed vegetation: an internal factor in lake ecosystem succession. *Am. Naturalist*, **118**, 372–83.

Carpenter, S. R. and Adams, M. S. (1977) The macrophyte nutrient pool of a hardwater eutrophic lake: implications for macrophyte harvesting. *Aquat. Bot.*, **3**, 239–55.

Carpenter, S. R., Kitchell, J. F. and Hodgson, J. R. (1985) Cascading trophic interactions and lake productivity. *Bioscience*, **35**, 634–9.

Carpenter, S. R., Kitchell, J. F., Hodgson, J. R., Cochran, P. A., Elser, J. J., Elser, M. M., Lodge, D. M., Kretchmer, D., He, X. and von Ende, C. N. (1987) Regulation of lake primary productivity by food web structure. *Ecology*, **68**, 1863–76.

Carr, J. F. and Hiltunen, J. K. (1965) Changes in the bottom fauna of western Lake Erie from 1930 to 1961. *Limnol. Oceanogr.*, **10**, 551–69.

Chadwick, G. G. and Brocksen, R. W. (1969) Accumulation of dieldrin by fish and selected fish-food organisms. *J. Wildlife Management*, **33**, 693–700.

Chambers, P. A. and Kalff, J. (1985) Depth distribution and biomass of submersed aquatic macrophyte communities in relation to Secchi depth. *Can. J. Fish. Aquat. Sci.*, **42**, 701–9.

Chambers, P. A., Prepas, E. E., Bothwell, M. L. and Hamilton, H. R. (1989) Roots versus shoots in nutrient uptake by aquatic macrophytes in flowing waters. *Can. J. Fish. Aquat. Sci.*, **46**, 435–9.

Chandler, J. R. (1970) A biological approach to water quality management. *Water Pollut. Control* **69**, 415–21.

Chapra, S. C. (1975) Comment on 'An empirical method of estimating retention of phosphorus in lakes' by W. B. Kirchner and P. J. Dillon. *Water Resources Res.*, **11**, 1033–4.

Chapra, S. C. and Reckhow, K. H. (1979) Expressing the phosphorus loading concept in probabilistic terms. *J. Fish. Res. Board Canada*, **36**, 225–9.

Chapra, S. C. and Tarapchak, S. J. (1976) A chlorophyll *a* model and its relationship to phosphorus loading plots for lakes. *Water Resources Res.*, **12**, 1260–4.

Charlson, R. J. and Rodhe, H. (1982) Factors controlling the acidity of natural rainwater. *Nature*, **295**, 683–5.

Chen, C. W. (1970) Concepts and utilities of an ecological model. *J. Sanit. Eng. Div.* Proc. Amer. Soc. Civil Eng., **96**, 1085–97.

Christensen, M. H. and Harremoes, P. (1972) *Biological Denitrification in Water Treatment*, Rep. 72–2, Department of Sanitary Engineers, Technical University of Denmark.

Chrystal, G. (1904) Some results in the mathematical theory of seiches. *Proc. R. Soc. Edinb.*, **25**, 328–37.

Chutter, F. M. (1969) The effects of silt and sand on the invertebrate fauna of streams and rivers. *Hydrobiolia*, **34**, 57–75.

Chutter, F. M. (1972) An empirical biotic index of the quality of water in South African streams and rivers. *Water Res.* **6**, 19–30.

Clubb, R. W., Gaufin, A. R. and Lords, J. L. (1975) Acute cadmium toxicity studies upon nine species of aquatic insects. *Environ. Research*, **9**, 332–41.

Cocking, A. W. (1959) The effects of high temperature on roach (*Rutilus rutilus*). II. The effects of temperature increasing at a known constant rate. *J. Exp. Biol.* **36**, 217–36.

Coffey, B. T. and McNabb, C. D. (1974) Eurasian water milfoil in Michigan. *Mich. Bot.*, **13**, 159–65.

Cole, R. A. (1973) Stream community response to nutrient enrichment. *J. Water Pollut. Control Fed.*, **45**, 1875–88.

Collins, G. B. and Weber, C. I. (1978) Phycoperiphyton (algae) as indicators of water quality. *Trans. Am. Microscop. Soc.*, **97**, 36–43.

Cooke, G. D. and Carlson, R. E. (1989) *Reservoir Management for Water Quality and THM Precursor Control*. AWWA Research Foundation and American Water Works Association, 387 pp.

Cooke, G. D., Heath, R. T., Kennedy, R. H. and McComas, M. R. (1978) The effect of sewage diversion and aluminum sulfate application on two eutrophic lakes. *Ecol. Res. Ser.* EPA 600/3-78-003, Cinn., OH, pp. 101.

Cooke, G. D., Heath, R. T., Kennedy R. H. and McComas, M. R. (1982) Change in lake trophic state and internal phosphorus release after aluminum sulfate application. *Water Res. Bull.*, **18**, 699–705.

Cooke, G. D., Welch, E. B., Peterson, S. A. and Newroth, P. R. (1986) *Lake and Reservoir Restoration*, Butterworths, Boston, USA.

Cornett, R. J. and Rigler, F. H. (1979) Hypolimnetic oxygen deficits: their predictions and interpretation. *Science*, **205**, 580–1.

Coutant, C. C. (1966) Alteration of the community structure of periphyton by heated effluents. Report to AEC for contract AT (45-1)-1830, Battelle Memorial Inst.

Coutant C. C. (1969) Temperature, reproduction, and behavior. *Chesapeake Sci.*, **10**, 261–74.

Cronberg, G., Gelin, C. and Larsson, K. (1975) The Lake Trummen restoration project. II. Bacteria, phytoplankton, and phytoplankton productivity. *Verh. Int. Verein. Limnol.*, **19**, 1088–96.

Csanady, G. T. (1970) Dispersal of effluents in the Great Lakes, in *Water Research*, Vol. 4, Pergamon Press, Elmsford, NY, pp. 79–114.

Cuker, B. E. (1987) Field experiment on the influences of suspended clay and P on the plankton of a small lake. *Limnol. Oceanogr.* **32**, 840–847.

Cuker, B. E., Gama, P. and Burkholder, J. M. (1990) Type of suspended clay influences lake productivity and phytoplankton community response to phosphorus loading. *Limnol. Oceanogr.* **35**, 830–839.

Cullen, P. and Forsberg, C. (1988) Experiences with reducing point sources of phosphorus to lakes. *Hydrobiologia*, **170**, 321–36.

Cummins, K. W. (1974) Structure and function of stream ecosystems. *Bioscience*,

24, 631–41.

Cummins, K. W., Klug, J. J., Wetzel, R. G., Petersen, R. C., Superkropp, K. F., Manny, B. A., Wuycheck, J. C. and Howard, F. O. (1972) Organic enrichment with leaf leachate in experimental lotic ecosystems. *Bioscience*, **22**, 719–22.

Curtis, E. J. and Harrington, D. W. (1971) The occurrence of sewage fungus in rivers of the United Kingdom. *Water Res.*, **5**, 281–90.

Cusimano, R. F., Brakke, D. F. and Chapman, G. A. (1986) Effects of pH on the toxicities of cadmium, copper and zinc to steelhead trout (*Salmo gairdneri*). *Can. J. Fish. Aquat. Sci.*, **43**, 1497–503.

Davis, G. E., Foster, J. and Warren, C. E. (1963) The influence of oxygen concentration on the swimming performance of juvenile Pacific salmon at various temperatures. *Trans. Am. Fish Soc.*, **92**, 111–24.

Davison, W., Reynolds, C. S. and Finlay, B. J. (1985) Algal control of lake geochemistry; redox cycles in Rostherne Mere, U.K. *Water Res.*, **19**, 265–67.

Delwiche, C. C. (1970) The nitrogen cycle. *Sci. Amer.*, **223**, 136–47.

DeSylva, D. P. (1969) Theoretical considerations of the effects of heated effluents on marine fishes, in *Biological Aspects of Thermal Pollution* (eds P. A. Krenkel and F. C. Parker) Vanderbilt Univ. Press, Nashville, TN, pp. 229–93.

Dillon, P. J. (1975) The phosphorus budget of Cameron Lake, Ontario: The importance of flushing rate relative to the degree of eutrophy of a lake. *Limnol. Oceanogr.*, **29**, 28–39.

Dillon, P. J. and Rigler, F. H. (1974a) A test of a simple nutrient budget model predicting the phosphorus concentration in lake water. *J. Fish. Res. Board Canada*, **31**, 1771–8.

Dillon, P. J. and Rigler, F. H. (1974b) The phosphorus-chlorophyll relationship in lakes. *Limnol. Oceanogr*, **19**, 767–73.

Dillon, P. J. and Rigler, F. H. (1975) A simple method for predicting the capacity of a lake for development based on lake trophic state. *J. Fish. Res. Board Canada*, **32**, 1519–31.

Dimond, J. B. (1967) *Pesticides and Stream Insects*. Bulletin 23, Maine Forest Service and the Conservation Foundation, Washington, DC.

Doherty, F. G. (1983) Interspecies correlations of acute aquatic median lethal concentrations for four standard testing species. *Environ. Sci. Technol*, **17**, 661–5.

Dominie, D. R. (1980) Hypolimnetic aluminum treatment of soft water Annabessacook Lake, in *Restoration of Lakes and Inland Waters* EPA-440/5-81-010, Cinn, OH, pp. 417–23.

Doudoroff, P. and Shumway, D. L. (1967) Dissolved oxygen criteria for the protection of fish, in *A Symposium on Water Quality Criteria to Protect Aquatic Life* American Fish Society, Special Publication No. 4, pp. 13–19.

Doudoroff, P., Leduc, G. L. and Schneider, C. R. (1966) Acute toxicity to fish of solutions containing complex metal cyanides in relation to concentrations of molecular hydrocyanic acid. *Trans. Am. Fish Soc.* **95**, 6–22.

Driscoll, C. T., Jr, Baker, J. P., Bisogni, J. J., Jr, and Schofield, C. L. (1980) Effect of aluminium speciation on fish in dilute acidified waters. *Nature*, **284**, 161–84.

Droop, M. R. (1973) Some thought on nutrient limitation in algae. *J. Phycol.*, **9**, 264–72.

Duarte, C. M. and Kalff, J. (1986) Littoral slope as a predictor of the maximum biomass of submerged macrophyte communities. *Limnol. Oceanogr.*, **31**, 1072–80.

Dugdale, R. C. (1967) Nutrient limitation in the sea: dynamics, identification, and significance. *Limnol. Oceanogr.*, **12**, 658–95.

Dunst, R. C. (1981) Dredging activities in Wisconsin's lake renewal program, in *Restoration of Lakes and Inland Waters* EPA-440/5-81-010, Washington, DC, pp. 86–8.

Dunst, R. C., Born, S. M., Uttormark, P. D., Smith, S. A., Nichols, S. A., Peterson, J. O., Knauer, D. R., Serns, S. L., Winter, D. R. and Wirth, T. L. (1974) *Survey of Lake Rehabilitation Techniques and Experiences* Technical Bulletin 75, Department of Natural Resources, Madison, WI, USA.

Edmondson, W. T. (1966) Changes in the oxygen deficit of Lake Washington. *Verh. Int. Verein. Limnol.* **16**, 153–8.

Edmondson, W. T. (1969) Eutrophication in North America, in *Eutrophication: Causes, Consequences, and Correctives*. National Academy of Science, Washington, DC.

Edmondson, W. T. (1970) Phosphorus, nitrogen, and algae in Lake Washington after diversion of sewage. *Science*, **169**, 690–1.

Edmondson, W. T. (1972) Nutrients and phytoplankton in Lake Washington, in *Nutrients and Eutrophication: The Limiting Nutrient Controversy* (ed G. E. Likens) Special Symposium, *Limnol. and Oceanogr.*, **1**, 172–93.

Edmondson, W. T. (1978) Trophic equilibrium of Lake Washington. *Ecol. Res. Ser.*, EPA-600/3-77-087, Cinn., OH, 36 pp.

Edmondson, W. T., Anderson, G. C. and Peterson, D. R. (1956). Artificial eutrophication of Lake Washington. *Limnol. Oceanogr.*, **1**, 47–53.

Edmondson, W. T. and Lehman, J. R. (1981) The effect of changes in the nutrient income on the condition of Lake Washington. *Limnol. Oceanogr.*, **26**, 1–28.

Edmondson, W. T. and Litt, A. H. (1982) *Daphnia* in Lake Washington. *Limnol. Oceanogr.*, **27**, 272–93.

Egloff, D. A. and Brakel, W. H. (1973) Stream pollution and a simplified diversity index. *J. Water Pollut. Control Fed.*, **45**, 2269–75.

Ehrlich, G. G. and Slack, K. V. (1969) *Uptake and Assimilation of Nitrogen in Microbiological Systems*, ASTM STP 488, Am. Soc. Test. Mat., Philadelphia, pp. 11–23.

Eichenberger, E. and Schlatter, A. (1978) Effect of herbivorous insects on the production of benthic algal vegetation in outdoor channels. *Verh. Int. Verein Limnol.*, **20**, 1806–10.

Eilers, J. M., Brakke, D. F. and Landers, D. N. (1988) Chemical and physical characteristics of lakes in the upper mid-west United States. *Environ. Sci. Technol.*, **22**, 164–72.

Einsele, W. (1936) Uber die Beziehungen des Eisenkreislaufs zum-phosphatkreislauf im entrophen see. *Arch. Hydrobiol.*, **29**, 664–86.

Ellis, M. M. (1937) Detection and measurement of stream pollution. Bulletin 22, US, Bureau of Fish, **48**, 365–537.

Elser, H. J. (1967) Observations on the decline of water milfoil and other aquatic plants: Maryland, 1962–7. Unpublished report., Department Chesapeake Bay Affairs, 14 pp.

Elwood, J. W., Newbold, J. D., Trimble A. F. and Stork, R. W. (1981) The limiting role of phosphorous in a woodland stream ecosystem: Effects of P enrichment on leaf decomposition and primary producers. *Ecology*, **62**, 146–58.

Enell, M. and Löfgren, S. (1988) Phosphorus in interstitial water: Methods and dynamics. *Hydrobiologia*, **170**, 103–32.

Eppley, R. W. (1972) Temperature and phytoplankton growth in the sea. *Fisheries Bull.*, **70**: 1063–85

Fallon, R. D. and Brock, T. D. (1984) Overwintering of *Microcystis* in Lake Mendota. *Freshwater Biol.*, **11**, 217–26.

Fisher, S. G. and Likens, G. E. (1972) Stream ecosystem: organic energy budget. *Bioscience*, **22**, 33–5.

Fitzgerald, G. P. (1964) The biotic relationships within water blooms, in *Algae and Man* (ed. D. F. Jackson), Plenum Press, New York, pp. 300–6.

Fitzgerald, G. P. and Nelson, T. C. (1966) Extractive and enzymatic analyses for limiting or surplus phosphorus in algae. *J. Phycol.*, **2**, 32–7.

Fjerdingstad, E. (1964) Pollution of streams estimated by benthal phytomicroorganisms. I. A saprobic system based on communities of organisms and ecological factors. *Int. Rev. Geol. Hydrobiol.*, **49**, 63–131.

Fogg, G. E. (1965) *Algal cultures and phytoplankton ecology*. Univ. Wisconsin Press, Madison, pp. 126.

Forel, F. A. (1895) *Le Leman: monographie limnologique. Tome 2, Mecanique, Chimie, Thermique, Optique, Acoustique*. F. Rouge, Lausanne, 651 pp.

Fox, H. M. (1950) Hemoglobin production in *Daphnia. Nature (Lond)*, **166**, 609–10.

Fox, H. M., Simmonds, B. G. and Washbourn, R. (1935) Metabolic rates of emphemerid nymphs from swiftly flowing and still waters. *J. Exp. Biol.*, **12**, 179–84.

Freeman, M. C. (1986) The role of nitrogen and phosphorus in the development of *Cladophora glomerata* (L.) Kutzing in the Manuwatu River, New Zealand. *Hydrobiologia*, **31**, 23–30.

Frodge, J. D. Thomas, G. L. and Pauley, G. B. (1987) Impact of triploid grass carp (*Ctenopharyngodon idella*) on water quality – evaluation of aquatic plants on water quality, in *An Evaluation of Triploid Grass Carp on Lakes in the Pacific Northwest* (eds Pauley, G. B. and G. L. Thomas) WA Coop. Fish. Unit, Univ. of Wash., Seattle, WA.

Fuhs, G. W., Demmerle, S. D., Canelli, E. and Chew, M. (1972) Characterization of phosphorus limited plankton, in *Nutrients and Eutrophication: The Limiting Nutrient Controversy*, (ed. G. E. Likens). Special Symposium, *Limnol. Oceanogr.*, **1**, 113–33.

Gabrielson, J. O., Perkins, M. A. and Welch, E. B. (1984) The uptake, translocation and release of phosphorus by *Elodea densa. Hydrobiologia*, **111**, 43–48.

Gächter, V. R. (1976) Die tiefenwasserableitung, ein weg zur sanierung von seen. *Zchweiz. Z. Hydro.*, **38**, 1–28.

Gächter, R. and Mares, A. (1985) Does settling seston release soluble reactive phosphorus in the hypolimnion of lakes? *Limnol. Oceanogr.*, **30**, 364–71.

Gächter, R., Meyer, J. S. and Mares, A. (1988) Contribution of bacteria to release and fixation of phosphorus in lake sediments. *Limnol. Oceanogr.*, **33**, 1542–58.

Galston, A. W. (1979) Herbicides: a mixed blessing. *Biol. Sci.*, **29**, 85–90.

Gameson, A. L. H., Barrett, M. J. and Shewbridge, J. S. (1973) The aerobic Thames estuary. *Adv. Water Pollut. Res.* Proc. Sixth Internat. Conf. (ed. S. H. Jenkins), Pergamon Press, Oxford and NY. 843–50.

Gammon, J. R. (1969) Aquatic life survey of the Wabash River, with special reference to the effects of thermal effluents on populations of macroinvertebrates and fish. Unpublished report, DePauw University, Greencastle, Ind. 65 pp.

Ganf, G. G. (1983) An ecological relationship between *Aphanizomenon* and *Daphnia pulex*. Aust. J. Mar. Freshwater Res., **34**, 755–73.

Gannon, J. E. and Beeton, A. M. (1971) The decline of the large zooplankton, *Limnocalanus macrurus* Sars (Copepoda: Calanoida), in Lake Erie. *Proc. 14th Conf. Great Lakes Res.* Internat. Assoc. Gt. Lakes Res., pp. 27–38.

Gaufin, A. R. (1958) The effects of pollution on a midwestern stream. *Ohio J. Sci.*, **58**, 197–208.

Gaufin, A. R. and Tarzwell, C. M. (1956) Aquatic macroinvertebrate communities as indicators of organic pollution in Lytle Creek. *Sewage Ind. Wastes*, **28**, 906–24.

Gerloff, G. C. (1975) Nutritional ecology of nuisance aquatic plants. *Ecol. Res. Ser.*, EPA-660/3-75-027, 78 pp.

Gerloff, G. C. and Krombholz, P. H. (1966) Tissue analysis to determine nutrient availability. *Limnol. Oceanogr.*, **11**, 529–37.

Gerloff, G. C. and Skoog, F. (1954) Cell contents of nitrogen and phosphorus as a measure of their availability for growth of *Microcystis aeruginosa*. *Ecology*, **35**, 348–53.

Gessner, F. (1959) *Hydrobotanik. Die Physiologischen Grundlagen der Pflanzenverbreitung im Wasser. II. Stoffhausholt.* VEB Deutscher Verlag der Wissenschaften, Berlin, 701 pp.

Gilmartin, M. (1964) The primary production of a British Columbia fjord. *J. Fish Res. Board Canada*, **21**, 505–38.

Gliwicz, Z. M. (1975) Effect of zooplankton grazing on photosynthetic activity and composition of phytoplankton. *Verh. Int. Verein. Limnol.*, **19**, 1490–7.

Gliwicz, Z. M. and Hillbricht-Ilkowska, A. (1973) Efficiency of the utilization of nanoplankton primary production by communities of filter-feeding animals measured *in situ*. *Verh. Int. Verein. Limnol.*, **18**, 197–212.

Godfrey, P. J. (1978) Diversity as a measure of benthic macroinvertebrate community response to water pollution. *Hydrobiologia*, **57**, 111–22.

Goldman, C. R. (1960a) Molybdenum as a factor limiting primary productivity in Castle Lake, Calif. *Science*, **132**, 1016–17.

Goldman, C. R. (1960b) Primary productivity and limiting factors in three lakes of the Alaskan Peninsula. *Ecol. Monogr.*, **30**: 207–30.

Goldman, C. R. (1962) Primary productivity and micronutrient limiting factors in some North American and New Zealand lakes. *Verh. Int. Verein. Limnol.*, **15**, 365–74.

Goldman, C. R. (1981) Lake Tahoe: two decades of change in a nitrogen deficient oligotrophic lake. *Verh. Int. Verein. Limnol.*, **21**, 45–70.

Goldman, C. R. and Carter, R. (1965) An investigation by rapid carbon-14 bioassay of factors affecting the cultural eutrophication of Lake Tahoe, California-Nevada. *J. Water Pollut. Control Fed.*, **37**, 1044–59.

Goldman, C. R. and Horne, A. J. (1983) *Limnology* McGraw-Hill, Inc., New York.

Goldman, C. R. and Wetzel, R. G. (1963) A study of the primary productivity of Clear Lake, Lake County, Calif. *Ecology*, **44**, 283–94.

Goldman, J. C. (1973) Carbon dioxide and pH: effect on species succession of algae. *Science*, **182**, 306–7.

Goldman, J. C. and Carpenter, E. J. (1974) A kinetic approach to the effect of temperature on algal growth. *Limnol. Oceanogr.*, **19**, 756–66.

Goldman, J. C., Porcella, D. B., Middlebrooks, E. J. and Toerien, D. F. (1971) *The Effect of Carbon on Algal Growth – Its Relationship to Eutrophication.* Occasional Paper 6, Utah State University, Water Research Lab. and College of Engineering, April, 56 pp.

Golterman, H L. (1972) Vertical movement of phosphate in fresh water, in *Handbook of Environmental Phosphorus* (eds E. J. Griffith, A. M. Beeton, J. Spencer and D. Mitchell) John Wiley & Sons, New York, pp. 509–37.

Goodman, D. (1975) The theory of diversity-stability relationships in ecology. *Q. Rev. Biol.* **50**, 237–66.

Grafius, E. and Anderson, N. H. (1972) *Literature Review of Foods of Aquatic*

Graham, J. M. (1949) Some effects of temperature and oxygen pressure on the metabolism and activity of the speckled trout, *Salvelinus fontinalis. Can. J. Res. D*, **27**, 270–88.

Grandberg, K. (1973) The eutrophication and pollution of Lake Päijäne, Central Finland. *Ann. Bot. Finn*, **10**, 267–308.

Grant, N. G. and Walsby, A. E. (1977) The contribution of photosynthate to turgor pressure rise in the planktonic blue-green alga *Anabaena flos-aquae. J. Exp. Bot.*, **28**, 409–15.

Grassle, J. F. and Grassle, J. P. (1974) Opportunistic life histories and genetic systems in marine benthic polychaetes. *J. Marine Res.*, **32**, 253–84.

Gray, N. F. and Hunter, C. A. (1985) Heterotrophic slimes in Irish rivers – evaluation of the problem. *Water Res.*, **6**, 685–91.

Graynoth, E. (1979) Effects of logging on stream environments and faunas in Nelson, N.Z. *N.Z.J. Marine Freshwater Res.*, **13**, 79–109.

Great Phosphorus Controversy, The (1970) *Environ. Sci. Technol.*, **4**, 725–6.

Green, G. H. and Hargrave, B. T. (1966) Primary and secondary production in Bas d'Or Lake, Nova Scotia, Canada. *Verh. Int. Verein. Limnol.*, **16**, 333–40.

Gross, M. L. (1983) Response of largemouth bass and black crappie to summer drawdown of Long Lake, Kitsap County, Washington, MS thesis, University of Washington.

Haines, T. A. (1973) Effects of nutrient enrichment and a rough-fish population (carp) on a game-fish population (smallmouth bass). *Trans. Am. Fish Soc.*, **102**, 346–54.

Håkanson, L. (1975) Mercury in Lake Vänern – present status and prognosis. Swedish Environ. Prot. Bd., NLU, Report No. 80, 121 pp.

Hall, D. J., Cooper W. and Werner, E. (1970) An experimental approach to the production dynamics and structure of freshwater animal communities. *Limnol. Oceanogr.*, **15**, 839–928.

Hall, J. D. and Lanz, R. L. (1969) The effects of logging on the habitat of coho salmon and cutthroat trout in coastal streams. *Symposium on Salmon and Trout in Streams* (ed. T. G. Northcote), H. R. MacMillan Lectures in Fisheries, University of British Columbia, Vancouver, pp. 355–75.

Halsey, T. G. (1968) Autumnal and overwinter limnology of three small eutrophic lakes with particular reference to experimental circulation and trout mortality. *J. Fish Res. Board Canada*, **25**, 81–99.

Hamilton, R. D. and Preslan, J. (1970) Observations on heterotrophic activity in the eastern tropical Pacific. *Limnol. Oceanogr.*, **15**, 395–401.

Hartman, R. T. and Brown, D. L. (1967) Changes in internal atmosphere of submerged vascular hydrophytes in relation to photosynthesis. *Ecology*, **48**, 252–8.

Hawkes, H. A. (1962) Biological aspects of river pollution, in *River Pollution. Two: Causes and Effects* (ed. L. Klein), Butterworths, London, pp. 311–432.

Hawkes H. A. (1969) Ecological changes of applied significance induced by the discharge of heated waters, in *Engineering Aspects of Thermal Pollution* (eds F. L. Parker and P. A. Krenkel), Vanderbilt University Press, Nashville, Tennessee, pp. 15–57.

Hays, F. R. and Phillips, J. E. (1958) Lake water and sediment, III: Radiophosphorus equilibrium with mud, plants, and bacteria under oxidized and reduced conditions. *Limnol. Oceanogr.*, **3**, 459–75.

Healy, F. P. (1978) Physiological indicators of nutrient deficiency in algae. *Mitt. Verh. Int. Verein. Limnol.*, **21**, 34–41.

Heinle, D. R. (1969a) Effects of elevated temperature on zooplankton, *Chesapeake Sci.*, **10**, 186–209.

Heinle, D. R. (1969b) *Thermal Loading and the Zooplankton Community.* Patuxent Thermal Studies, Supplementary reports, Nat. Res. Inst., Ref. No. 69-8, University of Maryland.

Hellawell, J. M. (1977) Change in natural and managed ecosystems: detection, measurement and assessment. *Proc. R. Soc. Lond. B*, **197**, 31–57.

Hellström, B. (1941) *Wind Effect on Lakes and Rivers*, Kungliga Tekniska Högskolan, Avhandling 26, Stockholm.

Hendrey, G. R. (1973) Productivity and growth kinetics of natural phytoplankton communities in four lakes of contrasting trophic state. PhD dissertation, University of Washington, Seattle.

Hendrey, G. R. and Welch, E. B. (1974) Phytoplankton productivity in Findley Lake. *Hydrobiologia*, **45**, 45–63.

Hendricks, A. C. (1978) Response of *Selenastrum capricornutum* to zinc sulfides. *J. Water Pollut Control Fed.*, **50**, 163–8.

Henrikson, A. (1980) Acidification of freshwaters – a large scale titration, in *Ecological Impact of Acid Precipitation* (eds. D. Drabløs and A. Tollen), SNF-project, NISK, 1432–Ås, pp. 68–74.

Henriksen, A. Skogheim, O. K. and Rosseland, B. O. (1984) Episodic changes in pH and aluminium – speciation kill fish in a Norwegian salmon river. *Vatten*, **40**, 255–60.

Henriksen, A., Lein, L., Traaen, T. S. *et al.* (1988) Lake acidification in Norway – Present and predicted chemical status. *Ambio*, **17**, 259–66.

Herbert, D., Elsworth, R. and Telling, R. C. (1956) The continuous culture of bacteria, a theoretical and experimental study. *J. Gen. Microbiol.*, **14**, 601–22.

Hester, F. E. and Dendy, J. B. (1962) A multiple-plate sampler for aquatic macroinvertebrates. *Trans. Am. Fish Soc.*, **91**, 420.

Hileman, B. (1988) The Great Lakes cleanup effort. *Chem Eng. News*, Am. Chem. Soc., Washington, DC, pp. 22–39.

Hillbricht-Ilkowska, A. I. (1972) Interlevel energy transfer efficiency in planktonic food chains. *International Biological Programme* – Section PH, 13 December 1972, Reading, England.

Hilsenhoff, W. L. (1977) *Use of Arthropods to Evaluate Water Quality of Streams.* Technical Bulletin No. 100, US Department of Nature Research, 16 pp.

Hodgson, R. H. and Otto, N. E. (1963) Pondweed growth and response to herbicides under controlled light and temperature. *Weed Sci.*, **11**, 232–7.

Hoehn, R. C., Stauffer, J. R., Masnik, M. T. and Hocutt, C. H. (1974) Relationships between sediment oil concentrations and the macroinvertebrates present in a small stream following an oil spill. *Environ. Lett.*, **7**, 345–52.

Holton, H., Brettum, P., Holton, G. and Kjellberg, G. (1981) *Kolbotnvatn med Fillop: Sammerstilling av Undersokelsesresultates 1978–1979.* Norsh Inst. for Vannfors., Norway, Report No. 0-78007.

Horne, A. J. and Goldman, C. R. (1972) Nitrogen fixation in Clear Lake, Calif. I. Seasonal variation and the role of heterocysts. *Limnol. Oceanogr.*, **17**, 678–92.

Horner, R. R. and Welch, E. B. (1981) Stream periphyton development in relation to current velocity and nutrients. *Can. J. Fish Aquat. Sci.*, **38**, 449–57.

Horner, R. R., Welch, E. B., Seeley M. R. and Jacoby, J. M. (1990) Responses of periphyton to changes in current velocity, suspended sediment and phosphorus concentration. *Freshwater Biol.*, **24**, 215–32.

Horner, R. R., Welch, E. B. and Veenstra, R. B. (1983) Development of nuisance periphytic algae in laboratory streams in relation to enrichment and velocity, in *Periphyton of Freshwater Ecosystems* (ed. R. G. Wetzel), *Dev. Hydrobiol.*, **17**, 121–34.

Horning, W. B., II, and Pearson, R. E. (1973) Growth temperature requirements and lower lethal temperatures for juvenile smallmouth bass (*Micropterus dolomieu*). *J. Fish Res. Board Canada*, **30**, 1226–30.

Howard, D. L., Frea, J. I., Pfister, R. M. and Dugan, P. R. (1970) Biological nitrogen fixation in Lake Erie. *Science*, **169**, 61–2.

Hrbacek, J., Dvorakova, M., Korinek, V. and Prochazkova, L. (1961) Demonstration of the effect of the fish stock on the species composition of zooplankton and the intensity of metabolism of the whole plankton assemblage. *Verh. Int. Verein. Limnol.*, **14**, 192–5.

Hughes, B. D. (1978) The influence of factors other than pollution on the value of Shannon's Diversity Index, for macroinvertebrates in streams. *Water Res.*, **12**, 359–64.

Hughs, J. C. and Lund, J. W. G. (1962) The rate of growth of *Asterionella formosa Hass.* in relation to its ecology. *Arch. Microbiol.*, **42**, 117–29.

Hulbert, S. H. (1971) The nonconcept of species diversity: a critique and alternative parameters. *Ecology*, **52**, 577–86.

Hutchinson, G. E. (1957) *A Treatise on Limnology*, Vol. 1, John Wiley & Sons, New York.

Hutchinson, G. E. (1967) *A Treatise on Limnology*, Vol. 2, John Wiley & Sons, New York.

Hutchinson, G. E. (1970a) The biosphere. *Sci. Am.*, **223**, 44–53.

Hutchinson, G. E. (1970b) The chemical ecology of three species of *Myriophyllum* (Angiospermae, Haloragaceae). *Limnol. Oceanogr.*, **15**, 1–5.

Hutchinson, G. E. and Bowen, V. T. (1950) Limnological studies in Connecticut. 9. A quantitative radiochemical study of the phosphorus cycle in Lindsley Pond. *Ecology*, **31**, 194–203.

Hynes, H. B. N. (1960) *The Biology of Polluted Water*. Liverpool University Press, Liverpool, England.

Infante, A. and Abella, S. E. B. (1985) Inhibition of *Daphnia* by *Oscillatoria* in Lake Washington. *Limnol. Oceanogr.*, **30**, 1046–52.

Infante, A. and Litt, A. H. (1985) Differences between two species of *Daphnia* in the use of 10 species of algae in Lake Washington. *Limnol. Oceanogr.*, **30**, 1053–9.

Isaac, G. W., Matsuda, R. I. and Welker, J. R. (1966) A limnological investigation of water quality conditions in Lake Sammamish. *Water Quality Series No. 2*, Municipality of Metropolitan Seattle, pp. 47.

Jacoby, J. M. (1985) Grazing effects on periphyton by *Theodoxis fluviatillis* (Gastropoda) in a lowland stream. *Freshwater Ecol.*, **3**, 265–74.

Jacoby, J. M. (1987) Alterations in periphyton characteristics due to grazing in a Cascade foothills stream. *Freshwater Biol.*, **18**, 495–508.

Jacoby, J. M., Bouchard, D. D. and Patmont, C. R. (1990) Response of periphyton to nutrient enrichment in Lake Chelan, WA. *Lake Reservoir Management*, **7**, 33–43.

Jacoby, J. M., Lynch, D. D., Welch, E. B. and Perkins, M. A. (1982) Internal phosphorus loading in a shallow, eutrophic lake. *Water Res.*, **16**, 911–19.

Jasper, S. and Bothwell, M. L. (1986) Photosynthetic characteristics of lotic periphyton. *Can. J. Fish. Aquat. Sci.*, **43**, 1960–9.

Jensen, K. W. and Snekvik, E. (1972) Low pH levels wipe out salmon and trout populations in southernmost Norway. *Ambio.*, **1**, 223–5.

Jensen, L. D. and Gaufin, A. R. (1964) Effects of ten organic insecticides on two species of stonefly naiads. *Trans. Am. Fish. Soc.*, **93**, 27–34.

Johnson, H. E. (1967) The effects of endrin on the reproduction of a freshwater fish (*Oryzias latipes*). PhD dissertation, University of Washington, Seattle, pp. 136.

Jones, C. A. and Welch, E. B. (1990) Internal phosphorus loading related to mixing and dilution in a dendritic, shallow prairie lake. *J. Water Pollut. Control Fed.*, **62**, 847–52.

Jones, J. G., Gardener, S. and Simon, B. M. (1983) Bacterial reduction of ferric iron in a stratified eutrophic lake. *J. Gen. Microbiol.*, **129**, 131–9.

Jones, J. R. and Bachmann, R. W. (1976) Prediction of phosphorus and chlorophyll levels in lakes. *J. Water Pollut. Control Fed.*, **48**, 2176–82.

Jones, J. R., Smart, M. M. and Burroughs, J. N. (1984) Factors related to algal biomass in Missouri Ozark streams. *Verh. Int. Verein. Limnol.*, **22**, 1867–75.

Jones, R. I. (1979) Notes on the growth and sporulation of a natural population of *Aphanizomenon flos-aquae*. *Hydrobiologia*, **62**, 55–8.

Jorgensen, S. E., Friis, M. B., Henrikson, J. *et al.* (1979) *Handbook of Environmental Data and Ecological Parameters*. Pergamon Press, Oxford.

Kaesler, R. L., Herricks, E. E. and Crossman, J. S. (1978) Use of indices of diversity and hierarchical diversity in stream surveys, in *Biological Data in Pollution Assessment: Quantitative and Statistical Analyses* (eds K. L. Dickson, J. Cairns, Jr., and R. J. Livingston), ASTM STP 652, Amer. Soc. Test. Mat., Philadelphia, pp. 92–112.

Kamp-Nielsen, C. (1974) Mud–water exchange of phosphate and other ions in undisturbed sediment cores and factors affecting the exchange rates. *Arch. Hydrobiol.* **73**, 218–37.

Kappers, F. I. (1976) Blue-green algae in the sediment of the lake Brielse Meer. *Hydrobiol. Bull.*, **10**, 164–71.

Karr, J. R., Fausch, K. D., Angermeier, P. L. *et al.* (1986) *Assessing biological integrity in running waters: a method and its rationale*. Illinois Natural History Survey, Champaigne, Ill., Special publ. 6.

Keating, K. I. (1977) Blue-green algal inhibition of diatom growth: transition from mesotrophic to eutrophic community structure. *Science*, **199**, 971–3.

Kerr, P. C., Paris, D. F. and Brockway, D. L. (1970) The interrelation of carbon and phosphorus in regulating heterotrophic and autotrophic populations in aquatic ecosystems. *EPA Report* 16060 FGS 07/70, Raleigh, NC.

Keup, L. E. (1966) Stream biology for assessing sewage treatment plant efficiency. *Water and Sewage Works*, **113**, 411–17.

Kilham, S. S. (1975) Kinetics of silicon-limited growth in the freshwater diatom *Asterionella formosa*. *J. Phycol.*, **11**, 396–9.

Kimmel, B. L. and Groeger, A. W. (1984) Factors controlling primary production in lakes and reservoirs: a perspective, in *Proc. Int. Conf. Lake and Reservoir Management*, North American Lake Management Society, EPA 440/5/84-001, pp. 277–81.

King, D. L. (1970) Role of carbon in eutrophication. *J. Water Pollut. Control Fed.*, **42**, 2035–51.

King, D. L. (1972) Carbon limitation in sewage lagoons, in *Nutrients and Eutrophication*, Special Symposium, Vol. 1, Amer. Soc. Limnol. Oceanogr., pp. 98–110.

King, D. L. and Novak, J. T. (1974) The kinetics of inorganic carbon-limited algal growth. *J. Water Pollut. Control Fed.*, **46**, 1812–16.

Kirchner, W. B. and Dillon, P. J. (1975) An empirical method of estimating the retention of phosphorus in lakes. *Water Resources Res.*, **11**, 182–3.

Klein, L. (1962) *River Pollution, Two: Causes and Effects*, Butterworths, London.

Klemer, A. R. (1973) Factors affecting the vertical distribution of a blue-green alga. PhD dissertation, University of Minnesota.

Klemer, A. R., Detenbeck, N., Grover, J. and Fung-Brasino, T. (1988) Macronutrient interactions involved in cyanobacterial bloom formation. *Verh. Int. Verein. Limnol.*, **23**, 1881–5.

Klemer, A. R., Feuillade, J. and Feuillade, M. (1982) Cyanobacterial blooms: carbon and nitrogen limitation have opposite effects on the buoyancy of *Oscillatoria*. *Science*, **215**, 1629–31.

Klemer, A. R. and Kanopka, A. E. (1989) Causes and consequences of blue-green algal (Cyanobacteria) blooms. *Lake Reservoir Management*, **5**, 9–20.

Klemer, A. R., Pierson, D. C. and Whiteside, M. C. (1985) Blue-green algal (cyanobacterial) nutrition, buoyancy and bloom formation. *Verh. Internat. Verein. Limnol.*, **22**, 2791–8.

Knapp, R. T. (1943) Density currents: their mixing characteristics and their effect on the turbulence structure of the associated flow. *Eng. Bull.* **27**, University Iowa Studies, Iowa City.

Knoechel, R. and Kalff, J. (1975) Algal sedimentation: the cause of a diatom-blue-green succession. *Verh. Int. Verein. Limnol.*, **19**, 745–54.

Knutzen, J. (1975) Course project in ecological effects of wastewater. 20 pp.

Kolkwitz, R. and Marsson, M. (1908) Ökologie der pflanzlichen Saprobien. *Berichte Deutsch. Bot. Gesellschaft*, **26a**, 505–19.

Kormondy, E. J. (1969) *Concepts of Ecology* Prentice-Hall, Englewood Cliffs, NJ.

Kuentzel, L. E. (1969) Bacteria, carbon dioxide, and algal blooms, *J. Water Pollut. Control Fed.*, **41**, 1737–47.

Kvarnäs, H. and Lindell, T. (1970) *Hydrologiska studier i Ekoln* Rapport over hydrologisk verksamhet inom Naturvårdsverkets Limnologiska Undersökning januari-augusti 1969. UNGI Rapport 3, Uppsala.

Lager, J. A. and Smith, W. G. (1975) *Urban Stormwater Management and Technology: An Assessment* EPA-670/2-74-040, Washington, DC.

Lake Erie Report (1968) *A Plan for Water-Pollution Control* Department of the Interior, Fed. Water Poll. Admin., 107 pp.

Lamberti, G. A., Ashkenas, L. R., Gregory, S. V. and Steinman, A. D. (1987) Effects of three herbivores on periphyton communities in laboratory streams. *J.N. Am. Benthol. Soc.*, **6**, 92–104.

Lamberti, G. A. and Resh, V. H. (1983) Stream periphyton and insect herbivores: an experimental study of grazing by a caddisfly population. *Ecology*, **64**, 1124–35.

Lampert, W. (1987) Laboratory studies on zooplankton cyanobacteria interactions. *N.Z. J. Marine Freshwater Res.*, **21**, 483–90.

Lampert, W. (1981) Toxicity of the blue-green *Microcystis aeruginosa*: Effective defense mechanism against grazing pressure by *Daphnia*. *Verh. Int. Verein. Limnol.*, **21**, 1436–40.

Landers, D. H. (1982) Effects of naturally senescing aquatic macrophytes on nutrient chemistry and chlorophyll *a* of surrounding water. *Limnol. Oceanogr.*, **27**, 428–39.

Landers, D. H., Overton, W. S., Linhurst, R. A. and Brakke, D. F. (1988) Eastern lake survey – regional estimates at lake chemistry. *Environ. Sci. Technol.*, **22**, 128–35.

La Point, T. W., Melancon, S. M. and Morris, M. K. (1984) Relationships among observed metal concentrations, criteria, and benthic community structural responses in 15 streams. *J. Water Pollut. Control Fed.*, **56**, 1030–8.

Larsen, D. P., Malueg, K. W., Schultz, D. W. and Brice, R. M. (1975) Response of eutrophic Shagawa Lake, Minnesota, USA, to point-source phosphorus reduction. *Verh. Int. Verein. Limnol.*, **19**, 884–92.

Larsen, D. P. and Mercier, H. T. (1976) Phosphorus retention capacity of lakes. *J. Fish Res. Board Canada*, **33**, 1742–50.

Laegrid, M., Alstad, J., Klareness, D. and Seip, H. M. (1983) Seasonal variation of cadmium toxicity toward the alga *Selenastrum capricornutum* Printz in two lakes with different humus content. *Environ. Sci. Technol.*, **17**, 357–61.

Larsen, D. P., Schultz, D. W. and Malueg, K. W. (1981) Summer internal phosphorus supplies in Shagawa Lake, Minnesota. *Limnol. Oceanogr.*, **26**, 740–53.

Larsen, D. P., Van Sickle, J., Malueg, K. W. and Smith, P. D. (1979) The effect of wastewater phosphorus removal on Shagawa Lake, Minnesota: phosphorus supplies, lake phosphorus and chlorophyll *a*. *Water Res.*, **13**, 1259–72.

Larson, D. W., Dahm, C. N. and Geiger, N. S. (1987) Vertical partitioning of the phytoplankton assemblage in ultraoligotrophic Crater Lake, Oregon, USA. *Freshwater Biol*, **18**, 429–42.

Lathrop, R. C. (1988) Phosphorus trends in the Yahara lakes since the mid-1960s. Research Mngt. Findings, Wisconsin, Department of Natural Resources, Madison, WI.

Lathrop, R.C. (1990) Response of Lake Mendota (Wisconsin, USA) to decreased phosphorus loadings and the effect on downstream lakes. *Ver. Internat. Verein. Limnol.*, **24**, 457–63.

Lawton, G. W. (1961) Limitation of nutrients as a step in ecological control. *Algae and Metropolitan Wastes*, R. A. Taft San. Eng. Center, Tech. Rept. W61–3.

Leach, J. M. and Thakore, A. N. (1975) Isolation and identification of constituents toxic to juvenile rainbow trout (*Salmo gairdneri*) in caustic extraction effluents from kraft pulp mill bleach plants. *J. Fish Res. Board Canada*, **32**, 1249–57.

Leivstad, H. and Muniz, I. P. (1976) Fish kills at low pH in a Norwegian river. *Nature.*, **259**, 391–2.

Lennox, L. J. (1984) Lough Ennell: laboratory studies on sediment phosphorus release under varying mixing, aerobic, anaerobic conditions. *Freshwater Biol.*, **14**, 183–7.

Likens, G. E. and Borman, F. H. (1974) Linkages between terrestrial and aquatic ecosystems. *Bioscience*, **24**, 447–56.

Lin, C. (1971) Availability of phosphorus for *Cladophora* growth in Lake Michigan. *Proc. 14th Conf. Great Lakes Res.*, Intern. Assoc. Great Lakes Des., pp. 39–43.

Lindell, T. (1975) Vänern, in *Vänern, Vättern, Mälaren, Hjälmaren – en översikt* Statens Naturvårdsverk, Publikationer 1976: 1, Stockholm.

Lindeman, R. L. (1942) Trophic dynamic aspect of ecology. *Ecology*, **23**, 399–418.

Lloyd, R. (1960) The toxicity of zinc sulfate to rainbow trout. *Ann. Appl. Biol.*, **48**, 84–94.

Loeb, S. L. (1986) Algal biofouling of oligotrophic Lake Tahoe: causal factors affecting production, in *Algal Biofouling* (eds L. V. Evans and K. D. Hoagland). Elsevier Science Publishers B. V., Amsterdam, The Netherlands, pp. 159–73.

Leob, S. L., Reuter, J. E. and Goldman, C. R. (1983) Littoral zone production of oligotrophic lakes, in *Periphyton of Freshwater Ecosystems* (ed. R. G. Wetzel) (*Developments in Hydrobiologia*, No. 17) Dr W. Junk Publishers, The Hague, pp. 161–7.

Lofgren, S. (1987) Phosphorus retention in sediments – implications for aerobic phosphorus release in shallow lakes. PhD dissertation, Uppsala University, Sweden.

Long, E. B. (1976) The interaction of phytoplankton and the bicarbonate system. PhD dissertation, Kent State University, Ohio.

Lorenzen, M. W. and Fast, A. W. (1977) *A Guide to Aeration/Circulation Techniques for Lake Management*, EPA-600 13-77-004, USEPA, Washington, DC.

Lund, J. W. G. (1950) Studies on *Asterionella formosa* Hass. II. Nutrient depletion and the spring maximum. *J. Ecol.*, **38**, 1–35.

Lynch, M. and Shapiro, J. (1981) Predation, enrichment, and phytoplankton community structure. *Limnol. Oceanogr.*, **26**, 86–102.

Maciolek, J. A., and Maciolek, M. G. (1968) Microseston dynamics in a simple Sierra Nevada lake-stream system. *Ecology*, **49**, 60–75.

Macon, T. T. (1974) *Freshwater Ecology* John Wiley & Sons, New York, 343 pp.

Mann, K. H. (1965) Heated effluents and their effects on the invertebrate fauna of rivers. *Proc. Soc. Water Treat. Exam.*, **14**, 45–53.

Marcuson, P. E. (1970) *Stream Sediment Investigation Progress Report* Mont. Fish and Game Dept., Helena.

Margalef, R. (1958) Temporal succession and spatial heterogeneity in phytoplankton. *Perspectives in Marine Ecology*. University of California Press, p. 323.

Margalef, R. (1969) Diversity and stability: a practical proposal and a model of interdependence. *Symposium on Diversity and Stability in Ecological Systems* Brookhaven National Lab., Upton, NY.

Markowski, S. (1959) The cooling water of power stations: new factor in the environment of marine and freshwater invertebrates. *J. Animal Ecol.*, **28**, 243–58.

Marshall, P. T. (1958) Primary production in the arctic. *J. Cons. Int. Explor. Mer.*, **23**, 173–7.

Martin, J. B., Jr., Bradford, B. N. and Kennedy, H. G. (1969) *Factors Affecting the Growth of Najas in Pickwick Reservoir*. TVA, National Fertilizer Development Center Report.

Mason, W. T., Jr., Anderson, J. B., Kreis, R. D. and Johnson, W. C. (1970) Artificial substrate sampling, macroinvertebrates in a polluted reach of the Klamath River, Oregon, *J. Water Pollut. Control Fed.*, **42**, R315–27.

Matthews, W. J. and Zimmerman, E. G. (1990) Potential effects of global warming on native fishes of the southern Great Plains and the southwest. *Fisheries* **15**, 26–32.

McBride, G. B. and Pridmore, R. D. (1988) Prediction of [chlorophylla-*a*] in impoundments of short hydraulic retention time: mixing effects. *Verh. Int. Verein. Limnol.*, **23**, 832–6.

McConnell, J. W. and Sigler, W. F. (1959) Chlorophyll and productivity in a mountain river. *Limnol. Oceanogr.*, **4**, 335–51.

McCormick, P. V. and Stevenson, R. V. (1989) Effects of snail grazing on benthic algal community structure in different nutrient environments. *J.N. Am. Benthol. Soc.* **8**, 162–72.

McFeters, G. A., Bond, P. J., Olson, S. B. and Tchan, Y. T. (1983) A comparison of microbial bioassays for the detection of aquatic toxicants. *Water Res.*, **17**, 1757–62.

McGauhey, P. H., Rohlich, G. A. and Pearson, E. P. (1968) *Eutrophication of Surface Waters – Lake Tahoe: Bioassay of Nutrient Sources*. First Progress Report, Lake Tahoe Area Council, 178 pp.

McIntire, C. D. (1966) Some effects of current velocity on periphyton communities in laboratory streams. *Hydrobiologia*, **37**, 559–70.

McIntire, C. D., Colby, J. A. and Hall, J. D. (1975) The dynamics of small lotic ecosystems: a modelling approach. *Verh. Int. Verein. Limnol.*, **19**, 1599–609.

McIntire, C. A. and Phinney, H. K. (1965) Laboratory studies of periphyton production and community metabolism in lotic environments. *Ecol. Monogr.*, **35**, 237–58.

McKnight, D. (1981) Chemical and biological processes controlling the response of a freshwater ecosystem to copper stress: A field study of the $CuSO_4$ treatment of Mill Pond Reservoir, Burlington, Massachusetts. *Limnol. Oceanogr.*, **26**, 518–31.

McMahon, J. W. and Rigler, F. H. (1965) Feeding rate of *Daphnia magna* Straus on different foods labeled with radioactive phosphorus. *Limnol. Oceanogr.*, **10**, 105–13.

McNaught, D. C. (1975) A hypothesis to explain the succession from calanoids to cladocerans during eutrophication. *Verh. Int. Verein. Limnol.*, **19**, 724–31.

McQueen, D. J. and Post, J. R. (1988) Limnocorral studies of cascading trophic interactions. *Verh. Int. Verein. Limnol.*, **23**, 739–47.

McQueen, D. J., Post, J. R. and Mills, E. L. (1986) Trophic relationships in fresh-water pelagic ecosystems. *Can. J. Fish. Aquat. Sci.*, **43**, 1571–81.

Menzel, D. W. and Ryther, J. H. (1964) The composition of particulate organic matter in the western North Atlantic. *Limnol. Oceanogr.*, **9**, 179–86.

Messer, J. and Brezonik, P. L. (1983) Comparison of denitrification rate estimation techniques in a large, shallow lake. *Water Res.*, **17**, 631–40.

METRO (1987) Municipality of Metropolitan Seattle, Seattle, WA.

Mihursky, J. (1969) *Patuxent Thermal Studies. Summary and Recommendations*. Nat. Res. Inst. Spec. Rept. No. 1, University of Maryland, College Park, MD.

Mihursky, J. and Kennedy, V. S. (1967) Water temperature criteria to protect aquatic life. In *Symposium on Water Quality Criteria* Am. Fish Soc. Spec. Pub. No. 4, pp. 20–32.

Milbrink, G. (1980) Oligochaete communities in pollution biology: the European situation with special reference to lakes in Scandinavia, in *Aquatic Oligochaete Biology* (eds R. O. Brinkhurst and D. G. Cook), Plenum Press, NY.

Miller, G. T., Jr. (1988) *Living in the Environment; an Introduction to Environmental Science* 5th edn, Wadsworth, Belmont, CA.

Miller, T. G., Melancon, S. M. and Janika J. J. (1983) Site specific water quality assessment, Prickley Pear Creek, Montana, US EPA 600/X-82-013, Las Vegas, NV.

Miller, W. E., Greene, J. C. and Shirogama, T. (1978) The *Salenastrum capricornutum* Printz algal assay bottle test. *US EPA Report* EPA-600-/9-78-018, Cincinnati, OH.

Miller, W. E., Maloney, T. E. and Greene, J. C. (1974) Algal productivity in 49 lake waters as determined by algal assays. *Water Res.*, **8**, 667–79.

Mills, E. L. and Forney, J. L. (1983) Impact on *Daphnia pulex* of predation by young yellow perch in Oneida Lake, New York. *Trans. Amer. Fish. Soc.*, **112**, 154–61.

Minshall, G. W. (1978) Autotrophy in stream ecosystems. *Bioscience*, **28**, 767–71.

Misra, R. D. (1938) Edaphic factors in the distribution of aquatic plants in the English Lakes. *J. Ecol.*, **26**, 411–51.

Moore, L. and Thornton, K., eds (1988) *Lake and Reservoir Restoration Guidance Manual* EPA 440/5-88-002.

Mortimer, C. H. (1941) The exchange of dissolved substances between mud and water in lakes (parts I and II). *J. Ecol.*, **29**, 280–329.

Mortimer, C. H. (1942) The exchange of dissolved substances between mud and water in lakes (parts III, IV, summary, and references). *J. Ecol.*, **30**, 147–301.

Mortimer, C. H. (1952) Water movements in lakes during summer stratification; evidence from the distribution of temperature in Lake Windermere, with an appendix by M. S. Longuet-Higgins. *Phil. Trans. Ser. B*, **236**, 355–404.

Mount, D. I. (1966) The effect of total hardness and pH on acute toxicity of zinc to fish. *Air Water Poll. Int. J.*, **10**, 49–56.

Mount, D. I. (1969) Developing thermal requirements for freshwater fishes, in *Biological Aspects of Thermal Pollution* (eds P. A. Krenkel and F. L. Parker), Vanderbilt University Press, Nashville, Tennessee, pp. 140–7.

Mount, D. I. and Norberg, T. J. (1984) A seven-day life-cycle cladoceran toxicity test. *Environ. Toxicol. Chem.*, **3**, 425–34.

Mulligan, H. F. and Barnowski, A. (1969) Growth of phytoplankton and vascular aquatic plants at different nutrient levels. *Verh. Int. Verein. Limnol.*, **17**, 302–10.

Muniz, I. P. and Leivestad, H. (1980) Acidification – effects on freshwater fish, in *Proc. Int. Conf. Ecol. Impact of Acid Precipitation* (eds. D. Drobløs and A. Tollam), Oslo, Norway, pp. 84–92.

Murphy, G. I. (1962) Effect of mixing depth and turbidity on the productivity of fresh-water impoundments. *Am. Fish. Soc.*, **91**, 69–76.

Nebeker, A. V. (1971a) Effect of water temperature on nymphal feeding rate, emergence, and adult longevity of the stone fly *Pteronarcys dorsata*. *J. Kans. Entomol. Soc.*, **44**, 21–6.

Nebeker, A. V. (1971b) Effect of high winter water temperatures on adult emergence of aquatic insects. *Water Res.*, **5**, 777–83.

Neil, J. H. and Owen, G. E. (1964) Distribution, environmental requirements, and significance of *Cladophora* in the Great Lakes. *Proc. 7th Conf. Great Lakes Res.*, **11**, 113–21.

Newbold, J. D., Elwood, J. W., O'Neill, R. V. and Van Winkle, W. (1981) Measuring nutrient spiralling in streams. *Can. J. Fish. Aquat. Sci.*, **38**, 860–3.

Newbold, J. D., Erman, D. C. and Roby, K. B. (1980) Effects of logging on macroinvertebrates in streams with and without buffer strips. *Can. J. Fish. Aquat. Sci.*, **37**, 1076–85.

Newroth, P. R. (1975) Management of nuisance aquatic plants. Unpublished report, B. C. Dept. of the Environ., 13 pp.

Nichols, D. S. and Keeney, D. R. (1976a) Nitrogen nutrition of *Myriophyllum spicatum*: variation of plant tissue nitrogen concentration with season and site in Lake Wingra. *Freshwater Biol.*, **6**, 137–44.

Nichols, D. S. and Keeney, D. R. (1976b) Nitrogen nutrition of *Myriophyllum spicatum*: uptake and translocation of ^{15}N by shoots and roots. *Freshwater Biol.*, **6**, 145–54.

Nichols, S. A. and Shaw, B. H. (1986) Ecological life histories of the three aquatic nuisance plants. *Myriophyllum spicatum, Potamogeton crispus* and *Elodea canadensis*. *Hydrobiologia*, **131**, 3–21.

Noonan, T. A. (1986) Water quality in Long Lake, Minnesota following Riplox sediment treatment. *Lake Reservoir Management*, **2**, 131–7.

Norberg, T. J. and Mount, D. I. (1985) A new fathead minnow (*Pimephales promelas*) subacute toxicity test. *Environ. Toxicol. Chem.*, **4**, 711–18.

Nūman, W. (1972) Predictions on the development of the salmonid communities in the European oligotrophic subalpine lakes during the next century. Unpublished manuscript. Staatliche Institut für Seenforschung und Seenbewirtschaftung, Langenargen Bodensee, Germany, 12 pp.

Nurnberg, G. K. (1984) The prediction of internal phosphorus load in lakes with anoxic hypolimnia. *Limnol Oceanogr.*, **29**, 111–24.

Nurnberg, G. K. (1987a) A comparison of internal phosphorus loads in lakes with anoxic hypolimnia: Laboratory incubation versus in situ hypolimnetic phosphorus accumulation. *Limnol. Oceanogr.*, **32**, 1160–64.

Nurnberg, G. K. (1987b) Hypolimnetic withdrawal as lake restoration technique. *J. Environ Eng.*, **113**, 1006–17.

O'Brien, W. J. (1979) The predator–prey interaction of planktivorous fish and zooplankton. *Am Sci.*, **67**, 572–81.

Odum, E. P. (1959) *Fundamentals of Ecology* W. B. Saunders, Philadelphia.

Odum, E. P. (1969) The strategy of ecosystem development. *Science*, **164**, 264–70.

Odum, H. T. (1956) Primary production in flowing waters. *Limnol. Oceanogr.*, **1**, 102–17.

OECD (1982) *Eutrophication of Waters. Monitoring, Assessment and Control* OECD, Paris, 154 pp.

Ohle, W. (1953) Phosphor als Initialfaktor der Gewässereutrophierung. *Vom Wasser*, **20**, 11–23.

Ohle, W. (1975) Typical steps in the change of a limnetic ecosystem by treatment with therapeutica. *Verh. Int. Verein. Limnol.*, **19**, 1250.

Olson, T. A. and Rueger, M. E. (1968) Relationship of oxygen requirements to index-organism classification of immature aquatic insects. *J. Water Pollut. Control Fed.*, **40**, 188–202.

Ormerod, J. G., Grynne, B. and Ormerod, K. S. (1966) Chemical and physical factors involved in heterotrophic growth response to organic pollution. *Verh. Int. Verein. Limnol.*, **16**, 906–10.

Osgood, R. A. (1988a) Lake mixis and internal phosphorus dynamics. *Arch. Hydrobiol.*, **113**, 629–38.

Osgood, R. A. (1988b) A hypothesis on the role of *Aphanizomenon* in translocating phosphorus. *Hydrobiologia*, **169**, 69–76.

Oskam, G. (1978) Light and zooplankton as algae regulating factors in eutrophic Biesbosch reservoirs. *Verh. Int. Verein. Limnol.*, **20**, 1612–18.

Paerl, H. W. and Ustach, J. F. (1982) Blue-green algal scums: An explanation for their occurrences during freshwater blooms. *Limnol. Oceanogr.*, **27**, 212–17.

Pagenkopf, G. K. (1983) Gill surface interaction model for trace-metal toxicity to fishes: role of complexation, pH and water hardness. *Environ. Sci. Technol.*, **17**, 342–7.

Parsons, T. R., Stephens, K. and Le Brasseuer, R. J. (1969) Production studies in the Strait of Georgia. Part I. Primary production under the Fraser River plume, February to May, 1967. *J. Exp. Mar. Biol. Ecol.*, **3**, 27–38.

Pastorak, R. A., Lorenzen, M. W. and Ginn, T. C. (1982) *Environmental Aspects of Artificial Aeration and Oxygenation of Reservoirs: A Review of Theory, Techniques and Experiences* Tech. Rept. No. E-82-3, U.S. Army Corps of Engineers.

Patalas, K. (1970) Primary and secondary production in a lake heated by a thermal power plant. *Proc. Inst. Environ. Sci. 16th Annual Tech. Meet.*, pp. 267–71.

Patalas, K. (1972) Crustacean plankton and the eutrophication of St Lawrence Great Lakes. *J. Fish. Res. Board Canada*, **29**, 1451–62.

Patten B. C., Egloff, D. A., Richardson, T. H. *et al.* (1975) Total Ecosystem model for a cove in Lake Taxoma, in *Systems Analysis and Simulation in Ecology* Vol. 3 (ed. B. C. Patten), Academic Press, NY.

Peakall, D. B. and Lovett, R. J. (1972) Mercury: its occurrence and effects in the ecosystem. *Bioscience*, **22**, 20–5.

Pearsall, W. H. (1920) The aquatic vegetation of the English Lakes. *J. Ecol.*, **8**, 163–99.

Pearsall, W. H. (1929) Dynamic factors affecting aquatic vegetation. *Proc. Int. Cong. Plant Sci.*, **1**, 667–72.

Pechlaner, R. (1978) Erfahrungen mit Restaurierungsmassnahmen an eutrophierten Badeseen Tirols. *Dester Wasserwertsch*, **30**, 112–19.

Pederson, G. L., Welch, E. B. and Litt, A. H. (1976) Plankton secondary productivity and biomass: their relation to lake trophic state. *Hydrobiologia*, **50**, 129–44.

Peltier, W. H. and Welch, E. B. (1969) Factors affecting growth of rooted aquatic plants in a river. *Weed Sci.*, **17**, 412–16.

Peltier, W. H. and Welch, E. B. (1970) Factors affecting growth of rooted aquatic plants in a reservoir. *Weed Sci.*, **18**, 7–9.

Percival, E. and Whitehead, H. (1929) A quantitative study of the fauna of some types of stream bed. *J. Ecol.*, **17**, 282–314.

Perkins, J. L. (1983) Bioassay evaluation of diversity and community composition indexes. *J. Water Pollut. Control Fed.*, **55**, 522–30.

Perrin, C. J., Bothwell, M. L. and Slaney, P. A. (1987) Experimental enrichment of a coastal stream in British Columbia: Effects of organic and inorganic additions on autotrophic periphyton production. *Can. J. Fish. Aquat. Sci.*, **44**, 1247–56.

Peters, J. C. (1960) *Stream Sedimentation Project Progress Report* Mont. Fish and Game Dept., Helena.

Peters, J. C. (1967) Effects on a trout stream of sediment from agricultural practices. *J. Wildlife Management*, **31**, 805–12.

Peterson, S. A., Miller, W. E., Greene, J. C. and Callahan, C. A. (1985) Use of bioassays to determine potential toxicity effects of environmental pollutants, in *Perspectives on Nonpoint Source Pollution* EPA 440/5-85-001.

Phaup, J. D. and Gannon, J. (1967) Ecology of *Sphaerotilus* in an experimental outdoor channel. *Water Res.*, **1**, 523–41.

Phillips, G. L., Einson, D. and Moss, E. (1978) A mechanism to account for macrophyte decline in progressively eutrophicated fresh waters. *Aquatic Bot.*, **3**, 239–55.

Phillips, J. E. (1964) The ecological role of phosphorus in waters with special reference to microorganisms, in *Principles and Applications in Aquatic Microbiology* (eds H. Heukelekian and N. C. Doudero), John Wiley & Sons, New York, pp. 61–81.

Phinney, H. K. and McIntire, C. D. (1965) Effect of temperature on metabolism of periphyton communities developed in laboratory streams. *Limnol. Oceanogr.*, **10**, 341–4.

Pickering, Q. H. (1968) Some effects of dissolved oxygen concentrations upon the toxicity of zinc to the bluegill *Lepomis macrochirus* Raf. *Water Res.*, **2**, 187–94.

Pickering, Q. H. (1988) Evaluation and comparison of two short-term fathead minnow tests for estimating chronic toxicity. *Water Res.*, **22**, 883–93.

Pomeroy, L. W. (1974) The ocean's food web, a changing paradigm. *Bioscience*, **24**, 499–503.

Pontasch, K. W. and Brusven, M. A. (1988) Diversity and community comparison indices: Assessing macroinvertebrate recovery following a gasoline spill. *Water Res.*, **22**, 619–26.

Pontasch, K. W., Smith, E. P. and Cairns, J. Jr. (1989) Diversity indices, community comparison indices and canonical discriminant analysis: interpreting the results of multispecies toxicity tests. *Water Res.*, **23**, 1229–38.

Porath, H. A. (1976) The effect of urban runoff on Lake Sammamish periphyton. MS thesis, University of Washington, Seattle, 74 pp.

Porcella, D. B., Peterson S. A. and Larsen, D. P. (1980) Index to evaluate lake restoration. *J. Environ. Eng.*, **106**, 1151–69.

Porter, G. W. (1977) The plant-animal interface in freshwater ecosystems. *Am. Sci.*, **65**, 159–69.

Porter, K. G. (1972) A method for the in situ study of zooplankton grazing effects on algal species composition and standing crop. *Limnol. Oceanogr.*, **17**, 913–17.

Porter, K. G. and McDonough, R. (1984) The energetic cost of response to blue-green algal filaments by cladocerans. *Limnol. Oceanogr.*, **29**, 365–9.

Post, J. R. and McQueen, D. J. (1987) The impact of planktivorous fish on the structure of a plankton community. *Freshwater Biol.*, **17**, 79–89.

Prairie, Y. T., Duarte, C. M. and Kalff, J. (1989) Unifying nutrient-chlorophyll relations in lakes. *Can. J. Fish. Aquat. Sci.*, **46**, 1176–82.

Prandtl, L. (1952) *Essentials of Fluid Dynamics*. Blackie & Son, Glasgow.

Prentki, R. T., Adams, M. S., Carpenter, S. R., Gasith, A., Smith, C. S. and Weiler, P. R. (1979) The role of submersed weedbeds in internal loading and interception of allochthonous materials in Lake Wingra, Wisconsin, USA. *Arch. Hydrobiol.*, **57**, 221–50.

Prescott, G. W. (1954) *How to Know the Freshwater Algae* Wm. C. Brown Co., Dubugue, Ia.

Pridmore, R. D. and McBride, G. B. (1984) Prediction of chlorophyll *a* concentrations in impoundments of shord hydraulic retention time. *J. Environ. Management*, **19**, 343–50.

Quinn, J. M. and McFarlane, P. N. (1987) Effects of slaughterhouse and dairy factory wastewaters on epilithon: a comparison in laboratory streams. *Water Res.*, **23**, 1267–73.

Raess, F. and Maly, E. J. (1986) The short-term effects of perch predation on a zooplankton prey community. *Hydrobiologia*, **140**, 155–60.

Rahel, F. J. and Magnuson, J. J. (1983) Low pH and the absence of fish species in naturally acidic Wisconsin lakes: inferences for cultural acidification. *Can. J. Fish Aquat. Sci.*, **40**, 3–9.

Rast, W. and Lee, G. F. (1978) Summary analysis of the North American (US portion) OECD eutrophication project: Nutrient Loading – lake response relationships and trophic state indices. *Ecol. Res. Ser.* EPA-600 13-78-008, 455 pp.

Rawson, D. S. (1945) The experimental introduction of smallmouth bass into lakes of the Prince Albert National Park, Saskatchewan. *Trans. Am. Fish Soc.*, **73**, 19–31.

Rawson, D. S. (1959) *Limnology and Fisheries of Cree and Wollaston Lakes in Northern Saskatchewan* Saskatchewan Dept. of Nat. Res., Fisheries Rept. No. 4.

Raymont, J. E. (1963) *Plankton and Productivity in the Oceans* Pergamon Press, Elmsford, NY.

Reckhow, K. H. and Chapra, S. C. (1983) *Data Analysis and Empirical Modeling Engineering Approaches for Lake Management* Vol. 1, Butterworths, Boston.

Redfield, A. C., Ketchum, B. H. and Richards, F. A. (1963) The influence of organisms on the composition of sea water, in *The Sea*, Vol. 2 (ed. N. Hill) Interscience, pp. 26–77.

Reinart, R. E. (1970) Pesticide concentrations in Great Lakes fish. *Pesticides Monitoring J.*, **3**, 233–40.

Repavich, W. M., Sonzogni, W. C., Standridge, J. H., Wedepohl, R. E. and Meisner, L. F. (1990) Cyanobacteria (blue-green algae) in Wisconsin waters: Acute and chronic toxicity. *Water Res.*, **24**, 225–31.

Resh, V. H. and Unzicker, J. D. (1975) Water quality monitoring and aquatic organisms: the importance of species identification. *J. Water Pollut. Control Fed.*, **47**, 9–19.

Revelle, P. and Revelle, C. (1988) *The Environment: Issues and Choices for Society*. Jones and Bartlett, Boston.

Reynolds, C. S. (1975) Interrelations of photosynthetic behavior and buoyancy regulation in a natural population of blue-green alga. *Freshwater Biol.*, **5**, 323–38.

Reynolds, C. S. (1984) *The Ecology of Freshwater Phytoplankton* Cambridge University Press, Cambridge.

Reynolds, C. S. (1987) Cyanobacterial water blooms, in *Advances in Botanical Research*, Vol. 3 (ed. P. Callow) Academic Press, London, pp. 67–143.

Reynolds, C. S., Jaworski, G. H. M., Cmiech, H. A. and Leedale, G. F. (1981) On the annual cycle of the blue-green alga *Microcystis aeruginosa* Kutz emend. Elenkin. *Phil. Trans. Soc. Lond. B* **293**, 419–77.

Reynolds, C. S., Oliver, R. L. and Walsby, A. E. (1987) Cyanobacterial dominance: The role of buoyancy regulation in dynamic lake environments. *N.Z. J. Marine Freshwater Res.*, **21**, 379–90.

Rhee, G. Y. (1978) Effects of N:P atomic ratios and nitrate limitations on algal growth, cell composition and nitrate uptake. *Limnol. Oceanogr.*, **23**, 10–25.

Rich, P. H. and Wetzel, R. G. (1978) Detritus in lake ecosystems. *Am. Mid. Nat.*, **112**, 57–71.

Rickert, D. A., Petersen, R. R., McKenzie, S. W., Hines, W. G. and Wille, S. A. (1977) *Algal Conditions and the Potential for Future Algal Problems in the Willamette River*, Oregon US Geological Circular 715-G, 39 pp.

Riley, E. T. and Prepase, E. E. (1984) Role of internal phosphorus loading in two shallow lakes in Alberta, Canada. *J. Fish. Aquat. Sci.*, **41**, 845–55.

Ringler, N. H. and Hall, J. D. (1975) Effects of logging on water temperature and dissolved oxygen in spawning beds. *Trans. Am. Fish Soc.*, **104**, 111–21.

Ripl, W. (1976) Biochemical oxidation of polluted lake sediments with nitrate – a new lake restoration method. *Ambio*, **5**, 132–5.

Ripl, W. (1986) Internal phosphorus recycling mechanisms in shallow lakes. *Lake Reservoir Management*, **2**, 138–42.

Robel, R. J. (1961) Water depth and turbidity in relation to growth of sago pondweed. *J. Wildlife Management*, **25**, 436–8.

Rock, C. A. (1974) The trophic status of Lake Sammamish and its relationship to nutrient income. PhD dissertation, University of Washington, Seattle.

Rodgers, G. K. (1966) *The Thermal Bar in Lake Ontario, Spring 1965 and Winter 1965–66* Great Lakes Res. Div., University of Michigan, Pub. No. 15, pp. 369–74.

Rodhe, W. (1948) Environmental requirements of freshwater plankton algae. *Experimental Studies in the Ecology of Phytoplankton* Symbol. Bot. Upsalien 10, 149 pp.

Rodhe, W. (1966) Standard correlations between pelagic photosynthesis and light, in *Primary Productivity in Aquatic Environments* (ed. C. R. Goldman), University of California Press, pp. 365–82.

Rodhe, W. (1969) Crystallization of eutrophication concepts in northern Europe. In *Eutrophication: Causes, Consequences, Correctives* National Academy of Science, Washington, DC.

Roelofs, T. D. and Ogelsby, R. T. (1970) Ecological observations on the planktonic cyanophyte *Gleotrichia echinulata*. *Limnol. Oceanogr.*, **15**, 224–9.

Rorslett, B., Berge, D. and Johansen, S. W. (1986) Lake enrichment by submersed macrophytes: a Norwegian whole-lake experience with *Elodea canadensis*. *Aquat. Bot.*, **26**, 325–40.

Rother, J. A. and Fay, P. (1977) Sporulation and the development of planktonic blue-green algae in two Salopian meres. *Proc. R. Soc. Lond. B*, **196**, 317–32.

Russell-Hunter, W. D. (1970) *Aquatic Productivity* Macmillan, New York.

Ryding, S.-O. (1985) Chemical and microbiological processes as regulators of the exchange of substances between sediments and water in shallow, eutrophic lakes. *Int. Rev. Ges. Hydrobiol.*, **70**, 657–702.

Ryding, S.-O. and Forsberg, C. (1976) Six polluted lakes: a preliminary evaluation of the treatment and recovery processes. *Ambio*, **5**, 151–6.

Ryther, J. H. (1960) Organic production by planktonic algae and its environmental control, in *The Ecology of Algae* Spec. Pub. No. 2, Pymatuning Lab. of Field Biol., University of Pittsburgh, pp. 72–83.

Ryther, J. H. and Dunstan, W. M. (1971) Nitrogen, phosphorus, and eutrophication in the coastal marine environment. *Science*, **171**, 1008–13.

Saether, O. A. (1979) Chironomid communities as water quality indicators. *Holarctic Ecol.*, **2**, 65–74.

Sakamoto, M. (1971) Chemical factors involved in the control of phytoplankton production in the Experimental Lakes Area, northwestern Ontario. *J. Fish Res. Board Canada*, **28**, 203–13.

Salonen, K., Jones, R. I. and Arvola, L. (1984) Hypolimnetic phosphorus retrieval by diel vertical migrations of lake phytoplankton. *Freshwater Biol.*, **14**, 431–8.

Sarnelle, O. (1986) Field assessment of the quality of phytoplanktonic food available to *Daphnia* and *Bosmina*. *Hydrobiologia*, **131**, 47–56.

Sas, H. (1989) *Lake Restoration by Reduction of Nutrient Loading: Expectations, Experiences, Extrapolations*. Academia-Verlag Richarz, St Augustin.

Savage, N. L. and Rabe, F. W. (1973) The effects of mine and domestic wastes on macroinvertebrate community structure in the Coeur d' Alene River. *Northwest Sci.*, **47**, 159–68.

Schell, W. R. (1976) *Biogeochemistry of Radionuclides in Aquatic Environments*. Annual progress report to Energy Research and Development Administration RLO-2225-T18-18.

Schelske, C. L. and Stoermer, E. F. (1971) Eutrophication, silica depletion, and predicted changes in algal quality in Lake Michigan. *Science*. **173**, 423–4.

Schelske, C. L., Stoermer, E. F., Fahnenstiel, G. L. and Haibach, M. (1986) Phosphorus enrichment, silica utilization and biogeochemical silica depletion in the Great Lakes, *Can. I. Fish. Aquat. Sci.*, **43**, 407–15.

Schiff, J. A. (1964) Protists, pigments, and photosynthesis, in *Principles and Applications in Aquatic Microbiology* (eds Heukelekian, H. and Doudero, N. C.), John Wiley & Sons, New York, pp. 298–313.

Schindler, D. W. (1971a) A hypothesis to explain the differences and similarities among lakes in the Experimental Lakes Area, northwestern Ontario. *J. Fish Res. Board Canada*, **28**, 295–301.

Schindler, D. W. (1971b) Carbon, nitrogen, and phosphorus and the eutrophication of freshwater lakes. *J. Phycol*, **7**, 321–9.

Schindler, D. W. (1974) Eutrophication and recovery in experimental lakes: implications for lake management. *Science*, **184**, 897–99.

Schindler, D. W. (1986) The significance of in-lake production of alkalinity. *Water Air Soil Pollut.* **18**, 259–71.

Schindler, D. W. and Fee, E. J. (1974) Experimental lakes area, whole lake experiments in eutrophication. *J. Fish Res. Board Canada*, **31**, 937–53.

Schmitt, M. R. and Adams, M. S. (1981) Dependence of rates of apparent photosynthesis on tissue phosphorus concentrations in *Myriophyllum spicatum* L. *Aquat. Bot.*, **11**, 379–87.

Schofield, C. L. and Trojnar, J. R. (1980) Aluminium toxicity to brook trout (*Salvelinus fontinalis*) in acidified waters, in *Polluted Rain* (eds T. Y. Toribare, M. W. Miller and P. E. Horrow), Plenum Press, New York, pp. 341–63.

Schultz, D. W. and Malueg, K. W. (1971) Uptake of radiophosphorus by rooted aquatic plants. *Proc. 3rd Natl Symposium on Radioecology* pp. 417–24.

Schumacher, G. J. and Whitford, L. A. (1965) Respiration and ^{22}P uptake in various species of freshwater algae as affected by current. *J. Phycol.*, **1**, 78–80.

Shannon, C. E. and Weaver, W. (1948) *The Mathematical Theory of Communication* University of Illinois Press, Urbana.

Shapiro, J. (1970) A statement on phosphorus. *J. Water Pollut. Control Fed.* **42**, 772–5

Shapiro, J. (1973) Blue-green algae: why they become dominant. *Science*, **179**, 382–4.

Shapiro, J. (1979) The importance of trophic level interactions to the abundance and species composition of algae in lakes. *Dev. Hydrobiol.*, **2**, 161–7.

Shapiro, J. (1984) Blue-green dominance in lakes: the role and management significance of pH and CO_2. *Int. Rev. Ges. Hydrobiol.*, **69**, 765–80.

Shapiro, J. (1990) Current beliefs regarding dominance by blue-greens: the case for the importance of CO_2 and pH. *Verh. Int. Verein Limnol.*, **24**, 38–54.

Shapiro, J., Forsberg, B., LaMarra, V., Lindmark, G., Lynch, M., Smeltzer, E. and Zoto, G. (1982) *Experiments and Experiences in Bio-Manipulation*. Interim Rept. No. 19, Limnological Res. Center, Univ. of Minnesota, EPA-600/3-82-096.

Shapiro, J., Lamarra, V. and Lynch, M. (1975) Biomanipulation: An ecosystem approach to lake restoration, in *Water Quality Management through Biological Control*, (eds Brezonik, P. L. and Fox, J. L.), Department of Environmental Engineering Sciences, University of Florida, Gainesville, pp. 85–96.

Shapiro, J. and Wright, D. I. (1984) Lake restoration by biomanipulation: Round Lake, Minnesota, the first two years. *Freshwater Biol.*, **14**, 371–83.

Sherman, R.E., Gloss, S. P. and Lion, L. W. (1987) A comparison of toxicity tests conducted in the laboratory and in experimental ponds using cadmium and the fathead minnow (*Pimephales promelas*). *Water Res.*, **21**, 317–23.

Shin, E. and Krenkel, P. A. (1976) Mercury uptake by fish and biomethylation mechanisms. *J. Water Pollut Control Fed.*, **48**, 473–501.

Shumway, D. L., Warren, C. E. and Doudoroff, P. (1964) Influence of oxygen concentration and water movement on the growth of steelhead trout and coho salmon embryos. *Trans. Am. Fish Soc.*, **93**, 342–56.

Shuster, J. I., Welch, E. B., Horner, R. R. and Spyridakis, D. E. (1986) Response of Lake Sammamish to urban runoff control. *Lake Reservoir Management*, **2**, 229–34.

Siefert, R. E. and Spoor, W. A. (1973) Effects of reduced oxygen on embryos and larvae of the white sucker, coho salmon, brook trout and walleye, in *The Early Life History of Fish* (ed. Baxter, J. H. S.), Proc. of Symposium, Marine Biol. Assn., Avon, Scotland, Springer-Verlag, Heidelberg, Germany. pp. 487–95.

Silvey, J. K., Henley, D. E. and Wyatt, J. T. (1972) Planktonic blue-green algae: growth and odor-production studies. *J. Am Water Works Assoc.*, **64**, 35–9.

Simons, T. J. (1973) *Development of Three-Dimensional Numerical Models of the Great Lakes* Environ. Canada Water Mgt., Scientific Series No. 12, Environment Canada, Ottawa.

Sivonen, K., Niemela, S. I., Niemi, R. M., Lepisto, L., Luoma, T. H. and Rasanen, L. A. (1990) Toxic cyanobacteria (blue-green algae) in Finnish fresh and coastal waters. *Hydrobiologia*, **190**, 267–75.

Skulberg, O. M. (1965) Algal cultures as a means to assess the fertilizing influence of pollution. *Adv. Water Pollut Res.*, **1**, 113–27.

Sládeček, V. (1973) The reality of three British biotic indices. *Water Res.*, **7**, 995–1002.

Sládečková, A. and Sládeček, V. (1963) Periphyton as indicator of the reservoir water quality. I. True periphyton. Prague, *Technol. Water*, **7**, 507–61.

Smayda, T. J. (1974) Bioassay of the growth potential of the surface water of lower Narragansett Bay over an annual cycle using the diatom *Thalassiosira pseudonana* (oceanic clone, 13–1). *Limnol. Oceanogr.*, **19**, 889–901.

Smith, C. S. and Adams, M. S. (1986) Phosphorus transfer from sediments by *Myriphyllum spicatum*. *Limnol. Oceanogr.*, **31**, 1312–21.

Smith, L. L., Jr., Kramer, R. H. and MacLeod, J. C. (1965) Effects of pulpwood fibers on fathead minnows and walleye fingerlings. *J. Water Pollut. Conf. Fed.*, **37**, 130–40.

Smith, M. W. (1969) Changes in environment and biota of a natural lake after fertilization. *J. Fish Res. Board Canada*, **26**, 3101–32.

Smith, S. V. (1984) Phosphorus versus nitrogen limitation in the marine environment. *Limnol. Oceanogr.*, **29**, 1149–60.

Smith, V. H. (1982) The nitrogen and phosphorus dependence of algal biomass in lakes: an empirical and theoretical analysis. *Limnol. Oceanogr.*, **27**, 1101–12.

Smith, V. H. (1983) Low nitrogen to phosphorus ratios favor dominance by blue green algae in lake phytoplankton. *Science*, **221**, 669–71.

Smith, V. H. (1986) Light and nutrient effects on the relative biomass of blue-green algae in lake phytoplankton. *Can. J. Fish. Aquat. Sci.*, **43**, 148–53.

Smith, V. H. (1990a) Nitrogen, phosphorus, and nitrogen fixation in lacustrine and estuarine ecosystems. *Limnol. Oceanogr.* **35**, 1852–9.

Smith, V. H. (1990b) Effects of nutrients and non-algal turbidity on blue-green algal biomass in four North Carolina reservoirs. *Lake Reservoir Management* **6**, 125–32.

Smith, V. H. and Shapiro, J. (1981) Chlorophyll–phosphorus relations in individual lakes: their importance to lake restoration strategies. *Environ. Sci. Technol.*, **15**, 444–51.

Smith, W. E. (1970) Tolerance of *Mysis relicta* to thermal shock and light. *Trans. Am. Fish Soc.*, **99**, 418–21.

Søballe, D. M. and Kimmel, B. L. (1987) A large-scale comparison of factors influencing phytoplankton abundance in rivers, lakes and impoundments. *Ecology* **68**, 1943–54.

Søballe, D. M. and Threlkeld, S. T. (1985) Advection, phytoplankton biomass, and nutrient transformations in a rapidly flushed impoundment. *Arch. Hydrobiol.*, **105**, 187–203.

Sodergren, A., Bengtsson, B. E., Jonsson, P., Lagergren, S., Larson, A., Olsson, M. and Renberg, L. (1988) Summary of results from the Swedish project 'Environment/Cellulose'. *Water Sci. Technol*, **20**, 49–60.

Solbe, J. F., de L. G. and Shurben, D. G. (1989) Toxicity of ammonia to early life stages of rainbow trout (*Salmo gairdneri*). *Water Res.*, **23**, 127–9.

Sonzogni, W. C. and Lee, G. F. (1974) Diversion of wastewaters from Madison lakes. *J. Environ. Eng. Div.*, **100**, 153–70.

Sorokin, C. (1959) Tabular comparative data for the low- and high-temperature strains of *Chlorella*. *Nature (Lond.)*, **184**, 613–14.

Sorokin, C. and Krauss, R. W. (1962) Effects of temperature and illuminance on chlorella growth uncoupled from cell division. *Plant Physiol.*, **37**, 37–42.

Spehar, R. L., Anderson, R. L. and Fiandt, J. T. (1978) Toxicity and bioaccumulation of cadmium and lead in aquatic invertebrates. *Environ. Pollut.*, **15**, 195–208.

Spence, D. H. N. (1967) Factors controlling the distribution of freshwater macrophytes with particular reference to the lochs of Scotland. *J. Ecol.*, **55**, 147–70.

Spencer, C. N. and King, D. L. (1984) Role of fish in regulation of plant and animal communities in eutrophic ponds. *Can. J. Fish Aquat. Sci.*, **41**, 1851–5.

Spencer, C. N. and King, D. L. (1987) Regulation of blue-green algal buoyancy and bloom formation by light, inorganic nitrogen, CO_2, and trophic interactions. *Hydrobiologia*, **144**, 183–92.

Spencer, D. F. (1990) Influence of organic sediment amendments on growth and tuber production by sago pondweed (*Potamogeton pectinatus* L.). *Freshwater Ecol.*, **5**, 255–63.

Spigel, R. H. and Imberger, J. (1987) Mixing processes relevant to phytoplankton dynamics in lakes. *N.Z. J. Marine and Freshwater Res.*, **21**, 361–77.

Stadelman, P. (1980) *Der Zustand des Rotsees bei Luzern, Kantonales amt fur Gewasserschutz*, Luzern.

Stanford, J. A. and Ward, J. V. (1988) The hyporheic habitat of river ecosystems. *Nature (Lond.)*, **335**, 64–6.

Stauffer, R. W. and Lee, G. F. (1973) The role of thermocline migration in regulating algal blooms, in *Modeling the Eutrophication Process*. (eds E. J. Middlebrooks, D. H. Falkenborg, and T. E. Maloney), Proc. Workshop at Utah State University, Logan, pp. 73–82.

Steele, J. H. (1962) Environmental control of photosynthesis in the sea. *Limnol. Oceanogr.*, **7**, 137–50.

Stensen, J. A. E., Bohlin, T., Henriksen, L., Nilsson, B. I., Oscarson, A. G. and Larsson, P. (1978) Effects of fish removal from a small lake. *Verh. Int. Verein. Limnol.*, **20**, 794–801.

Stefan, H. G. and Hanson, M. J. (1981) Phosphorus recycling in five shallow lakes. *J. Am. Soc. Civil Eng., EE Div.*, **107**, 713–30.

Stewart, K. M. (1976) Oxygen deficits, clarity, and eutrophication in some Madison Lakes. *Int. Rev. Ges. Hydrobiol.* **61**, 563–79.

Stewart, N. E., Shumway, D. L. and Doudoroff, P. (1967) Influence of oxygen concentration on the growth of juvenile largemouth bass. *J. Fish Res. Board Canada*, **24**, 475–94.

Stockner, J. G. (1972) Paleolimnology as a means of assessing eutrophication. *Verh. Int. Verein. Limnol.*, **18**, 1018–30.

Stockner, J. G. and Shortreed, K. R. S. (1978) Enhancement of autotrophic production by nutrient addition in a coastal rain-forest stream on Vancouver Island. *J. Fish Res. Board Canada*, **35**, 28–34.

Stockner, J. G. and Shortreed, K. S. (1985) Whole-lake fertilization experiments in coastal British Columbia lakes: empirical relationships between nutrient inputs and phytoplankton biomass and production. *Can. J. Fish. Aquat. Sci.*, **42**, 649–58.

Stockner, J. G. and Shortread, K. S. (1988) Response of *Anabaena* and *Synechococcus* to manipulation of nitrogen: phosphorus ratios in a lake fertilization experiment. *Limnol. Oceanogr.*, **33**, 1348–61.

Storr, J. F. and Sweeney, R. A. (1971) Development of a theoretical seasonal growth response curve of *Cladophora glomerata* to temperature and photoperiod. *Proc. 14th Conf. Great Lakes Res.*, **14**, 119–27.

Stumm, W. (1963) *Proc. Int. Conf. Water Poll. Res.* Vol. 2. Pergamon Press, Elmsford, NY.

Stumm, W. and Leckie, J. O. (1971) Phosphate exchange with sediments: its role in the productivity of surface waters. *Proc. 5th Int. Conf. Water Pollut. Res.* III-26/1-16, Pergamon Press, Elmsford, NY.

Sundborg, A. (1956) *The River Klarälven. A Study of Fluvial Processes.* Geografiska Annaler Häfte 2–3/1956, University of Uppsala, Sweden, pp. 127–316.

Sverdrup, H. U. (1953). On conditions for the vernal blooming of phytoplankton. *J. Conseil.*, **18**, 287–95.

Sverdrup, H. U., Johnson, M. W. and Fleming, R. H. (1942) *The Oceans, their Physics, Chemistry, and General Biology* Prentice Hall, Englewood Cliffs, NJ.

Szumski, D. S., Barton, D. A., Putnam, H. D. and Polta, R. C. (1982) Evaluation of EPA un-ionized ammonia toxicity criteria. *J. Water Pollut. Control Fed.*, **54**, 281–91.

Talling, J. F. (1957a) Photosynthetic characteristics of some freshwater plankton diatoms in relation to underwater radiation. *New Phytol.*, **56**, 29–50.

Talling, J. F. (1957b) The phytoplankton population as a compound photosynthetic system. *New Phytol.*, **56**, 133–49.

Talling, J. F. (1971) The underwater light climate as a controlling factor in the production ecology of freshwater phytoplankton. *Mitt. Int. Verein. Limnol.*, **19**, 214–43.

Taylor, E. W. (1966) *Forty-Second Report on the Results of the Bacteriological Examinations of the London Waters for the Years 1965–6* Metropolitan Water Board, New River Head, London.

Templeton, W. L., Becker, C. D., Berlin, J. D., Coutant, C. C., Dean, J. M., Fujihara, M. P., Nakatani, R. E., Olson, P. A., Prentice, E. F. and Watson, D. G. (1969) *Biological Effects of Thermal Discharges. Progress Report for 1968* AEC Research and Development Report, BNWL-1050 reprint, Battelle Northwest Lab, Richland, WA.

Tett, P., Gallegos, C., Kelly, M. G., Hornberger, G. M. and Cosby, G. J. (1978) Relationships among substrate, flow and benthic microalgal pigment density in the Mechuma River, Virginia. *Limnol. Oceanogr.*, **3**, 785–97.

Thomas, E. A. (1969) The process of eutrophication in Central European lakes, in *Eutrophication: Causes, Consequences and Correctives.* Natl Acad. Sci., Washington, DC.

Thornton, J. A. (1987a) Aspects of eutrophication management in tropical/sub-tropical regions. *J. Limnol. Soc. S. Africa*, **13**, 25–43.

Thornton, J. A. (1987h) A review of some unique aspects of limnology of shallow southern African man-made lakes. *Geojournal*, **14**, 339–52.

Thut, R. N. (1969) A study of the profundal bottom fauna of Lake Washington. *Ecol. Monogr.* **39**, 79–100.

Traaen, T. S. and Lindström, E. A. (1983) Influence of current velocity on periphyton distribution, in *Periphyton of Freshwater Ecosystems* (ed. R. G. Wetzel) *Developments in Hydrobiology*, Vol. 17, pp. 97–9.

Trembly, F. J. (1960) *Research Project on Effects of Condenser Discharge Water on Aquatic Life*. Progress Report 1956–9. Inst. of Research, Lehigh University, Bethlehem, PA.

Trimbee, A. M. and Harris, G. P. (1984) Phytoplankton population dynamics of a small reservoir: use of sedimentation traps to quantify the loss of diatoms and recruitment of summer bloom-forming blue-green algae. *J. Plankton Res.* **6**, 897–918.

Trimbee, A. M. and Prepas, E. E. (1988) The effect of oxygen depletion on the timing and magnitude of blue-green algal blooms. *Verh. Int. Verein. Limnol.*, **23**, 220–6.

Tsai, C. (1973) Water quality and fish life below sewage outfalls. *Trans. Am. Fish Soc.*, **102**, 281–92.

Turback, S. C., Olson, S. B. and McFeters, G. A. (1986) Comparison of algal assay systems for detecting waterborne herbicides and metals. *Water Res.* **20**, 91–6.

USEPA (1982) Secondary treatment information, in *Code of Federal Regulations* Vol. 40, US Government Printing Office, Washington, DC.

USEPA (1983) *Assessing Water Quality in Streams* Environmental Monitoring Systems Laboratory, Las Vegas, NV.

USEPA (1987) *Clean Lakes Program* 1987 Annual Report, North American Lake Management Society for USEPA.

Uhlmann, D. (1971) Influence of dilution, sinking and grazing rate on phytoplankton populations of hyperfertilized ponds and microecosystems. *Mitt. Inte. Verein. Limnol.* **19**, 100–24.

Usinger, R. L. (1956) *Aquatic Insects of California*. University of California Press, Berkeley.

Vallentyne, J. (1972) Nutrients and Eutrophication. Special Symposium, *Limnol. Oceanog.*, **1**, 107.

Vance, B. D. (1965) Composition and succession of cyanophycean water blooms. *J. Phycol.*, **1**, 81–6.

Vannote, R. L., Minshall, G. W., Cummins, K. W., Sedell, J. R. and Cushing, C. E. (1980) The river continuum concept. *Can. J. Fish. Aquat. Sci.*, **37**, 130–7.

Van't Hoff, J. H. (1884) *Etudes de Dynamique Chimique*. Amsterdam.

Viessman, W. Jr. and Welty, C. (1985) *Water Management: Technology and Institutions* Harper & Row, New York.

Vollenweider, R. A. (1968) *Scientific Fundamentals of the Eutrophication of Lakes and Flowing Waters, with Particular Reference to Nitrogen and Phosphorus as Factors in Eutrophication* Paris, Rept. Organization for Economic Cooperation and Development. DASCSI/68.27, p. 182.

Vollenweider, R. A. (1969a) Possibilities and limits of elementary models concerning the budget of substances in lakes. *Arch. Hydrobiol.*, **66**, 1–36.

Vollenweider, R. A. (1969b) A manual on methods for measuring primary production in aquatic environments. *Int. Biol. Program Handbook 12* Blackwell Scientific Publications, Oxford, 213 pp.

Vollenweider, R. A. (1975) Input-output models with reference to the phosphorus-loading concept in limnology. *Schweizerische Zeitschrift Hydrol*, **37**, 53–84.

Vollenweider, R. A. (1976) Advances in defining critical loading levels for phosphorus in lake eutrophication. *Mem. Inst. Ital. Idrobiol.*, **33**, 53–83.

Vollenweider, R. A. and Dillon, P. J. (1974) *The Application of the Phosphorus-Loading Concept to Eutrophication Research* National Res. Council, Canada. Tech. Rept. 13690, 42 pp.

Wagner, K. J. (1986) Biological management of a pond ecosystem to meet water use objectives. *Lake Reservoir Management*, **2**, 54–9.

Walker, W. W. Jr. (1977) Some analytical methods applied to lake water quality problems. PhD dissertation, Harvard University.

Walker, W. W. Jr. (1981) *Empirical Methods for Predicting Eutrophication in Impoundments*. Report 1, Phase 1: Data base development. Technical Report E-81-9, EWQOS, US Army Corps of Engineers, Vicksburg, MS.

Walker, W. W. Jr. (1985) Statistical bases for mean chlorophyll *a* criteria. *Lake Reservoir Management*, **1**, 57–67.

Walker, W. W. Jr. (1987a) Empirical methods for predicting eutrophication in impoundments. Report 4, Phase III: Applications Manual. Tech. Rept. E-81-9. US Army Engineer Waterways Exp. Sta., Vicksburg, MS.

Walker, W. W. Jr. (1987b) Phosphorus removal by urban runoff detention basins. *Lake Reservoir Management*, **3**, 314–26.

Walshe, B. M. (1948) The oxygen requirements and thermal resistance of chironomed larvae from flowing and from still waters. *J. Exp. Biol.*, **25**, 35.

Walton, S. P. (1990) Effects of grazing by *Dicosmoecus gilvipes* (caddisfly) larvae and phosphorus enrichment on periphyton. MS thesis, University of Washington.

Warnick, S. L. and Bell, H. L. (1969) The acute toxicity of some heavy metals to different species of aquatic insects. *J. Water Pollut. Control Fed.*, **41**, 280–4.

Warren, C. E. (1971) *Biology and Water Pollution Control*. W. B. Saunders, Philadelphia.

Warren, C. E., Wales, J. H., Davis, G. E. and Doudoroff, P. (1964) Trout production in an experimental stream enriched with sucrose. *J. Wildlife Management* **28**, 617–60.

Warwick, W. F. (1980) Paleolimnology of the Bay of Quinte, Lake Ontario: 2800 years of cultural influence. *Can. Bull. Fish. Aquat. Sci.* **206**, 118 pp.

Washington Department of Wildlife (1990) *Grass Carp Use in Washington* Fish Management Division, FM No. 90–4.

Washington, H. G. (1984) Diversity, biotic and similarity indices, a review. *Water Res.* **18**, 653–94.

Weibel, S. R. (1969) Urban drainage as a factor in eutrophication, in *Eutrophication: Causes, Consequences and Correctives*. Natl Acad. Sci., Washington, DC.

Welch, E. B. (1969a) Factors controlling phytoplankton blooms and resulting dissolved oxygen in Duwamish River estuary, Washington, US Geol. Survey Water Supply Paper 1873-A, 62 pp.

Welch, E. B. (1969b) Discussion of ecological changes of applied significance induced by the dischage of heated waters, in *Thermal Pollution, Engineering Aspects* (eds F. L. Parker and P. A. Krenkel), Vanderbilt University Press, Nashville, Tennessee, pp. 58–68.

Welch, E. B. (1988) In-lake eutrophication control: progress and limitations, in *Water Pollution Control in Asia* (eds T. Panswad, C. Polprasert and K.

Yamamoto) Pergamon Press, New York, pp. 37–44.

Welch, E. B., Brenner, M V. and Carlson, K. L. (1984) Control of algal biomass by inflow NO_3, in *Lake and Reservoir Management* EPA 440/5/84/001, pp. 493–7.

Welch, E. B., Bush, R. M., Spyridakis, D. E. and Saikewicz, M. B. (1973) *Alternatives for Eutrophication Control in Moses Lake, Washington* Dept of Civil Engineering, University of Washington, Seattle, 102 pp.

Welch, E. B., Buckley, J. A. and Bush, R. M. (1972) Dilution as an algal bloom control. *J. Water Pollut Control Fed.*, **44**, 2245–65.

Welch, E. B., DeGasperi, C. L. and Spyridakis, D. E. (1988a) Sources for internal P loading in a shallow lake. *Verh. Int. Verein. Limnol.*, **23**, 307–14.

Welch, E. B., DeGasperi, C. L., Spyridakis, D. E. and Belnick, T. J. (1988b) Internal phosphorus loading and alum effectiveness in shallow lakes. *Lake Reservoir Management*, **4**, 27–34.

Welch, E. B., Horner, R. R. and Patmont, C. R. (1989b) Prediction of nuisance periphytic biomass: a management approach. *Water Res.* **23**, 401–5.

Welch, E. B., Jacoby, J. M., Horner, R. R. and Seeley, M. R. (1988c) Nuisance biomass levels of periphyton algae in streams. *Hydrobiologia* **157**, 161–8.

Welch, E. B. and Jones, C. A. (1990) Predicting phosphorus in a dendritic, shallow prairie lake following dilution and diversion. *Verh. Int. Verein. Limnol.*, **24**, 427.

Welch, E. B., Jones, C. A. and Barbiero, R. P. (1989a) *Moses Lake Quality: Results of Dilution, Sewage Diversion and BMPs – 1977 through 1988* Dept. of Civil Eng., University of Washington, Water Resources Series Tech. Rept. No. 118.

Welch, E. B. and T. S. Kelly (1990) Internal phosphorus loading and macrophytes: an alternative hypothesis. *Lake Reservoir Management* **6**, 43–8.

Welch, E. B. and Patmont, C. R. (1980) Lake restoration by dilution: Moses Lake, Washington *Water Res.*, **14**, 1316–25.

Welch, E. B. and Perkins, M. A. (1979) Oxygen deficit rate as a trophic state index. *J. Water Pollut Control Fed.*, **51**, 2823–28.

Welch, E. B., Quinn, J. M. and Hickey, C. W. (1992) Periphyton biomass related to point-source nutrient enrichment in seven New Zealand streams. *Water Res.* **26**, 669–75.

Welch, E. B., Rock, C. A., Howe, R. C. and Perkins, M. A. (1980) Lake Sammamish response to wastewater diversion and increasing urban runoff. *Water Res.*, **14**, 821–8.

Welch, E. B., Spyridakis, D. E., Shuster, J. I. and Horner, R. R. (1986) Declining lake sediment phosphorus release and oxygen deficit following wastewater diversion. *J. Water Pollut. Control Fed.*, **58**, 92–6.

Welch, E. B. and Wojtalik, T. A. (1968) *Some Aspects of Increased Water Temperature on Aquatic Life* Tennessee Valley Authorily, Chattanooga, TN, 48 pp.

Wetzel, R. G. (1964) A comparative study of the primary productivity of higher aquatic plants, periphyton, and phytoplankton in a large shallow lake. *Int. Rev. Geol. Hydrobiol.*, **49**, 1–64.

Wetzel, R. G. (1972) The role of carbon in hard-water marl lakes, in *Nutrients and Eutrophication: The limiting nutrient controversy* (ed. G. E. Likens), Sec. Symp. Amer. Soc. Limnol. Oceanogr., **1**, 84–91.

Wetzel, R. G. (1975, 1983) *Limnology* W. B. Saunders, Philadelphia.

Wetzel, R. G. and Hough, R. A. (1973) Productivity and role of aquatic macrophytes in lakes: an assessment. *Pol. Arch. Hydrobiol.* **20**, 9–19.

Wezernak, C. T., Lyzenga, D. R. and Polcyn, F. C. (1974) *Cladophora* distribution in Lake Ontario. *Ecol. Res. Ser.* EPA-660/3-74-022, 84 pp.

White, E. (1983) Lake eutrophication in New Zealand – a comparison with other countries of the Organization for Economic and Co-operative Development. *N.Z. J. Marine Freshwater Res.*, **17**, 437–44.

Whitord, L. A. (1960) The current effect and growth of freshwater algae. *Trans. Am. Microcop. Soc.*, **79**, 302–9.

Wiederholm, T. (1974) Bottom fauna and eutrophication in the large lakes of Sweden. PhD dissertation abstract, University of Uppsala, Sweden, 11 pp.

Wiederholm, T. (1980) Use of benthos in lake monitoring. *J. Water Pollut. Control Fed.*, **52**, 537–47.

Wildman, R. B., Loescher, J. H. and Winger, C. L. (1975) Development and germination of akinetes of *Aphanizomenon flos-aquae*. *J. Phycol.*, **11**, 96–104.

Wile, I. (1975) Lake restoration through mechanical harvesting of aquatic vegetation. *Verh. Int. Verein. Limnol.*, **19**, 660–71.

Wilhm, J. (1972) Graphical and mathematical analyses of biotic communities in polluted streams. *Annu. Rev. Entomol.*, **17**, 223–52.

Wilhm, J. and Dorris, T. (1968) Biological parameters for water quality criteria. *Bioscience*, **18**, 477–81.

Williams, J. D. H., Syers, J. K. and Walker, T. W. (1967) Fractionation of soil inorganic phosphate by a modification of Chang and Jackson's procedure. *Soil Sci. Soc. Am.*, **31**, 736–9.

Williams, L. G. and Mount, D. I. (1965) Influence of zinc on periphyton communities. *Am. J. Bot.*, **52**, 26–34.

Wilson, B. and Musick, T. A. (1989) *Lake Assessment Program 1988, Shagawa Lake, St. Louis County, Minnesota* unpublished manuscrpt Minnesota Pollution Control Agency, 85 pp.

Winner, R. W., Boesel, M. W. and Farrell, M. P. (1980) Insect community structure as an index of heavy-metal pollution in lotic ecosystems. *Can. J. Fish. Aquat. Sci.*, **37**, 647–55.

Winner, R. W., Scott Van Dyke, J., Caris, N. and Farrel, M. P. (1975) Response of the macroinvertebrate fauna to a copper gradient in an experimentally-polluted stream. *Verh. Int. Verein. Limnol.*, **19**, 2121–7.

Winter, D. F., Banse, K. and Anderson, G. C. (1975) The dynamics of phytoplankton blooms in Puget Sound, a fjord in Northwestern United States. *Marine Biol.*, **29**, 139–76.

Winter, T. C. (1981) Uncertainties in estimating the water balance of lakes. *Water Res. Bull.*, **17**, 82–115.

Wissmar, R. C., Richey, J. E. and Spyridakis, D. E. (1977) The importance of allochthonous particulate carbon pathways in a subalpine lake. *J. Fish Res. Board Canada.*, **43**, 1410–18.

Witting, R. (1909) Zur Kenntnis des vom Winde erzeugten Oberflächenstromes. *Ann. Hydrogr. Berlin.*, **37**, 193–203.

Woelke, C. E. (1968) Application of shellfish bioassay results to the Puget Sound Pulp Mill pollution problem. *Northwest Sci.*, **42**, 125–33.

Wojtalik, T. A., Hall, T. F. and Hill, L. O. (1971) Monitoring ecological conditions associated with wide-scale applications of DMA 2, 4-D to aquatic environments. *Pesticides Monitoring J.*, **4**, 184–203.

Woodwell, G. M. (1970) The energy cycle of the biosphere. *Sci. Am.*, **223** (3), 64–97.

WQC (1968) Water Quality Criteria. Tech. Advisory Comm. to EPA, US Government Printing Office.

414   REFERENCES

WQC (1973) Water Quality Criteria-1972. Tech Advisory Comm., Nat. Acad of Sci. and Acad of Engineers. US Government Printing Office.

WQC (1986) *Quality criteria for Water – 1986* EPA 440/5-86-001.

Wrenn, W. B. (1980) Effects of elevated temperature on growth and survival of smallmouth bass. *Trans. Am. Fish. Soc.*, **109**, 617–25.

Wright, J. C. (1958) The limnology of Canyon Ferry Reservoir. *Limnol. Oceanogr.*, **3**, 150–9.

Wright, R. F. (1983) *Predicting Acidification of North American Lakes* Report 4, Norwegian Inst. Water Res., Oslo, 165 pp.

Wright, R. F., Dale, T. Gjessing, E. J., Hendrey, G. R., Henriksen, A., Johannessen, M. and Muniz, I. P. (1976) Impact of acid precipitation on freshwater ecosystems in Norway. *Water, Air Soil Poll.*, **6**, 483–99.

Wright, R. F. and Henriksen, A. (1983) Restoration of Norwegian lakes by reduction in sulfur deposition. *Nature*, **303**, 422–4.

Wright, R. F. and Snekvir, E. (1978) Acid precipitation: chemistry and fish populations in 700 lakes in southern-most Norway. *Verh. Int. Verein. Limnol.*, **20**, 765–75.

Young, T. C. and King, D. L. (1980) Interacting limits to algal growth: light, phosphorus, and carbon dioxide availability. *Water Res.*, **14**, 409–12.

Younos, T. M. and Weigmann, D. L. (1988) Pesticides: a continuing delemma. *J. Water Pollut. Control Fed.*, **60**, 1199–205.

Zand, S. M. (1976) Indexes associated with information theory in water quality. *J. Water Pollut. Control Fed.*, **48**, 2026–31.

Zevenbroom, W. and Mur, L. R. (1980) N_2-fixing cyanobacteria: why they do not become dominant in Dutch hypereutrophic lakes. *Dev. Hydrobiol.*, **2**, 123–30.

INDEX